Lecture Notes in Artificial Intelligence   5077

Edited by R. Goebel, J. Siekmann, and W. Wahlster

Subseries of Lecture Notes in Computer Science

T0241055

Petra Perner (Ed.)

Advances in Data Mining

8th Industrial Conference, ICDM 2008
Leipzig, Germany, July 16-18, 2008
Proceedings

 Springer

Volume Editor

Petra Perner
Institute of Computer Vision
and Applied Computer Sciences, IBaI
Arno-Nitzsche-Str. 43
04277 Leipzig, Germany
E-mail: pperner@ibai-institut.de

Library of Congress Control Number: Applied for

CR Subject Classification (1998): I.2.6, I.2, H.2.8, K.4.4, J.3, I.4, J.6, J.1

LNCS Sublibrary: SL 7 – Artificial Intelligence

ISSN   0302-9743
ISBN-10   3-540-70717-4 Springer Berlin Heidelberg New York
ISBN-13   978-3-540-70717-2 Springer Berlin Heidelberg New York

Springer is a part of Springer Science+Business Media

springer.com

© Springer-Verlag Berlin Heidelberg 2008
Printed in Germany

Typesetting: Camera-ready by author, data conversion by Scientific Publishing Services, Chennai, India
Printed on acid-free paper   SPIN: 12441785   06/3180   5 4 3 2 1 0

Preface

ICDM / MLDM Medaillie (limited edition)
Meissner Porcellan, the "White Gold" of King
August the Strongest of Saxonia

ICDM 2008 was the eighth event of the Industrial Conference on Data Mining held in Leipzig (www.data-mining-forum.de).

For this edition the Program Committee received 116 submissions from 20 countries.

After the peer-review process, we accepted 36 high-quality papers for oral presentation, which are included in these proceedings. The topics range from aspects of classification and prediction, clustering, Web mining, data mining in medicine, applications of data mining, time series and frequent pattern mining, and association rule mining.

Thirteen papers were selected for poster presentations that are published in the ICDM Poster Proceeding Volume.

In conjunction with ICDM there were three workshops focusing on special hot application-oriented topics in data mining. The workshop Data Mining in Life Science DMLS 2008 was held the third time this year and the workshop Data Mining in Marketing DMM 2008 ran for the second time this year. Additionally, we introduced an International Workshop on Case-Based Reasoning for Multimedia Data CBR-MD.

We were pleased to give out the best paper award for ICDM for the third time this year. Announcements were made at www.data-mining-forum.de. The final decision was made by the Best Paper Award Committee based on the presentation by the authors and the discussion with the auditorium. The ceremony took place under the guidance of Isabelle Bichindaritz from the University of Washington, USA. This prize was sponsored by IBaI solutions (www.ibai-solutions.de), one of the leading data mining companies in data mining for marketing, Web mining and e-commerce.

The conference was rounded up by an overview of new challenging topics in data mining before the Best Paper Award Ceremony.

The Conference Summary Volume summarizes the vision of the conference and workshops as well as the paper presentations and also presents "Who Is Who" in Machine Learning and Data Mining by giving each author the chance to present himself.

We thank the members of the Institute of Applied Computer Sciences, Leipzig, Germany (www.ibai-institut.de), who handled the conference as secretariat. We appreciate the help and understanding of the editorial staff at Springer, and in particular Alfred Hofmann, who supported the publication of these proceedings in the LNAI series.

Last, but not least, we wish to thank all the speakers and participants who contributed to the success of the conference.

July 2008   Petra Perner

Industrial Conference on Data Mining, ICDM 2008

Chair

Petra Perner   IBaI Leipzig, Germany

Committee

Klaus-Peter Adlassnig	Medical University of Vienna, Austria
Andrea Ahlemeyer-Stubbe	ENBIS, Amsterdam, The Netherlands
Klaus-Dieter Althoff	University of Hildesheim, Germany
Chid Apte	IBM Yorktown Heights, USA
Eva Armengol	IIA CSIC, Spain
Isabelle Bichindaritz	University of Washington, USA
Leon Bobrowski	Bialystok Technical University, Poland
Marc Boullé	France Télécom, France
Henning Christiansen	Roskilde University, Denmark
Frans Coenen	University of Liverpool, UK
Juan M. Corchado	Universidad de Salamanca, Spain
Da Deng	University of Otago, New Zealand
Antonio Dourado	University of Coimbra, Portugal
Peter Funk	Mälardalen University, Sweden
Brent Gordon	NASA Goddard Space Flight Center, USA
Gary F. Holness	Quantum Leap Innovations Inc., USA
Eyke Hüllermeier	University of Marburg, Germany
Piotr Jedrzejowicz	Gdynia Maritime University, Poland
Mehmed Kantardzic	University of Louisville, USA
Ron Kenett	KPA Ltd., Israel
Mineichi Kudo	Hokkaido University, Japan
Eduardo F. Morales	INAOE, Ciencias Computacionales, Mexico
Stefania Montani	Università del Piemonte Orientale, Italy
Eric Pauwels	CWI Utrecht, The Netherlands
Ashwin Ram	Georgia Institute of Technology, USA
Rainer Schmidt	University of Rostock, Germany
Yuval Shahar	Ben Gurion University, Israel
David Taniar	Monash University, Australia
Stijn Viaene	KU Leuven, Belgium
Rob A. Vingerhoeds	Ecole Nationale d'Ingénieurs de Tarbes, France
Claus Weihs	University of Dortmund, Germany
Terry Windeatt	University of Surrey, UK

Additional Reviewers

Kerstin Bach	University of Hildesheim, Germany
Javier Bajo	University of Salamanca, Spain
Cezary Boldak	Bialystok Technical University, Poland
Fabrice Clerot	France Telecom, France
Jan-Oliver Deutsch	University of Hildesheim, Germany
Sylvain Ferrandiz	France Telecom, France
Francoise Fessant	France Telecom, France
Alexandre Hanft	University of Hildesheim, Germany
Marek Kretowski	Bialystok Technical University, Poland
Vincent Lemaire	France Telecom, France
Régis Newo	University of Hildesheim, Germany
Robin Nunkesser	University of Dortmund, Germany
Elzbieta Olejarczyk	IBIB PAN, Poland
Nils Raabe	University of Dortmund, Germany
Meike Reichle	University of Hildesheim, Germany
Sara Rodríguez	University of Salamanca, Spain
Martin Schaaf	University of Hildesheim, Germany
Julia Schiffner	University of Dortmund, Germany
Gero Szepannek	University of Dortmund, Germany
Dante Tapia	University of Salamanca, Spain

Table of Contents

Lifescience and Biotechnological Applications for Data Mining

Clustering and Classification

Association Rule Mining

E-Mail, WebMining

Information Retrieval

Industrial Applications

Frequent Item Set, Sequence Mining

Aspects of Data Mining

Prototypes for Medical Case-Based Applications

Rainer Schmidt[1], Tina Waligora[1], and Olga Vorobieva[1,2]

[1] Institut für Medizinische Informatik und Biometrie, Universität Rostock, Germany
rainer.schmidt@medizin.uni-rostock.de
[2] Sechenov Institute of Evolutionary Physiology and Biochemistry, St.Petersburg, Russia

Abstract. Already in the early stages of Case-Based Reasoning prototypes were considered as an interesting technique to structure the case base and to fill the knowledge gap between single cases and general knowledge. Unfortunately, later on prototypes never became a hot topic within the CBR community. However, for medical applications they have been used rather regularly, because they correspond to the reasoning of doctors in a natural way. In this paper, we illustrate the role of prototypes by application programs, which cover all typical medical tasks: diagnosis, therapy, and course analysis.

1 Introduction

Cases are the most specialised form of knowledge representation. The knowledge of physicians consists of general knowledge they have read in medical books and of their experiences in form of cases they have treated themselves or colleagues have told them about. Not all cases are of the same importance. Some are typical while others are rather exceptional, e.g. a paediatrician does not remember all his patients with measles, but maybe those with serious complications or those where his measles diagnosis was surprisingly wrong. Doctors consider differences between their current patient and typical or known exceptional cases.

We believe that medical Case-Based Reasoning (CBR) systems should take the reasoning of doctors into account [1]. Such systems should not only consist of general medical domain knowledge plus a flat case base, but the case base should be structured by typical case generalisations called prototypes [2].

Though the use of prototypes had been early introduced in the CBR community [3, 4], their use is still rather seldom. Later on it fell into oblivion and was brought up again by Bergman in form of generalised cases [5]. His notion of generalised cases is similar but not identical to our idea of prototypes. While generalised cases are general or abstract in contrast to concrete cases, prototypes contain the typical features of a set of cases.

However, since doctors reason with typical cases anyway, in medical CBR systems prototypes are a rather common knowledge form, they are used in a variety of applications, e.g. for diabetes [6], for eating disorders [7], and for pulmonology [8]. Prototypical images that can be transformed after certain image processing steps in prototypes are used for the diagnosis of medical images [9].

A Prototype is generalised from a set of single cases. The cases in this set are very similar to each other or they belong in some other specific way together and form a

P. Perner (Ed.): ICDM 2008, LNAI 5077, pp. 1–15, 2008.

sort of class. For example, in a diagnostic system all patients diagnosed as measles patient might be grouped together. Usually, prototypes have the same structure as cases but have less and more general features, namely just the typical ones. Sometimes prototypes are defined by medical experts, sometimes they can be found in literature (e.g. the typical symptoms for measles), and sometimes they are computed. The use of case oriented generalised knowledge presents the opportunity to structure case bases. Cases can be clustered into groups, prototypical diseases or schema. Clancey [10] distinguishes between prototypes that represent specific expressions of diseases or therapies and schema that contain essential features of diseases or therapies. As Selz [11] characterises a schema as a description of an entity where at least one part remains vague, the distinction between prototypes and schema seems to be fluid. We only use the term prototype and refer to a hierarchy of prototypes where the most general prototypes that contain the most common features are situated on top and the most specific ones are placed at the bottom. This notion of prototypes differs from the usual notion of classes and clusters [12] in many ways. Prototypes are not the result of a classification process. Whether a case belongs to a prototype, is determined by its features or defined by an expert. There may be a hierarchy of prototypes but there are not relations (similarity, is-a and so on), and the set of cases belonging to a prototype is not represented by its most representative case but by the prototype. The main purpose of such generalised knowledge is to guide the retrieval and sometimes to decrease the amount of storage by erasing redundant cases. In domains with rather weak domain theories another advantage of case-oriented techniques is their ability to learn from cases. Only gathering new cases may improve the systems ability to find suitable similar cases for current problems, but it does not elicit the intrinsic knowledge of the stored cases. To learn the knowledge contained in cases a generalisation process is necessary. Generally speaking, prototypes fill the knowledge gap between the specificity of single cases and abstract knowledge usually expressed as rules.

In this paper we present three systems we developed during the last years and focus on the role of prototypes within them. A more detailed presentation on this topic can be found in [13]. We start with a prototype-based system for diagnosis of dysmorphic syndromes. Subsequently we present a system for course analysis and prognosis of the kidney function and finally we present ISOR, a system that deals with therapeutic problems in the endocrine domain.

2 Prototype-Based Diagnosis of Dysmorphic Syndromes

In this application, retrieval does not search for former single cases but only for prototypes. Each prototype represents and characterises one specific diagnosis. We assume that this idea is rather typical for diagnostic tasks, because it seems to be reasonable to search for a general description of a disease instead of searching for single patients.

When a child is born with dysmorphic features or with multiple congenital malformations or if mental retardation is observed at a later stage, finding the correct diagnosis is extremely important. Knowledge of the nature and the etiology of the disease enables the pediatrician to predict the patient's future course. So, an initial goal for medical specialists is to diagnose a patient to a recognised syndrome. Genetic counselling and a course of treatments may then be established.

A dysmorphic syndrome describes a morphological disorder and it is characterised by a combination of various symptoms, which form a pattern of morphologic defects. An example is Down Syndrome which can be described in terms of characteristic clinical and radiographic manifestations such as mental retardation, sloping forehead, a flat nose, short broad hands, and generally dwarfed physique [14].

The main problems of diagnosing dysmorphic syndromes are [15]:

- more than 200 syndromes are known,
- many cases remain undiagnosed with respect to known syndromes,
- usually many symptoms are used to describe a case (between 40 and 130),
- every dysmorphic syndrome is characterised by nearly as many symptoms.

Furthermore, knowledge about dysmorphic disorders is continuously modified, new cases are observed that cannot be diagnosed (it exists even a journal that only publishes reports of newly observed interesting cases [16]), and sometimes even new syndromes are discovered. Usually, even experts of paediatric genetics only see a small count of dysmorphic syndromes during their lifetime.

So, we have developed a diagnostic system that uses a large case base. Starting point to build the case base was a large case collection of the paediatric genetics of the University of Munich, which consists of nearly 2,000 cases and 229 prototypes. A prototype (prototypical case) represents a dysmorphic syndrome by its typical symptoms. Most of the dysmorphic syndromes are already known and have been defined in literature. And nearly one third of the prototypes were determined by semiautomatic knowledge acquisition, where an expert selected cases that should belong to same syndrome and subsequently a prototype, characterised by the most frequent symptoms of his cases, was generated. To this database we have added rare dysmorphic syndromes, namely from "clinical dysmorphology" [16] and from the London dysmorphic database [17].

2.1 Prototypicality Measures

In CBR usually cases are represented as attribute-value pairs. In medicine, especially in diagnostic applications, this is not always the case, instead often a list of symptoms describes a patient's disease. Sometimes these lists can be very long, and often their lengths are not fixed but vary with the patient. For dysmorphic syndromes usually between 40 and 130 symptoms are used to characterise a patient.

Furthermore, for dysmorphic syndromes it is unreasonable to search for single similar patients (and of course none of the systems mentioned above does so) but for more general prototypes that contain the typical features of a syndrome. To determine the most similar prototype for a given query patient instead of a similarity measure a prototypicality measure is required. One speciality is that for prototypes the list of symptoms is usually much shorter than for single cases.

The result should not be just the one and only most similar prototype, but a list of them – sorted according to their similarity. So, the usual CBR retrieval methods like indexing or nearest neighbour search are inappropriate. Instead, rather old measures for dissimilarities between concepts [18, 19] are appropriate.

As humans look upon cases as more typical for a query case as more features they have in common [19], distances between prototypes and cases usually mainly consider the shared features. In some experiments we have compared the measure proposed by Tversky [18] with a similar measure proposed by Mervis and Rosch [19]. The measure proposed by Tversky (2.1) counts the number of matching symptoms of the query patient (X) and a prototype (Y). Subsequently, two numbers are subtracted, namely the number of symptoms that are observed for the patient but are not used to characterise the prototype (X-Y), and secondly the number of symptoms that are used for the prototype but are not observed for the patient (Y-X). Finally, the result is normalised by the number of symptoms used for the prototype (Y). In (2.1) f is a general function, however in our application it is just a counting function. The prototypicality measure proposed by Rosch and Mervis [19] differs from Tversky's measure only in one point: the factor X-Y is not considered.

$$D (X,Y) = \frac{f (X + Y) - f (X-Y) - f (Y-X)}{f (Y)} \tag{2.1}$$

2.2 The System

Our system process consists of three steps. At first the user has to select the symptoms that characterise a new patient. This selection is a long and very time consuming process, because we consider more than 800 symptoms. However, diagnosis of dysmorphic syndromes is not a task where the result is very urgent, but it usually requires thorough reasoning and subsequently a long-term therapy has to be started. In the second step, a prototypicality measure is sequentially applied on all prototypes (syndromes). Since the syndrome with maximal similarity is not always the right diagnosis, the 20 syndromes with best similarities are listed in a menu. In the third and final step, the user can optionally choose to apply adaptation rules on the syndromes. Such a rule states that a specific combination of symptoms favours or disfavours specific dysmorphic syndromes. However, how shall the adaptation rules alter the results? Our first idea was that the adaptation rules should increase or decrease the similarity scores for favoured and disfavoured syndromes. But the question is how. Of course no medical expert can determine values to manipulate the similarities by adaptation rules and any general value for favoured or disfavoured syndromes would be arbitrary. So, instead the result after applying adaptation rules is a menu that contains up to three lists (figure 1).

On top the favoured syndromes are depicted, then those neither favoured nor disfavoured, and at the bottom the disfavoured ones. Additionally, the user can get information about the specific rules that have been applied on a particular syndrome.

In the example presented in figure1, the correct diagnosis is Lenz-syndrome. The computation of the prototypicality measure provided Lenz-syndrome as the most similar but one syndrome. After application of adaptation rules, the ranking is not obvious. Two syndromes have been favoured, the more similar one is the right one. However, Dubowitz-syndrome is favoured too (by a completely different rule), because a specific combination of symptoms makes it probable, while other observed symptoms indicate a rather low similarity.

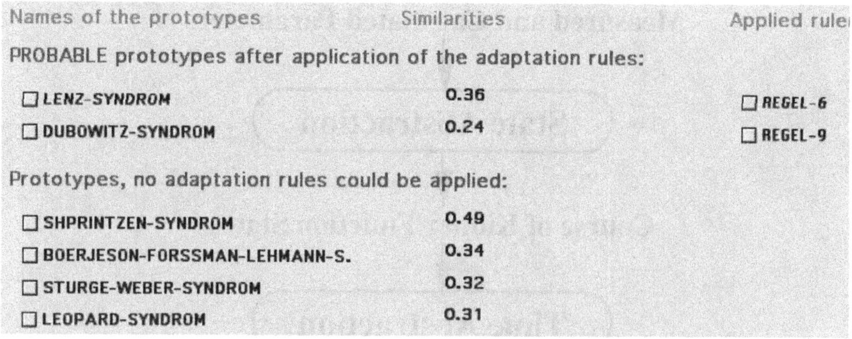

Names of the prototypes	Similarities	Applied rule
PROBABLE prototypes after application of the adaptation rules:		
☐ LENZ-SYNDROM	0.36	☐ REGEL-6
☐ DUBOWITZ-SYNDROM	0.24	☐ REGEL-9
Prototypes, no adaptation rules could be applied:		
☐ SHPRINTZEN-SYNDROM	0.49	
☐ BOERJESON-FORSSMAN-LEHMANN-S.	0.34	
☐ STURGE-WEBER-SYNDROM	0.32	
☐ LEOPARD-SYNDROM	0.31	

Fig. 1. Top part of the listed prototypes after additionally applying adaptation rules

3 Prognosis of Kidney Function Courses

Up to 60% of the body mass of an adult person consists of water. The electrolytes dissolved in body water are of great importance for an adequate cell function. The human body tends to balance the fluid and electrolyte situation. But intensive care patients are often no longer able to maintain adequate fluid and electrolyte balances themselves due to impaired organ functions, e.g. renal failure, or medical treatment, e.g. parenteral nutrition of mechanically ventilated patients. Therefore physicians need objective criteria for the monitoring of fluid and electrolyte balances and for choosing therapeutic interventions as necessary.

At our ICU, physicians daily get a printed renal report from the monitoring system NIMON [20] which consists of 13 measured and 33 calculated parameters of those patients where renal function monitoring is applied. The interpretation of all reported parameters is quite complex and needs special knowledge of the renal physiology. The aim of our knowledge based system ICONS [21] is to give an automatic interpretation of the renal state to elicit impairments of the kidney function on time and to give early warnings against forthcoming kidney failures. That means, we need a time course analysis of many parameters without any well-defined standards. However, in the domain of fluid and electrolyte balance, neither a prototypical approach in ICU settings is known nor exists complete knowledge about the kidney function. Especially, knowledge about the behaviour of the various parameters on time is yet incomplete. So, we combined the idea of RÉSUMÉ [22] to abstract many parameters into one single parameter with the idea of Haimowitz and Kohane [23] to compare many parameters of current courses with well-known standards. Since well-known standards were not available, we used former similar cases instead.

3.1 Prognostic Method

The method that was developed for ICONS is shown in figure 2. The steps are explained in this section.

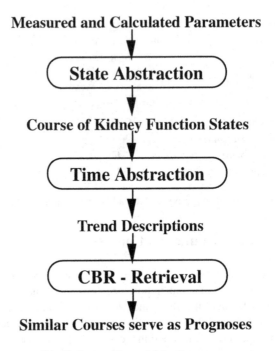

Fig. 2. Prognostic model for time course

State Abstraction. For the data abstraction we use states of the renal function, which determine states of increasing severity beginning with a normal renal function and ending with a renal failure. Based on the kidney function states, characterised by required and optional conditions for selected renal parameters, we first check the required conditions. For each state that satisfies the required conditions we calculate a similarity value concerning the optional conditions. We use a variation of Tversky's [18] measure of dissimilarity between concepts. Only if two or more states are under consideration, ICONS presents them to the user sorted according to their similarity values together with information about the satisfied and not satisfied optional conditions.

The user can accept or reject a presented state. When a suggested state has been rejected, ICONS selects another one. Finally, we determine the central state of occasionally more than one states the user has accepted. This central state is the closest one towards a kidney failure. Our intention is to find the state indicating the most profound impairment of the kidney function.

Temporal Abstraction. First, we have fixed five assessment definitions for the transition of the kidney function state of one day to the state of the respectively next day. These assessment definitions are related to the grade of renal impairment: steady, increasing, sharply increasing, decreasing, and sharply decreasing. These assessment definitions are used to determine the state transitions from one qualitative value to another. Based on these state transitions, we generate three trend descriptions. Two trend descriptions especially consider the current state transitions.

short-term trend:=	current state transition; Abbreviation: T1
medium-term trend:=	looks recursively back from the current state transition to the one before and unites them if they are both of the same direction or one of them has a "steady" assessment; Abbreviation: T2
long-term trend:=	characterises the considered course of at most seven days; Abbreviation: T3

For the long-term trend description we additionally introduced four new assessment definitions (alternating, oscillating, fluctuating, and nearly steady). If none of the five former assessments fits the complete considered course, we attempt to fit one of these four additional definitions.

Only if there are several courses with the same trend descriptions, we use a minor fourth trend description T4 to find the most similar among them. We assess the considered course by adding up the state transition values inversely weighted by the distances to the current day. Together with the current kidney function state these four trend descriptions form a course depiction, that abstracts the sequence of the kidney function states.

Example. The following kidney function states may be observed in this temporal sequence (figure 3):

selective tubular damage, reduced kidney function, reduced kidney function, selective tubular damage, reduced kidney function, reduced kidney function, sharply reduced kidney function

So we get these six state transitions:

decreasing, steady, increasing, decreasing, steady, decreasing

with these trend descriptions:

current state: sharply reduced kidney function
T1: decreasing, reduced kidney function, one transition
T2: decreasing, selective tubular damage, three transitions
T3: fluctuating, selective tubular damage, six transitions
T4:1.23

Retrieval. We use the parameters of the four trend descriptions and the current kidney function state to search for similar courses. As the aim is to develop an early warning system, we need a prognosis. For this reason and to avoid a sequential runtime search along the entire cases, we store a course of the previous seven days and a maximal projection of three days for each day a patient spent on the intensive care unit.

Since there are many different possible continuations for the same previous course, it is necessary to search for similar courses and for different projections. Therefore, we divided the search space into nine parts corresponding to the possible continuation directions. Each direction forms an own part of the search space. During the retrieval these parts are searched separately and each part may provide at most one similar

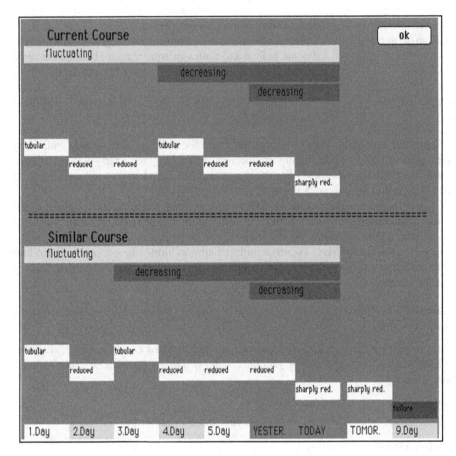

Fig. 3. Comparative presentation of a current and a similar course. In the lower part of each course the (abbreviated) kidney function states are depicted. The upper part of each course shows the deduced trend descriptions.

case. The similar cases of these parts together are presented in the order of their computed similarity values.

Before the main retrieval, we search for a prototypical case (see section 3.2) that matches most of the trend descriptions. Below this prototype the main retrieval starts. First we search with an activation algorithm concerning qualitative features. Only if two or more courses are selected in the same projection part, we use the sequential similarity measure TSCALE [24], which goes back to Tversky [18], concerning the quantitative features in a second step.

Continuation of the example. For the example above, the following similar course (figure 3) with these transitions is retrieved:

decreasing, increasing, decreasing, steady, steady, decreasing

with these trend descriptions:

current state: sharply reduced kidney function

T1: decreasing, reduced kidney function, one transition
T2: decreasing, selective tubular damage, four transitions
T3: fluctuating, selective tubular damage, six transitions
T4: 1.17

After another day with a "sharply reduced kidney function" the patient belonging to the similar course had a kidney failure. The physician may notice this as a warning and it is up to him to interpret it.

3.2 Learning a Tree of Prototypes

Prognosis of multi-parametric courses of the kidney function for ICU patients is a domain without a medical theory. Moreover, we can not expect such a theory to be formulated in the near future. So we attempt to learn prototypical course pattern. Therefore, knowledge on this domain is stored as a tree of generalised cases (prototypes) with three levels and a root node (figure 4).

Except for the root, where all not yet clustered courses are stored, each level corresponds to one of the trend descriptions T1, T2 or T3. As soon as enough courses that share another trend description are stored at a prototype, a new prototype with this trend is created. At a prototype at level 1, we cluster courses that share T1, at

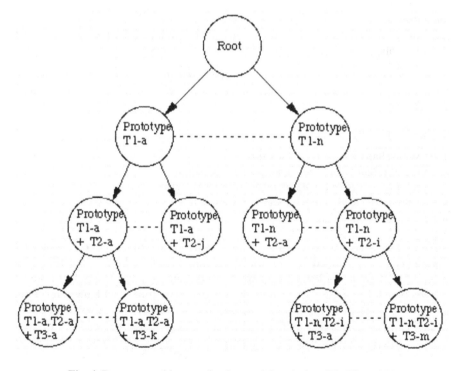

Fig. 4. Prototype architecture for the trend descriptions T1, T2, and T3

level 2, courses that share T1 and T2 and at level 3, courses that share all three trend descriptions T3. This can be done, because regarding their importance, the short-, medium- and long-term trend descriptions T1, T2 and T3 refer to hierarchically related time periods. T1 is more important than T2 and T3, and so forth.

The retrieval starts with a search for a prototype that has most of the trend descriptions in common with the query course. The search begins at the root node with a check for a prototype with the same short-term trend description T1. If such a prototype can be found, the search goes on below this prototype for a prototype that has the same trend descriptions T1 and T2, and so forth. If no prototype with a further trend in common can be found, we search for a course at the last accepted prototype.

If no prototype exists that has the same T1 as the query course, the search starts at the root node, where links to all courses in the case base exist.

Continuation of the example. In the example above, we can create just one prototype at level one, because at the second level the query course and the similar one, called "similar-1" differ in their length. Although the long-term trend description T3 is equal for both courses, we can not create a prototype at level three because of the strictly hierarchical organisation of the prototype tree. However, learning a prototypical description "fluctuating in seven days from a selective tubular damage to sharply reduced kidney function" which does not consider any more similarities or deviations within this time period would be too general and therefore too impracticable.

Assuming we find another similar course, called "similar-2", for the current case of the example above with the following kidney function states:

reduced kidney function, reduced kidney function, selective tubular damage, selective tubular damage, reduced kidney function, reduced kidney function, sharply reduced kidney function

with these trend descriptions:

current state: sharply reduced kidney function
T1: decreasing, reduced kidney function, one transition
T2: decreasing, selective tubular damage, four transitions
T3: oscillating, reduced kidney function, six transitions
T4: 1.33

The current query course, "similar-1", and "similar-2" will be clustered at level 1 to prototype T1-a, defined by T1 as "decreasing, reduced kidney function, one transition". Afterwards at level 2 the current course and "similar-2" will be clustered to a prototype T1-a + T2-a, defined by T1 as "decreasing, reduced kidney function, one transition" plus by T2 as "decreasing, selective tubular damage, four transitions". The attempt to create another prototype at level 3 fails, because the trend descriptions T3 have different assessments and different start states. The result, a tree of prototypes learned from the three courses is shown in figure 5.

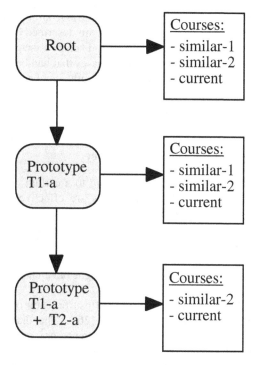

Fig. 5. Generated prototype tree from three example courses

4 ISOR

ISOR is a Case-Based Reasoning system for long-term therapy support in the endocrine domain [25]. It performs typical therapeutic tasks, such as computing initial therapies, initial dose recommendations, and dose updates. Apart from these tasks ISOR deals especially with situations where therapies become ineffective. Causes for inefficacy have to be found and better therapy recommendations should be computed. In addition to the typical Case-Based Reasoning knowledge, namely former already solved cases, ISOR uses further knowledge forms, especially medical histories of query patients themselves and prototypical cases (prototypes).

ISOR uses prototypes in two ways, namely in form of guidelines for dose calculations and as generalised solutions for therapy inefficacy.

4.1 Computing Initial Doses: Guidelines as Prototypes

For hypothyroidism only one drug exists, namely Levothyroxine. The problem is to calculate effective initial doses. Firstly, a couple of prototypes exist. These are recommendations that have been defined by expert commissions. Though we are not sure whether they are officially accepted, we call them guidelines. The assignment of a patient to a fitting guideline is obvious because of the way the guidelines have been defined. With the help of these guidelines a range for good doses can be calculated.

To compute a dose with best expected impact, we retrieve similar cases whose initial doses are within the calculated ranges. Since cases are described by few attributes and since our case base is rather small, we use Tversky's sequential measure of dissimilarity [18]. On the basis of those retrieved cases that had best therapy results an average initial therapy is calculated. Best therapy results can be determined by values of a blood test after two weeks of treatment with the initial dose. The opposite idea to consider cases with bad therapy results does not work here, because bad results can also be caused by various other reasons.

4.2 Generalised Solutions for Therapy Inefficacy

In medical practice, therapies prescribed according to a certain diagnosis sometimes do not give desired results. Sometimes therapies are effective for some time but suddenly stop helping any more. There are many different reasons. A diagnosis might have been erroneous, the state of a patient might have changed completely or the state might have changed just slightly but with important implications for an existing therapy. Furthermore, a patient might have caught an additional disease, some other complication might have occurred, or a patient might have changed his/her lifestyle (e.g. started a diet) and so on.

For diagnosis and therapy of hypothyroidism (insufficiency of the thyroid gland) complete and elaborate knowledge exists (e.g. [26]). The diagnosis is based on analyses of laboratory results. Hypothyroidic patients are treated with hormonal therapy in form of levothyroxine. With a proper dose of levothyroxine the states of hypothyroidic patients should become stable. If the achieved stability is disturbed, it is necessary to ascertain the reason of the disturbance and to eliminate it. Usually, a hypothyroidic patient should visit his/her doctor twice a year. Between these visits certain changes may occur in his/her condition. Some of them the doctor can find out during the standard interrogatory of the patient. These include drugs prescribed by another physician, new diseases, physiological conditions, pregnancy etc. However, there are changes that are much more difficult to find out, because a patient often does not attach significance to some changes in his/her diet and lifestyle. Therefore substantial facts may escape the attention of the doctor. The task of ISOR is to help catching these seemingly insignificant facts. Furthermore, it shall indicate reasons that prevent the expected effects of the drug. ISOR first attempts to find causes for inefficacy and subsequently computes new therapy recommendations that should perform better than those administered before. Apart from a case base ISOR consists of a knowledge base, which contains therapies, conflicts, instructions and so on, of medical histories, and of prototypes. For details about the architecture of ISOR see [25]. In ISOR, prototypes are generalized cases with general solutions. They play a particular role, because they help to select a proper solution from the list of probable or available solutions. A prototype may help to point out a reason of inefficacy or it may support the doctor's choice of a drug.Since ISOR is problem oriented, cases have appropriate attributes, namely three main ones: problem code, diagnosis and therapy. Each combination of these attribute values has its specific additional attributes. The combination A4 (problem code), hypothyroidism (diagnosis), and levothyroxine (therapy), for example, has the additional attributes age, sex, prescribed drugs, and supervisor (yes/no). Sets of cases with equal main attribute values and with similar additional attribute values build a prototype. For the combination mentioned above, one prototype P1 is, for example, characterized by these additional values:

P-1: Age between 15 and 19, female, no prescribed drugs, and no supervisor.

The prototypes do not have unique but four general solutions:

A. irregular levothyroxine intake.
B. changes in the hormonal condition of the patient.
C. uncontrolled intake of any stuff that inhibits levothyroxine absorption.
D. Intake of prescribed drugs that can inhibit levothyroxine absorption.

The additional attribute values determine the order in which these solutions are probable. For the prototype P-1 the order is B, C, and A, whereas solution D is eliminated. Some of the general solutions are specialized into some more specific solutions.

5 Summary: The Role of Prototypes

The presented systems have one thing in common that distinguishes them from most CBR systems: They use prototypes as a form of knowledge representation that fills the gap between specific cases and general rules. The main purpose of such generalised knowledge is to structure the case base, to guide the retrieval process, and sometimes to decrease the amount of storage by erasing redundant cases. In domains with very poor domain theories they may help to learn general knowledge. In domains with rather weak domain theories another advantage of case-oriented techniques is their ability to learn from cases. Only gathering new cases may improve the system's ability to find suitable similar cases for current problems, but it does not elicit the intrinsic knowledge of the stored cases. To learn the knowledge contained in cases a generalisation process is necessary. In our early warning system concerning the kidney function, apart from guiding the retrieval and structuring the case base prototypes mainly serve to learn typical course pattern, because just the relevant kidney parameters are known but no knowledge about their temporal course behaviour exists.

For diagnosis of dysmophic syndromes prototypes correspond directly to the physician's sense of prototypes. As comparisons with single cases are unable to identify typical features, in this application the use of prototypes is not only sensible, but even necessary.

In ISOR, the prototypes for dose calculation are guidelines and the prototypes for therapy inefficacy are similar to those for diagnosis of dysmorphic syndromes. The main difference is that in ISOR all prototypes are defined by medical experts.

Summarising our experiences we would like to make quite clear that the role of prototypes depends on the application and the task. For medical diagnoses they even seem to be necessary because of their correspondence to medical prototypes which guide the physicians diagnoses.

In this paper, we have just presented and discussed prototypes within our own case-based systems. Recently Isabelle Bichindaritz [27] has presented her experience with prototypes for medical CBR systems. However, she focuses on their role for knowledge maintenance, which is an interesting topic but it lays beyond the scope of this paper.

References

1. Strube, G., Janetzko, D.: Episodisches Wissen und fallbasiertes Schließen: Aufgaben für die Wissensdiagnostik und die Wissenspsychologie. Schweizerische Zeitschrift für Psychologie 49(4), 211–221 (1990)
2. Swanson, D.B., Feltovich, P.J., Johnson, P.E.: Psychological Analysis of Physician Expertise: Implications for Design of Decision Support Systems. In: Shires, D.B., Wolf, H. (eds.) Proc MEDINFO 1977, pp. 161–164. North-Holland, Amsterdam (1977)
3. Schank, R.C.: Dynamic Memory: a theory of learning in computer and people. Cambridge University Press, New York (1982)
4. Bareiss, R.: Exemplar-based knowledge acquisition. Academic Press, San Diego (1989)
5. Maximini, K., Maximini, R., Bergmann, R.: An Investigation of Generalized Cases. In: Ashley, K.D., Bridge, D.G. (eds.) ICCBR 2003. LNCS, vol. 2689, pp. 261–275. Springer, Heidelberg (2003)
6. Bellazzi, R., Montani, S., Portinale, I.: Retrieval in a prototype-based case library: a case study in diabetes therapy revision. In: Smyth, B., Cunningham, P. (eds.) EWCBR 1998. LNCS (LNAI), vol. 1488, pp. 64–75. Springer, Heidelberg (1998)
7. Bichindaritz, I.: Case-based reasoning adaptive to several cognitive tasks. In: Aamodt, A., Veloso, M.M. (eds.) ICCBR 1995. LNCS, vol. 1010, pp. 391–400. Springer, Heidelberg (1995)
8. Turner, R.: Organizing and Using Schematic Knowledge for Medical Diagnosis. In: Kolodner, J. (ed.) Proc of a Workshop on CBR, Florida, pp. 435–446 (1988)
9. Perner, P.: A Comparative Study of Catalogue-Based Classification. In: Roth-Berghofer, T.R., Göker, M.H., Güvenir, H.A. (eds.) ECCBR 2006. LNCS (LNAI), vol. 4106, pp. 301–308. Springer, Heidelberg (2006)
10. Clancey, W.J.: Heuristic Classification. Artificial Intelligence 27, 289–350 (1985)
11. Selz, O.: Über die Gesetze des geordneten Denkverlaufs. Stuttgart (1913)
12. Perner, P.: Are case-based reasoning and dissimilarity-based classification two sides of the same coin? Journal Engineering Applications of Artificial Intelligence 15/2, 205–216 (2004)
13. Schmidt, R., Waligora, T., Vorobieva, O.: Prototypes and Case-Based Reasoning for Medical Applications. In: Perner, P. (ed.) Case-Based Reasoning on Images and Signals, pp. 285–317. Springer, Heidelberg (2008)
14. Taybi, H., Lachman, R.S.: Radiology of Syndromes, Metabolic Disorders, and Skeletal Dysplasia. Year Book Medical Publishers, Chicago (1990)
15. Gierl, L., Stengel-Rutkowski, S.: Integrating Consultation and Semi-automatic Knowledge Acquisition in a Prototype-based Architecture: Experiences with Dysmorphic Syndromes. Artificial Intelligence in Medicine 6, 29–49 (1994)
16. Clinical Dysmorphology, http://www.clindysmorphol.com
17. Winter, R.M., Baraitser, M., Douglas, J.M.: A computerised data base for the diagnosis of rare dysmorphic syndromes. Journal medical genetics 21(2), 121–123 (1984)
18. Tversky, A.: Features of Similarity. Psychological Review 84(4), 327–352 (1977)
19. Rosch, E., Mervis, C.B.: Family Resemblance: Studies in the Internal Structures of Categories. Cognitive Psychology 7, 573–605 (1975)

20. Wenkebach, U., Pollwein, B., Finsterer, U.: Visualization of large datasets in intensive care. In: Proc. Annu. Symp. Comput. Appl. Med. Care, pp. 18–22 (1992)
21. Schmidt, R., Pollwein, B., Gierl, L.: Medical multiparametric time course prognoses applied to kidney function assessments. International Journal in Medical Informatics 53(2-3), 253–264 (1999)
22. Shahar, Y.: Timing is Everything: Temporal Reasoning and Temporal Data Maintenance in Medicine. In: Horn, W., Shahar, Y., Lindberg, G., Andreassen, S., Wyatt, J. (eds.) AIMDM 1999. LNCS (LNAI), vol. 1620, pp. 30–46. Springer, Heidelberg (1999)
23. Haimowitz, I.J., Kohane, I.S.: Automated trend detection with alternate temporal hypotheses. In: Proc. of the 13th International Joint Conference on Artificial Intelligence, pp. 146–151. Morgan Kaufmann Publishers, San Mateo (1993)
24. DeSarbo, W.S., et al.: TSCALE: A new multidemensional scaling procedure based on Tversky's contrast model. Psychometrika 57, 43–69 (1992)
25. Schmidt, R., Vorobieva, O.: Case-Based Reasoning Investigation of Therapy Inefficacy. Knowledge-Based Systems 19(5), 333–340 (2006)
26. DeGroot, L.J.: Thyroid Physiology and Hypothyroidsm. In: Besser, G.M., Turner, M. (eds.) Clinical endocrinilogy, ch. 15. Wolfe, London (1994)
27. Bichindaritz, I.: Prototypical Cases for Knowledge Maintenance for Biomedical CBR. In: Weber, R.O., Richter, M.M. (eds.) ICCBR 2007. LNCS (LNAI), vol. 4626, pp. 492–506. Springer, Heidelberg (2007)

Hopfield Networks in Relevance and Redundancy Feature Selection Applied to Classification of Biomedical High-Resolution Micro-CT Images*

Benjamin Auffarth**, Maite López, and Jesús Cerquides

Volume Visualization and Artificial Intelligence research group,
Departament de Matemàtica Aplicada i Anàlisi (MAIA), Universitat de Barcelona,
C/Gran Via, 585, 08007 Barcelona, Spain
{benjamin,maite,jcerquide}@maia.ub.es

Abstract. We study filter–based feature selection methods for classification of biomedical images. For feature selection, we use two filters — a relevance filter which measures usefulness of individual features for target prediction, and a redundancy filter, which measures similarity between features. As selection method that combines relevance and redundancy we try out a Hopfield network. We experimentally compare selection methods, running unitary redundancy and relevance filters, against a greedy algorithm with redundancy thresholds [9], the min-redundancy max-relevance integration [8,23,36], and our Hopfield network selection. We conclude that on the whole, Hopfield selection was one of the most successful methods, outperforming min-redundancy max-relevance when more features are selected.

Keywords: feature selection, image features, pattern classification.

1 Introduction

Computerized Tomography (CT) is a technique of producing a 3-dimensional image from a large series of X-ray images taken around a single axis of rotation. The 3-D image is cut into sections (Greek tomos = cutting), so the data is an array of 2-D images which together constitute a volume. The information of density of cell tissue is given in gray intensity levels.

Volume visualization of such CT–slices can help experts in biomedical analysis, such as e. g. inspection of cell tissue and of anatomical structures, or in gaining a better understanding of cell growth (more general: Computer-Assisted Diagnosis). For this purpose, first, a transfer function maps from possible voxel values to RGBA space, defined by colors and opacity (red, green, blue, alpha).

---

* This research was supported by the Spanish MEC Project "3D Reconstruction, classification and visualization of temporal sequences of bioimplant Micro-CT images" (MAT-2005-07244-C03-03).

** Corresponding author.

P. Perner (Ed.): ICDM 2008, LNAI 5077, pp. 16–31, 2008.

Using volume visualization techniques, 2–dimensional projections on different planes can then be displayed.

The opacity of voxels depends on cell tissue that the voxels represent. Therefore, distinguishing between different tissues can enhance the volume visualization. Hence, we break the transfer function between intensity values and optical properties into two parts: i) classification function, and ii) transfer function from tissue to optical properties.

Classification, in the context of this work, refers to the process of distinguishing between different kinds of data, here biomaterial and non-biomaterial, and the result of this process. Our data consist of slices in a 3-D volume taken from CT of bones, in which was artificially introduced a biomaterial for tracing purposes[1]. The introduced biomaterial is the target class and relatively small as compared to the non-target class. The data set originally was of dimensions $423 \times 486 \times 562$, but because 403 slices did not contain any biomaterial, in the current study, dimensionality was reduced to $423 \times 486 \times 159$.

Earlier, working on these data, we introduced a pipeline process for classification and subsequent volume rendering [22]. In classification, instead of a unique single-run classifier, as in most approaches, we applied a learning pipeline consisting of three steps. After initial Gentle Boost [13] classification based on image properties, a conditional random field [20] on an image of reduced scale works on spatial characteristics of uncertain pixels (output of Gentle Boost), and finally we refined the result in a last step. This article aims at extending the framework with a feature selection step.

For organic tissues, as in our case, distributions of intensities overlap considerably [33]. In order to produce a reliable classification model, we extracted characteristics (features) – a process called "feature extraction" (see 4.1) – from images by integrating image intensity within a window around each pixel. With high number of features, classifiers become slow and tend to produce unstable models with low generalization performance, so our problem was then selecting from a number of features, compiling a set of features that would give good performance.

Each of the extracted features can have merits on its own and merits when used in combination with a selection of other features and we did not know beforehand, which of the features to use. "Feature selection" refers to methods dedicated to finding a set of features that together can be more successful than others. Feature selection within the context of pattern classification will be the focus of most of this work.

The outline of this article is as follows: First, in section 2, we explain the concepts of relevance and redundancy filters, briefly survey related research on feature selection, and line out two heuristics for combination of the information from the two filters. In section 3 we present a novel method that uses a Hopfield network with the idea of taking into account more complex redundancy relations than other methods. In section 4 we describe our experimental benchmarks of

---

[1] Samples from the data set are available on one of the authors' homepage: `http://www.maia.ub.es/~maite/out-slice-250-299.arff`

several feature selection methods, and interpret conclusions based on the results regarding the best method for feature selection, quality of selection, and finally the best features. Lastly, we draw conclusions in section 5 and outline future work in section 6.

2   Relevance and Redundancy Feature Selection

Feature selection in biomedical research is still often done manually by experts, however due to great quantities of data it is becoming increasingly automatized. A comparison of methods over articles by different authors is difficult, because of incompatible performance indicators, often unknown significance, and different data sets methods are applied to. Saeys et al. [28] review research in feature selection in application to biological data.

Sets of features can be evaluated by either filters, which measure statistical properties or information content, or a performance score of a classifier ("wrapper approach"). There exist many heuristics for choosing subsets of features. Two standard iterative search strategies are forward selection and backward selection. Forward selection, starts from the empty set and adds at each step a feature, which gives the most performance improvement. Backward selection starts from all features, eliminating at each iteration one or several features. Forward-backward algorithms make an initial guess of a useful feature set and then refine the guess by eliminating variables and adding new ones.

In the context of this work, we define the feature selection task as follows: given a selection criterion (error function) $\varepsilon(\cdot)$ and an initial feature set X with m features we want to find a subset $X^* \subseteq X$ such that $|X^*| = s$ (s for number of selected features) and $X^* = \arg\min_{\bar{X} \subseteq X, |\bar{X}| = s} \varepsilon(\bar{X})$.

Many approaches to feature selection in bioinformatics are either based on ranks ("univariate filter paradigm") and thereby do not take into account relationships between features, or are wrapper approaches which require high computational costs. We chose a filter-based feature selection approach for being fast and giving good results, which other computation-heavy methods are not guaranteed to achieve (cf. [16]). Filter-based have the additional advantage of providing a clearer picture of why a certain feature subset is chosen through the use of scoring methods in which inherent characteristics of the selected set of variables is optimized. This is contrary to wrapper-based approaches which treat selection as a "black-box" optimizing the prediction ability according to a chosen classifier.

Multivariate filter-based feature selection with the idea to have a set of features of maximal relevance to the target, which are least redundant has enjoyed increased popularity [28]. It has been shown that the best subset of features may not be the set of the best individual features (e. g. [6]). The idea behind combining redundancy and relevance information is simple: you should take the features that together have the highest value for prediction and not the ones which alone have highest prediction value.

Relevance criteria determine how well a variable discriminates between the classes. They are a measure between a feature and the class.

Redundancy criteria should capture similarities of mappings from attributes to classes, i.e. given a predictor function $f \in F : \mathbb{R} \to C$ then our intuition is that for two non-redundant features X_k and X_l, $f(X_k)$ should be different to $f(X_l)$ (and hopefully provide complementary information). Formally the redundancy between features X_1 and X_2 given class targets $Y \in C^n = \{c_1, \ldots, |C|\}^n$ can be written as

$$\mathrm{Red}(X_1, X_2, Y) = \frac{1}{|C|} \sum_{i=1}^{|C|} \Delta (X_1|Y = c_i, X_2|Y = c_i), \tag{1}$$

where $X_1|Y = c_i$ denotes the distribution of feature 1, given class i (i. e. $\{X_1^l|\forall l, Y^l = c_i\}$, and Δ one of the distributional similarity measures that we applied. Given a relevance measure $Rel()$, features X_1 and X_2, and targets $Y \in C^n$, we can define $\mathrm{Red}(X_1, X_2, Y) = \frac{1}{|C|} \sum_{\forall i \in [1,|C|]} \mathrm{Rel}(X_1|c_i, X_2|c_i)$.

Ding, Peng, et al. [8,23,36] select features in a framework they call "min-redundancy max-relevance" (here short: mRmR) that integrates relevance and redundancy information of each variable into a single scoring mechanism to automatically annotate the fruitfly's embryonic tissue.

Knijnenburg [19] presented a cluster-based approach where variables are first hierarchically (complete linkage) clustered and then from each cluster the most relevant feature is selected. Relevance and redundancy were measured by Pearson Correlation Coefficients. He concluded that cluster-based selection could not improve upon greedy ranking-based selection, but a second approach that integrated relevance and redundancy into a single score (in a way similar to mRmR [8]) did so.

Duch et al. [9] presented an algorithm that proceeds at each step including variables starting from highest relevance and excluding variables that are redundant. Their heuristics is simple, straightforward, and seemed to work.

In the next subsection we will describe the mRmR approach and thereafter describe Duch and Biesiada's [9] greedy heuristics with threshold.

2.1   Minimum Redundancy Maximum Relevance

Ding, Peng et al. [8,23,36] presented minimum redundancy maximum relevance feature selection. The method boils down to a forward scheme[2] maximizing one of two formulas for combination of redundancy and relevance information (mutual information in both cases) by subtraction and division, respectively. These formulas are:

– $\arg \max_i \mathrm{rel}(i, c) - \frac{\sum_j \mathrm{red}(i,j)}{m}$, with i and j being two features, c the matching target, and m the numbers of competing features at each step

---

[2] Peng et al. [23] also discuss and test a backward scheme but it is given less importance than the forward scheme.

– $\arg \max_i \dfrac{\text{rel}(i,c)}{\frac{\sum_j \text{red}(i,j)}{m}}$

Peng et al. [23] use mutual information as measure for relevance and redundancy and they refer to the first formula as mutual information difference (MID), to the second as mutual information quotient (MIQ). We will refer to (dropping the reference to mutual information) $mRmRD$ and $mRmRQ$, respectively. We implemented the mRmR forward search and integrated it with our redundancy and relevance methods. The algorithm works as lined out in algorithm 1. $best()$ is the selection formula, i.e. either quotient or difference. Features are $X_i, i \in [1, \dots, m]$.

Algorithm 1. mRmR feature selection

Input: rel $\in \mathbb{R}^m$: relevance scores ;
red $\in \mathbb{R}^{m^2}$ redundancy scores ;
s: number of features that need to be selected (assumed $k \geq 1$)
Initialize set $D = \{X_1, \dots, X_m\}$;
for $i \leftarrow 1; i \leq s;\ i{+}{+}$ **do**
    $S_i \leftarrow best(D)$;
    $D \leftarrow D \setminus S(i)$;
end
Output: S: s features ordered by mRmR

2.2 Greedy Algorithm with Redundancy Threshold

Duch et al. [9] in their forward scheme used the Kolmogorov–Smirnov test for measuring redundancy and set the cut–off threshold to a p-value> 0.95.

Algorithm 2 lines out the workings of the implemented algorithm. The Greedy selection scheme could be extended to select a specific number of variables, however this would mean giving up on the strict thresholding and to introduce arbitrariness into feature selection.

Algorithm 2. Greedy feature selection algorithm with Thresholding

Input: m: number of features, ;
rel $\in \mathbb{R}^m$: relevance scores, ;
red $\in \mathbb{R}^{m^2}$: redundancy scores, ϵ: threshold
Initialize sets: set $S \leftarrow \emptyset$, and $C \leftarrow \{c_1, \dots, c_m\} = \{1, \dots, m\}$;
while $|C| > 0$ **do**
    $S \leftarrow S \cup \arg_i \max \text{rel}_{c_i}$;
    $C \leftarrow C \setminus \{i\}$;
    $C \leftarrow C \setminus \{j | \exists s_i \in S, \text{red}_{s_i, c_j} \geq \epsilon\}$;
end
Output: Selected features are in S

3   Hopfield Network for Relevance and Redundancy Feature Selection

The spaces of feature combinations and the corresponding space of their energy or error functions has numerous local optima, which iterative algorithms intrinsically have difficulties dealing with. This inspired us to think of other graph based methods as a manner of partitioning and selecting the best features.

We can form complete graphs from the redundancy matrices, if we think of them as a proximity matrices $D = \mathbb{R}^{m^2}$, where m is the number of variables. We had the idea of the features as nodes, and to add as additional dimension relevance, redundancy constituting inhibitory connections between the features.

The lateral connections represented by the redundancies could create an attractor network that forms basins of attractions, where redundancies are lowest and relevancies are highest. In this manner, the choice of features could come up as an emergent pattern within the configuration space of the network arising from the connections and activations.

A recurrent attractor network with well-studied convergences is a Hopfield network[17,18,31]. A Hopfield network has the advantage of being able of generating arbitrary shapes and providing insight into the number of variables without prior knowledge.

In the simplest form of the Hopfield network, we formalize connections (having an appropriate normalization) as symmetric, real-valued connections w_{ij}, units $S_i \in [0,1]$ and corresponding bias units I_i. The input to each unit S is

$$n_i \leftarrow \sum_j w_{ij} S_j + I_i \ , \tag{2}$$

where I_i is a bias term of unit i.

In the classical (bipolar) formalization, nodes can be asynchronously (serially) updated at each time step t:

$$S_i(t+1) \leftarrow \begin{cases} 1 & \text{if } n_i(t) > 0 \\ 0 & \text{if } n_i(t) < 0 \\ S_i(t) & \text{otherwise} \end{cases} \tag{3}$$

The energy function of such a network is

$$E = -\frac{1}{2} \sum_{ij} S_i S_j - \sum_i I_i S_i + \sum_i \int_0^{S_i} g_\lambda^{-1}(S) dS, \tag{4}$$

where g_λ is a sigmoidal funcion, often the sigmoid function $\equiv \frac{1}{1+e^{-\lambda x}}$, and λ is its gain, which guarantees the convergence to a continuous local minimum of the energy function over time and the synchronization of clusters of units.

We tried many different parameters, normalizations of activations and connections. Parameters included annealing with different rates, including using e.g. Rprop[25]. Finally, we chose a simple implementation for continuous (graded) activation (and responses) and asynchronous updating in discrete time-steps[3] with the hyperbolic tangent $\tanh x = \frac{\sinh x}{\cosh x}$ as our sigmoidal function. Weights and activations were normalized in the range $[-1, 0]$ and $[0, 1]$ respectively, with the diagonal of the weight matrix set to 0. We set the noise parameter $u_0 = 0.015$ (cf. [18]) and we fixed the learning rate $\lambda = 0.1$.

The update of the activation S of a neuron i at time step t is then

$$S_i(t + 1) \leftarrow (1 - \lambda)\, S_i(t) + \lambda \left(1 + \tanh\left(\frac{u_i}{u_0} \right) /2 \right), \text{ where} \tag{5}$$

$$u_i = \sum_j w_{ij} \times S_j(t). \tag{6}$$

For application to feature selection, at the end we choose the most highly activated units, thresholding the unit activations:

$$S_i \leftarrow \begin{cases} 1 \text{ if } \sum_j w_{ij} a_j > \theta, \\ 0 \quad \text{otherwise.} \end{cases} \tag{7}$$

In the coming section (4) we will submit the presented feature selection schemes to a testing procedure using information of relevance and redundancies (all combinations). Afterwards we will present results and compare methods.

4   Experiments and Results

We conducted experiments in order to find out which selection schemes and which relevance and redundancy measures perform best. For the experiments we need to extract a set of features from the images and compute measures of redundancy and relevance. After describing methods corresponding to these, we come to our experimental design, and methods of statistical validation. After this we look at results.

4.1   Feature Extraction

A standard method for compact image encoding is a method called Laplacian pyramids [3]. For their computation, an image is iteratively smoothed by computing averages in constant windows as low-pass filters. The bottom level of this representation (g_0) is the original image. The Laplacian pyramid is then the sequence of difference maps between two levels at the pyramid $L_n = g_n - g_{n+1}$, for $0 \leq n < N$, with N denoting the number of levels in the smoothed pyramid.

Gabor filters (e.g. [12] have received considerable attention because the characteristics of simple cells in the primary visual cortex of some mammals can be

---

[3] Our attempts at converging at a good implementation were streamlined considerably by Hervé Abdi [1].

approximated by these filters. They are used a lot in pattern recognition and texture segmentation. Gabor filters, in contrast to the other features presented here, incorporate orientation information.

In eye-tracking studies, Reinagel and Zador [24] gave evidence for increased luminance contrast in fixated regions as compared to control points (fixated points on different images). They defined luminance contrast (LC) as the variance of luminance within a patch (a rectangular patch for practical purposes) divided by the mean intensity of the image. Given a patch P of pixel intensity from image I and around a pixel (x, y): $LC_P = \frac{\delta_P}{\mu_P}$. Another texture function from neuropsychological research is texture contrast [10]. The texture contrast (TC) of a patch is the standard deviation of the luminance contrast values in the patch standardized by the luminance contrast mean of the image. More formally, given a patch $\bar{P}$ from LC_I: $TC_{\bar{P}} = \frac{\delta_P}{\mu_P} = LC_{\bar{P}}$.

We extracted 10 features from the Laplacian Pyramid, 100 Gabor features (10 orientations at 10 scales), 9 features from luminance contrast, 7 features from texture contrast, and intensity. We added 50 probes which have a function in performance assessment; a good feature selection method should eliminate most of these probes. 25 probes were standard normal distributed, 24 uniformly distributed in the interval $(0, 1)$. The last probe was a variable of zeros.

4.2 Relevance and Redundancy Criteria

Due to the uncertainty of the true distribution underlying data, we prefer non-parametric and model-free metrics. Non-parametric tests have less power (i. e. the probability that they reject the null hypothesis is smaller) but are more robust to outliers than parametric tests.

The four relevance criteria that we used in experiments are: Symmetric Uncertainty (SU), Spearman Rank Correlation Coefficient (CC), Value Difference Metric (VDM), and Fit Criterion (FC). In [2], we showed how a measure of probability difference, presented before as the "value difference metric" [29], can be adapted as a relevance criterion. We also use a measure, which we call "fit criterion", presented in [2].

As for redundancy criteria, we used seven measures: Kolmogorov-Smirnov test on class-conditional distributions (KSC), Kolmogorov-Smirnov test ignoring classes (KSD), Value Difference Metric adapted to redundancy (RVDM) [2], Redundancy Fit Criterion (RFC) [2], Spearman Rank Correlation Coefficients (CC), Jensen-Shannon Divergence (JS), and the Sign-test (ST).

As for discretization, we use histograms. Conforming to Cromwell's rule of avoiding probabilities of 1 and 0 (except for logical true and false), we apply the Laplacian rule of succession by calculating the probabilities of bin i with frequency count n_i as $\tilde{p}(i) = \frac{n_i + 1}{k + \sum_{j=1}^{k} n_j}$. In order to avoid any problems with optimization of a bandwidth or bin number and because of impracticality of mixture modeling, we chose a rigid bin number of 100.

4.3   Experimental Design

We benchmarked first each relevance and redundancy criterion on its own ("unitary filters"), then all 28 combinations of mentioned relevance and redundancy measures with the selection methods mRmRQ, mRmRD, Greedy, and Hopfield. For the threshold in Greedy we used all thresholds possible in combination with the redundancy measure. As for unitary filters, for relevance measures, the s highest relevant features were used and for redundancy measures, at each step the most redundant feature with all the remaining features is removed until the desired numbers of features s are left. We also introduced a baseline of random selection.

We selected feature sets of sizes $[4, 8, 12, 16, 20, 30, 45, 60, 80, 100]$. We emphasized feature sets of sizes ≤ 30 because that was were they were the greatest differences between the different methods. The reported experiments and comparisons are based on the set of 177 features and their respective relevance measures and mutual redundancies. We used three classifiers for benchmarking: Naïve Bayes, GentleBoost, and a linear Support Vector Machine.

As for Naïve Bayes we relied on our own implementation for multi-valued attributes using 100 bins for discretization. Given m features $X_1, \ldots, X_m$ and corresponding targets $Y \in C^n$, classifying a pattern $x = \{x_1, \ldots, x_m\}$ by Naïve Bayes means $\mathrm{argmax}_{c \in C} \ p(c) \prod_{i=1}^m p(x_i | c = Y)$. As for GentleBoost we used Antonio Torralba's matlab toolbox [27] and fixed the iterations to 50, which seemed to be a good trade–off between speed and performance. As for SVM [5] we used libsvm 2.84 [4], accessed from within MATLAB using an interface by Michael Vogt [30] from Technical University Darmstadt. We made the cost function to compensate for unequal class priors, by setting the weight of the less frequent class to $\max\left(\frac{\#(Y=c_2)}{\#(Y=c_1)}, \frac{\#(Y=c_1)}{\#(Y=c_2)}\right)$. Further, we set the SVM complexity parameter C to 1 which seemed to be a good choice and in the right order of magnitude. We tried out several normalization methods. Comparisons showed that classification performance being approximately equal, there was a notable loss of speed with classification after normalization according to Graf et al. [15], with z-normalization performing faster than normalization between $[0, 1]$ or $k[-1, 1]$. Consequently, features were z-normalized.

The whole set of experimental conditions can be obtained by combining selection schemes with corresponding relevance and redundancy measures, classifiers, and numbers of features. Greedy, Hopfield, mRmRQ, and mRmRD, were tried out with the 28 redundancy and relevance combinations, all classifier, at each number of features. Unitary filters with their redundancy or relevance measures, were combined with a classifier and a number of features. Random selection ran with each classifier at each number of features. In total we had 3700 experimental conditions.

In order to have many validations at acceptable speed – we made 10 random samplings of size $n/10$ and for each sampling we did 5-fold cross-validation. As for random feature selection, we did 10 random samplings of the data of size $n/10$ and tested 10 random selections of features in 5-fold cross-validation.

4.4   Statistical Evaluation

As performance measure, we used the area under the curve (AUC) throughout the analysis and — following the recommendations of Janez Demšar [7], who surveyed the state of the art of comparing classifiers — we did not base our statistics on performances of single folds but took averages (medians[4]) over folds.

4.5   Results

Statistics were extracted from performance vectors and are given over all three classifiers (Naïve Bayes, GentleBoost, and SVM). For feature selection, what is the "best" method depends on how many features there are, which is the application, and what computational resources are available.

We will focus on three questions:

1. Which is the best feature selection scheme?
   (a) In particular, are there differences with respect to numbers of features?
2. Are the best methods the ones with fewest probes?
3. What is the best feature set?

Question 1 includes feature selection schemes, measures of redundancy and relevance (short: RR measures), and combinations of relevance and redundancy. Apart from an overall winner according to our experimental setup, we will look at which selection scheme gives the best results. We will have to look whether there are differences between the methods with respect to RR measures.

As for question 1.a, we will analyze, if relative performance of the different methods is the same when the number of attributes selected increases. We will have to decide which is a good feature size for our classification task and with respect to this decision decide on the best selection scheme.

As for question 2, we look at probe frequency and see whether a selection scheme with good performance is automatically one with few probes.

Question 3 deals with the final result of our feature selection: which are the best features for our classification task?

In table 1, the first column gives the name of the method (the selection scheme followed by redundancy and relevance measures), ordering (column two) follows the mean rank of performance (third column), win–loss statistics (W/L) from statistical tests based on ranks at all feature numbers respectively show

---

[4] According to the central limit theorem, any sum (such as e. g. a performance benchmark), if finite variance, of many independent identically distributed random features will converge to a Gaussian distribution. This is however not necessarily to expect for only 5 values, i. e. from 5-folds of cross-validations. After finding partly huge differences between means (which are usually taken) and medians over cross-validations, in pre-trial runs, we decided to take the more robust median (which in case of normal distributions is equal to the arithmetic mean anyway). As for the error-bar, we plot the interquartile range (short: IQR), which is the difference between values at the first (25%) and the third quartile (75%).

Table 1. Ranking of all Selection Schemes

	index	mean rank	F/N W/L	SR W/L
mRmRD	1	2.10	3/0	4/0
Hopfield	2	2.85	2/0	2/0
Red	3	3.50	1/0	2/0
mRmRQ	4	3.70	1/0	1/1
Rel	5	3.95	1/1	1/1
Greedy	6	5.00	1/2	1/3
rand	7	6.90	0/6	0/6

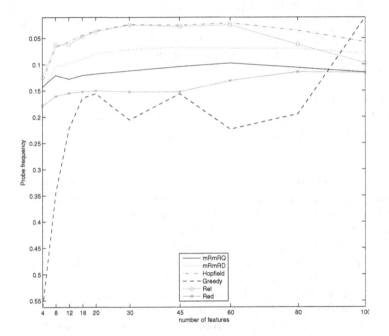

Fig. 1. Normalized Probe Frequencies of all Selection Schemes

in column four and five: Friedman test with Nemenyi post-hoc test (F/N) and Wilcoxon Signed Rank Test (SR, also called the *Mann-Whitney U test*).[5]

According to table 1, mRmRD is overall winner followed by Hopfield. mRmRQ is by Wilcoxon Signrank worse than mRmRD. Hopfield and unitary redundancy filters are not statistically worse than mRmRD. Random feature selection is clearly (and statistically significantly) worse than all other selection schemes. Greedy is the worst non-random scheme. Unitary redundancy filters come high up in third place.

Fig. 2 shows changes with different numbers of features. Because of the complications with the Greedy scheme, rankings of selection methods at all numbers of features were normalized by the total number of competing methods with their

---

[5] We included the Greedy schemes using a threshold of $\frac{|s_{\text{design}} - s_{\text{Greedy}}|}{s_{\text{design}}} \leq 0.1$.

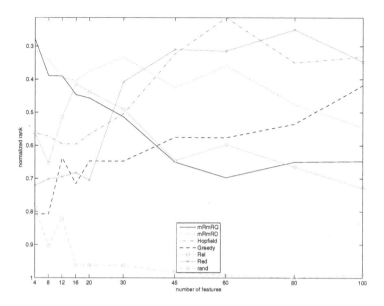

Fig. 2. Normalized Median Rankings of all Selection Schemes

different combinations of methods. The medians for each selection scheme are depicted in fig. 2. We see that with more features all selection schemes become better than random choice because of the inclusion of less probes. MRmRQ starting as best at 4 variables, adding more features improves relatively less than most other selection schemes. Hopfield, unitary redundancy filters, and Greedy see best improvements as compared to other methods as compared to mRmRQ/D and unitary relevance filters. We observe that Hopfield copes better with higher feature spaces than other methods and constitutes one of the best methods from 45 features on.

Fig. 1 shows the expected frequency of probes in the selection of schemes. Frequencies are normalized by numbers of features in the selected set. We see that Greedy, the worst selection scheme, was very resistant to probes, however chosen features could obviously not have been the most useful ones. The same is true for (unitary) redundancy measures. Redundancy and Greedy curves show an increasing probe tolerance (as was to expect), the Greedy curve exhibiting a steep rise from 80 to 100 features. Unitary relevance filters and Hopfield let in most probes as compared to the other measures. mRmRD/Q were in the mid-field.

Over all classifiers, Spearman correlations of normalized ranks and inverse (subtracting from 1) normalized probe frequencies over all RR combinations at each number of features ranged between −0.08 and 0.6. This suggests to us that low probe frequencies are not sufficient for good classifier performance. The correlations follow a curious pattern: they start at medium range with 4 features (0.5), go down (until −0.08) at 45 features, and climb up again. This suggests

Table 2. Best Selection Methods for each Number of Features

#features	selection method	LP	Gabors	LC	TC	Int.	RN	RU	Zeros
4	mRmRQ, CC+CC	4.42	0.89	0.00	0.00	0.00	0.00	0.00	44.25
8	mRmRD. CC+VDM	6.64	1.11	0.00	0.00	0.00	0.00	0.00	0.00
12	mRmRD. CC+VDM	7.38	0.89	0.00	2.11	0.00	0.00	0.00	0.00
16	Redundancy Filter VDM	0.00	1.33	4.92	0.00	0.00	0.00	0.00	0.00
20	Redundancy Filter VDM	0.00	1.33	4.92	0.00	0.00	0.00	0.00	0.00
30	mRmRQ RVDM FC	5.31	1.00	1.97	0.84	0.00	0.00	0.00	0.00
45	mRmRQ, RVDM+FC	3.54	1.14	2.62	0.56	0.00	0.00	0.00	0.00
60	Hopfield. KSC+CC	0.00	1.24	2.95	2.95	2.95	0.00	0.00	2.95
80	Hopfield, KSD+FC	0.89	1.28	2.21	2.21	2.21	0.00	0.00	2.21
100	Hopfield, KSD+FC	1.77	1.27	1.77	1.77	1.77	0.00	0.00	1.77

low probe frequencies being relatively important at low numbers of variables and when there are few to choose from (but not in-between). This explains for higher numbers of features the success of redundancy filters and the revival of Greedy.

As we will see below in table 2, the probe of zeros (feature index 177), enjoyed some popularity. This seems to be a problem that comes from the skewed class-distributions in our data, which makes that 0 can be to 90% associated with one class.

Table 2 lists from 4 to 100 numbers of features (first column) the selection method (second column) that provided the best performing feature set. At the end of the second column we put the redundancy and relevance measures (cf. 4.2). From column 3-10 you see normalized frequencies of features from Laplacian Pyramid (LP), Gabor filters, luminance contrast (LC), texture contrast (TC), intensity (Int.), random normal probes (RN), random uniform probes (RU), and the zero probe. The frequency of each feature type in the selected set was divided by the frequency expected from prior probabilities. E.g. as for Gabor filters, the a-priori probability is $100/177 \approx 0.56$. For 100 features, the expected number of Gabor features is $0.56 \times 100 = 56$. The figures corresponding to each number of features and feature type tell how much the frequencies found for the type exceed or fall behind expectations. E.g. for 100 features from Gabor filters there were 1.27 times more than expected.

Laplacian Pyramids, intensity, texture contrast, and luminance contrast appear prominently (relative to their proportion in the feature set). There are many Gabor filters present. It is remarkable that only few probes are selected (however the zeros each time).

5   Conclusions

In this paper, we presented a new method for feature selection based on redundancy and relevance of features. Approaches that use neural networks for feature selection (e.g. [35]) or that use feature selection before feeding data into neural networks exist in manifold (e.g. [32]. However, to our knowledge, a non-supervised neural network has not been used before for feature selection in the minimal redundancy and maximal relevance framework. As a recurrent artificial neural network attractor model, the Hopfield network [17,18], shares phenomenology with the associative memory function of the cortex. Similarly in both, if several patterns are presented simultaneously, a rivalry process leads to competition, from which stable states result in perceptual groupings. In the brain, this process possibly functions by synchronization of neural cell oscillation [14].

On the whole, for all the tested features we saw that mRmRD was the best combination scheme. Curiously, the selections with least numbers of probes are not necessarily the best ones. We observed a log-shaped performance curve over number of features, unsaturated until 100 for some methods. Therefore we decided that the best selection came from the best method at 100 features, a Hopfield network using the Kolmogorov-Smirnov test as redundancy measure and FC relevance. Selection with Hopfield networks showed improvements for higher-dimensioned feature sets.

Not explained yet, but what deserves mention is that the SVM classifier was the best classifier, followed by GentleBoost and Naïve Bayes. Performance estimations obtained could be optimistically biased, because we used only one data set for estimation and the methods were partly chosen for the expected aptness in the domain.

6 Future Work

We identify four lines of future work. These concern relevance and redundancy measures, discretization methods, extension to multi-class classification, and more numbers of features.

Though not reported in this article, due to space limitations[6], we obtained some interesting results comparing different relevance and redundancy measures. As for redundancy, Jensen–Shannon Divergence, was best, followed by VDM, and the sign–test. As for relevance measures, VDM and FC were good. Some measures had difficulties with the zero–probe and this should be taken care of. Future research could try out other relevance and redundancy measures and try them out on several data sets.

Liu et al. [21] systematize and test several methods of discretization and again found the minimum description length best performing. Ding, Peng et al. [23] chose to discretize data using $mean \pm \alpha\sigma$, with $\alpha \in [0, 2, 0.5]$. We have not tried out this method, neither did we try out minimum description length.

Although this study was limited to binary classification, its methods are not and it remains to be seen how our feature selection scales up from a two-class problem to multi-class domains with thousands of features.

In our experimental design we emphasized few numbers of features (70% below 50), which turned out favorably for mRmRD. The Hopfield network selection seems to perform well for high-dimensional feature spaces and could be used in analysis of complex data. It stands out to test the methods in higher dimensional feature spaces. Studies in this direction for the NIPS feature selection challenge are in preparation.

---

[6] For relevance and redundancy measures check [2] for details.

References

1. Abdi, H.: Les reseaux de neurones. Presses Universitaires de Grenoble (1994)
2. Auffarth, B.: Classification of biomedical high-resolution micro-ct images for direct volume rendering. Master's thesis, University of Barcelona, Barcelona, Spain (2007)
3. Burt, P.J., Adelson, E.H.: The laplacian pyramid as a compact image code. IEEE Trans. Communications 31, 532–540 (1983)
4. Chang, C.-C., Lin, C.-J.: LIBSVM: a library for support vector machines (2001), http://www.csie.ntu.edu.tw/~cjlin/libsvm
5. Cortes, C., Vapnik, V.: Support-vector network. Machine Learning 20, 273–297 (1995)
6. Cover, T.M.: The best two independent measurements are not the two best. IEEE Transactions on Systems, Man, and Cybernetics 4, 116–117 (1974)
7. Demsar, J.: Statistical comparisons of classifiers over multiple data sets. Journal of Machine Learning Research 7, 1–30 (2006)
8. Ding, C., Peng, H.: Minimum redundancy feature selection from microarray gene expression data. In: Second IEEE Computational Systems Bioinformatics Conference, pp. 523–529 (2003)
9. Duch, W., Biesiada, J.: Feature selection for high-dimensional data: A kolmogorov-smirnov correlation-based filter solution. In: Kurzynski, M., Puchala, E., Wozniak, M., Zolnierek, A. (eds.) Advances in Soft Computing, pp. 95–104. Springer, Heidelberg (2005)
10. Einhäuser, W., Kruse, W., Hoffman, K.-P., König, P.: Differences of monkey and human overt attention under natural conditions. Vision Research 46(8-9), 1194–1209 (2006)
11. Fawcett, T.: Roc graphs: Notes and practical considerations for researchers. technical report, HP Laboratories, Palo Alto (2004)
12. Fogel, I., Sagi, D.: Gabor filters as texture discriminator. BioCyber 61, 102–113 (1989)
13. Friedman, J., Hastie, T., Tibshirani, R.: Additive logistic regression: a statistical view of boosting, Tech. report, Department of Statistics, Stanford University (1998)
14. Fries, P., Reynolds, J.H., Rorie, A.E., Desimone, R.: Modulation of oscillatory neuronal synchronization by selective visual attention. Science 291, 1560–1563 (2001)
15. Graf, A.B.A., Borer, S.: Normalization in support vector machines. In: Radig, B., Florczyk, S. (eds.) DAGM 2001. LNCS, vol. 2191, pp. 277–282. Springer, Heidelberg (2001)
16. Guyon, I., Gunn, S., Ben-Hur, A., Dror, G.: Result analysis of the NIPS feature selection challenge, vol. 17, pp. 545–552. MIT Press, Cambridge (2004)
17. Hopfield, J.J.: Neural networks and physical systems with emergent collective computational abilities. In: Proceedings of the National Academy of Science, vol. 79, pp. 2554–2558 (1982)
18. Hopfield, J.J.: Neurons with graded responses have collective computational properties like those of two-state neurons. Proceedings of the National Academy of Sciences 81, 3088–3092 (1984)
19. Knijnenburg, T.A.: Selecting relevant and non-relevant features in microarray classification applications. Master's thesis, Delft Technical University, Faculty of Electrical Engineering, 2628 CD Delft (2004)
20. Lafferty, J., McCallum, A., Pereira, F.: Conditional random fields: Probabilistic models for segmenting and labeling sequence data. In: Proc. 18th International Conf. on Machine Learning, pp. 282–289. Morgan Kaufmann, San Francisco (2001)

21. Liu, H., Hussain, F., Tan, C.L., Dash, M.: Discretization: An enabling technique. Data Min. Knowl. Discov. 6(4), 393–423 (2002)
22. López-Sánchez, M., Cerquides, J., Masip, D., Puig, A.: Classification of biomedical high-resolution micro-ct images for direct volume rendering. In: Proceedings of IASTED International Conference on Artificial Intelligence and Applications (AIA 2007), Austria, pp. 341–346. IASTED (2007)
23. Peng, H., Long, F., Ding, C.: Feature selection based on mutual information: criteria of max-dependency, max-relevance, and min-redundancy. IEEE Transactions on Pattern Analysis and Machine Intelligence 27(8), 1226–1238 (2005)
24. Reinagel, P., Zador, A.: Natural scene statistics at center of gaze. Network: Comp. Neural Syst. 10, 341–350 (1999)
25. Riedmiller, M., Braun, H.: Rprop – description and implementation details. Technical report, Universitat Karlsruhe (1994)
26. Rossi, F., Lendasse, A., François, D., Wertz, V., Verleysen, M.: Mutual information for the selection of relevant variables in spectrometric nonlinear modelling. Chemometrics and Intelligent Laboratory Systems 2(80), 215–226 (2006)
27. Russell, B.C., Torralba, A., Murphy, K.P., Freeman, W.T.: Labelme: a database and web-based tool for image annotation. MIT AI Lab Memo AIM-2005-025, MIT CSAIL (September 2005)
28. Saeys, Y., Inza, I., Larrañaga, P.: A review of feature selection techniques in bioinformatics. Bioinformatics (August 24, 2007)
29. Stanfill, C., Waltz, D.: Toward memory-based reasoning. Communications of the ACM 29(12), 1213–1228 (1986)
30. Vogt, M. (accessed 9-October-2007), http://pc228.rt.e-technik.tu-darmstadt.de/~vogt/de/software.html
31. Tank, D.W., Hopfield, J.J.: Simple "neural" optimization networks: An a/d converter, signal decision circuit, and a linear programming circuit. ieeetcas 33, 533–541 (1986)
32. Valenzuela, O., Rojas, I., Herrera, L.J., Guillén, A., Rojas, F., Marquez, L., Pasadas, M.: Feature selection using mutual information and neural networks. Monografias del Seminario Matematico Garcia de Galdeano 33, 331–340 (2006)
33. Xu, D., Lee, J., Raicu, D.S., Furst, J.D., Channin, D.: Texture classification of normal tissues in computed tomography. In: The 2005 Annual Meeting of the Society for Computer Research (2005)
34. Yu, L., Liu, H.: Feature selection for high-dimensional data: A fast correlation-based filter solution. In: ICML, pp. 856–863 (2003)
35. Yu, L., Liu, H.: Redundancy based feature selection for microarray data. In: ACM SIGKDD 2004, pp. 737–742 (2004)
36. Zhou, J., Peng, H.: Automatic recognition and annotation of gene expression patterns of fly embryos. Bioinformatics 23(5), 589–596 (2007)

Modelling Medical Time Series Using Grammar-Guided Genetic Programming

Fernando Alonso, Loïc Martínez, Aurora Pérez, Agustín Santamaría,
and Juan Pedro Valente

Facultad de Informática. Universidad Politécnica de Madrid. Campus de Montegancedo.
28660 Boadilla del Monte. Madrid. Spain
{falonso,loic,aurora,jpvalente}@fi.upm.es
Agustin.Santamaria@Sun.com

Abstract. The analysis of time series is extremely important in the field of medicine, because this is the format of many medical data types. Most of the approaches that address this problem are based on numerical algorithms that calculate distances, clusters, reference models, etc. However, a symbolic rather than numerical analysis is sometimes needed to search for the characteristics of time series. Symbolic information helps users to efficiently analyse and compare time series in the same or in a similar way as a domain expert would. This paper describes the definition of the symbolic domain, the process of converting numerical into symbolic time series and a distance for comparing symbolic temporal sequences. Then, the paper focuses on a method to create the symbolic reference model for a certain population using grammar-guided genetic programming. The work is applied to the isokinetics domain within an application called I4.

Keywords: Time series characterization, isokinetics, symbolic distance, information extraction, reference model, text mining.

1 Introduction

An important domain for the application of time series analysis in the medical field is physiotherapy and, more specifically, muscle function assessment based on isokinetics data.

Isokinetics data is retrieved by an isokinetics machine (Fig. 1a), on which patients perform exercises using any of their joints (knee, elbow, ankle, etc.) at maximum strength. To assure that the patient exercises at constant speed, the machine puts up the required resistance to the strength the patient exerts. As our patients are chiefly sportspeople, we decided to focus on knee exercises (extensions and flexions) since most of the data and knowledge gathered by sports physicians is related to this joint. The data takes the form of a strength curve with additional information on the angle of the knee (Fig. 1b). The positive values of the curve represent extensions (knee angle from 90° to 0°) and the negative values represent flexions (angle from 0° to 90°).

P. Perner (Ed.): ICDM 2008, LNAI 5077, pp. 32–46, 2008.

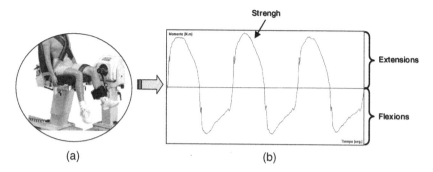

Fig. 1. Isokinetics machine (a) and collected data (b)

After observing experts at work, we found that they apply their knowledge and expertise to focus on certain sections of the isokinetics curve and ignore others. Therefore, we looked for a way of bringing this machine's output closer to the information sports physicians deal with in their routine work, since they demand a representation related to their own way of thinking and operating. Hence, symbolic series have been used as an option that more closely resembles an expert's conceptual mechanisms.

To do this, our research focused primarily on the design of the symbols extraction method that translates numerical time series into symbolic temporal series. Then, we designed an isokinetics symbolic distance measure to indicate how similar two symbolic time series are [1]. This way, symbolic sequences can be automatically compared to detect similarities, class patients, etc. Finally, we applied data mining (DM) techniques based on grammar-guided genetic programming (GGGP) to create reference models useful for defining population groups.

Section 2 of the paper gives an overview of the I4 (Intelligent Interpretation of Isokinetics Information) system, of which this research is part. Section 3 describes the symbol extraction method. Section 4 explains the symbolic distance measure. Section 5 describes the use of genetic algorithms to create symbolic models from symbolic time sequences. Finally, section 6 presents the research results and section 7 outlines some conclusions, and mentions future lines of research.

2 I4 System Overview

The I4 System provides sports physicians with a set of tools to analyze patient strength data output by an isokinetics machine. It is composed of a data preparation subsystem, a Knowledge-Based System (KBS), a numerical Knowledge Discovery in Databases (nKDD) subsystem, a symbolic Knowledge Discovery in Databases (sKDD) subsystem (objective of this paper), and a Visualization module (Fig. 2). The data preparation subsystem manages the tasks of translating, formatting, cleaning and pre-processing the time series obtained from the isokinetics data. These tasks use expert knowledge and generate a database in which data is homogeneous, consistent and noise free. The KBS module analyses this data to make it easier for novice users and also blind physiotherapists to interpret the isokinetics curves. The nKDD performs

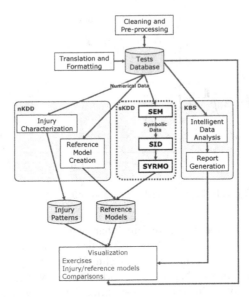

Fig. 2. I4 System Overview

data mining on the numerical isokinetics data to define reference models [2] for patient groups and to identify injury patterns. Finally, the Visualization module displays exercises, injury patterns, reference models, etc.

Many of these functionalities are used on a daily basis by specialized physicians to assess the potential of their patients (mostly top-competition sportsmen and women), diagnose injuries and analyse what progress patients have made in injury recovery. The system is reliable and outputs equivalent results to what an expert would. However, it has failed to gain experts' total confidence. This is because the information the expert receives from the I4 system does not highlight the significant aspects of the isokinetics series in a language that they can easily understand. This has led to the need to build a symbolic Knowledge Discovery in Databases subsystem (sKDD) to solve this problem. The sKDD subsystem contains: a Symbolic Extraction Method (SEM) to extract the symbolic sequence from a numerical series; a Symbolic Isokinetics Distance (SID) module to get a similarity measure between two symbolic isokinetics sequences; and a SYmbolic Reference MOdels (SYRMO) method to create a reference model from a set of isokinetics exercises. The sKDD subsystem should produce results that are equally reliable as the nKDD subsystem, and it should also give a reasonable explanation of the results in terms of the domain under study. This paper focuses on the design of the sKDD.

3   Conversion of Numerical into Symbolic Time Sequences

To be able to develop a symbolic comparison method it is necessary to translate the numerical time sequences output by the isokinetics machine into symbolic time sequences. The first option was to use SAX (Symbolic Aggregate approXimation)

[3,4] to do this translation. This method is able to reduce the temporal sequence dimensionality and assures that the symbolic distance used is less than or equal to distance between the original two sequences. However, the symbols output automatically using the Piecewise Aggregate Approximation (PAA) and the Gaussian distribution do not have expert semantic content, that is, they would not be equivalent to the symbols that the expert identifies when analysing the sequence. To do this it is necessary [1] to find out what symbol alphabet the isokinetics expert uses to analyse the temporal sequences.

After the first few interviews, the expert stated that there were two visually distinguishable regions in every exercise: knee extension and flexion. Both had a similar morphology (the shape shown in Fig. 3), from which we were able to identify the following symbols:

- *Ascent*: part where the patient gradually increases the strength applied.
- *Descent*: part where the patient gradually decreases the strength applied.
- *Peak*: a prominent part in any part of the sequence.
- *Trough*: a depression in any part of the sequence.
- *Curvature*: the upper section of a region.
- *Transition*: the changeover from extension to flexion (or vice versa).

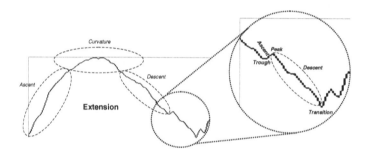

Fig. 3. Symbols of an isokinetics curve

Each isokinetics symbol can have different shapes, that is, different types that are taken into account when translating a numerical temporal sequence into a symbolic series. The types were also elicited from the expert as he analyzed test cases that constituted a significant sample of the whole database. As the expert separated an extension from a flexion, each symbol had to be labelled with its type and also with the keyword "Ext" or "Flex". The set of symbols, types and regions form an alphabet called ISA (Isokinetics Symbols Alphabet), shown in Table 1.

This ISA is used to get symbolic sequences from numerical temporal sequences. The Symbolic Extraction Method (SEM), shown in Fig. 4, was designed to make this transformation. First, a pre-prepared numerical sequence is put through the domain-independent module (DIM), which outputs a set of domain independent features, that is, peaks and troughs. Both the features output by the DIM (peaks and troughs) and the actual numerical sequence data will be used as input for the domain-dependent

Table 1. Isokinetics Symbols Alphabet

Region	Symbol	Types		
EXT	*Ascent*	Sharp		Gentle
	Descent	Sharp		Gentle
	Trough	Big		Small
	Peak	Big		Small
	Curvature	Sharp	Flat	Irregular
FLEX	*Transition*	-		

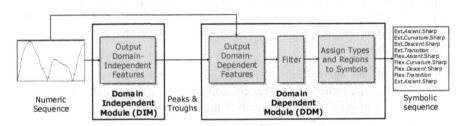

Fig. 4. Architecture of SEM

module (DDM). The DDM outputs all the domain-dependent data of the sequence. This module is divided into three submodules:

- *Output of domain-dependent features.* The aim is to get all the symbols that characterize the given numerical sequence. This module selects the relevant peaks and troughs and identifies the ascents, descents and curvatures.
- *Filter.* The set of symbols output by the above submodule is put through a filtering stage. Apart from other filtering processes, this filter checks that there are no consecutive symbols that are equal. For example, it makes no sense to have two ascents one after the other, because they would really be just one ascent.
- *Assign types to symbols.* The goal of this submodule is to label each symbol with a type. This process is based on a set of rules that use a number of thresholds to define the symbol type in each case.

A graphical interface has been designed to easily work with the SEM (see Fig. 5). An exercise is selected as input to the SEM. The original temporal sequence of the exercise is displayed at the top of the interface. The central part displays the translation of the temporal sequence into symbols, illustrating all the SEM stages. The first stage outputs the domain-independent features, as is shown on the left under the heading "FEATURES". This list contains all the information related to each peak/trough and is formatted as follows:

```
<feature>.  Grad:<gradient_value> Start:<initial_value>
End:<final_value>   Ampl:<amplitude_value>
Dur:<duration_value>    <value_of_the_point>
```

The next stage of the method is to output the domain dependent symbols, which are shown under the heading "DOMAIN-DEPENDENT SYMBOLS". The threshold parameters that are used to output these symbols are listed under "FILTERING PARAMETERS".

The result of the last stage of the SEM is set out on the right side of the interface, under the heading "DOMAIN-DEPENDENT TYPED SYMBOLS". It is the type characterisation of each symbol. The threshold parameters used are shown as "TYPOLOGY PARAMETERS".

The curve reconstructed from the symbols is shown at the bottom.

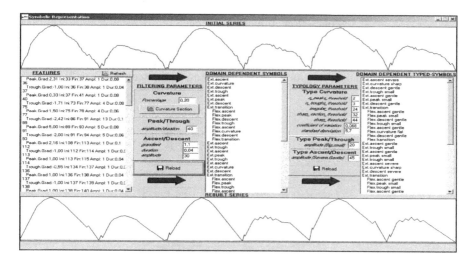

Fig. 5. Symbolic representation interface

As stated by the expert, SEM is an important aid for physicians in writing reports, examining the evolution of an athlete's joint, diagnosing injuries or controlling treatment after a medical diagnosis.

4 Comparison of Isokinetics Symbolic Sequences

Our next goal is to find a similarity measure that can be used to compare symbolic isokinetics sequences and perform data mining tasks.

After a thorough study to select the best similarity measure for the medical field of isokinetics, we reached the conclusion [1] that a new measure needed to be designed. This measure is based on edit distances, which are the best fit for the isokinetics domain, as they take into account the order of the components and the morphology of the sequence. However, none of the distances we examined exactly fits our problem, because the symbols used in the isokinetics domain also have an associated type that needs to be taken into account to calculate the distances.

This led us to propose a variation on the Needleman-Wunch distance [5]. The suggested distance, the Symbolic Isokinetics Distance (SID), allocates a variable cost to the *insert* and *delete* operations depending on the symbol and symbol type to be inserted or deleted. It also allocates a variable cost to the *substitute* operations depending on the symbol and type that are substituted.

Fig. 6 shows the three steps required to calculate the SID of two symbolic sequences: calculate the distance between each pair of subsequences, normalize these distances and calculate the arithmetic mean to get the total distance.

The researched isokinetics sequences are composed of three repetitions, and each repetition is composed of an extension and a flexion. Therefore, an isokinetics sequence contains six parts, each of which is represented by the notation shown in (1).

$$<Zone><Repetition><Sequence> \qquad (1)$$

where <Zone> can take the value E (for Extension) or F (for Flexion), <Repetition> can take the value R^1, R^2 or R^3 depending on whether it is repetition 1, 2 or 3, and<Sequence> can take the value S^1 or S^2 depending on the sequence 1 or 2.

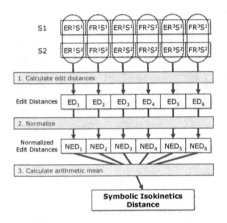

Fig. 6. Computing the Symbolic Isokinetics Distance

The SID between two series, S1, of length n, and S2, of length m, is calculated by building a matrix of mxn elements. This matrix includes the accumulated costs of the *insert, delete* or *substitute* operations, always calculating the best alignment between the two symbolic sequences for comparison. This prevents trapping in local minima. The value of each matrix element is calculated using equation (2): element (i,j) indicates the SID between S1' and S2' (the subsequences —prefixes— of S1 and S2 ending in elements j and i, respectively); element (m,n) indicates the final SID between S1 and S2. This way, the SID can be used to get the least costly edit command sequence (delete, insert and substitute) for transforming S_1 into S_2.

$$D(i,j) = \min \begin{cases} D(i-1, j-1) & \text{if } s_i = t_j & //copy \\ D(i-1, j-1) + SubstituteGapCost & \text{if } si != tj & //substitute \\ D(i-1, j) + InsertGapCost & & //insert \\ D(i, j-1) + DeleteGapCost & & //delete \end{cases} \qquad (2)$$

Due partly to qualitative aspects (each symbol has a different structural weight) and partly to quantitative issues, not all the operations or all the symbols can be allocated an identical gapcost in the isokinetics field. For example, curvatures are symbols that are part of any repetition, whereas peaks and troughs are circumstantial symbols, usually induced by minor patient injuries and, therefore, may or may not appear. Additionally, the presence of a large peak cannot be considered the same as there being a small peak. Therefore, each symbol has to be allocated a different weight, and a distinction has to be made depending on the symbol type.

We had to define both the cost of substituting one symbol type by another and the cost of inserting or deleting a particular symbol type. This was done with the help of an isokinetics expert. The *insert* and *delete* costs were unified to assure that the comparison of two series is symmetric.

For the *substitute* cost, several possibilities were weighed up. Initially, we designed a tabular structure, where the table rows and columns included all the symbol types, and the cell (i,j) represented the cost of substituting the symbol type i by the symbol type j. However, this table was hard work for the expert to build. For instance, $(nxm)^2/2-(nxm)$ values are needed if the number of symbols is n and the mean number of types per symbol is m (the table is symmetric and the cost is 0 along the main diagonal). Additionally, this table is not very open to the entry of any change in the symbols alphabet, as the expert would have to put in a lot of work to reformulate the table to accommodate the changes.

To overcome these two problems, we opted for a graph structure, where the principal cost of substituting two symbols is determined mainly by the symbol, whereas the symbol type serves to further specify that cost. Fig. 7b shows this substitution graph. The expert will have to define $n^2/2-n + nxm$ values, which is clearly fewer than for the table. Additionally, this structure is much more open to the entry of any changes in the symbols alphabet, and it is also more self-explanatory for the expert.

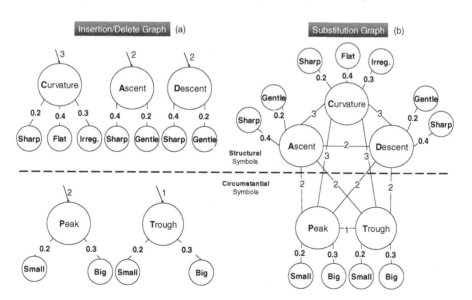

Fig. 7. Insertion/Deletion and Substitution Graph

For the sake of coherency, we have used a similar representation for the *insert* and *delete* costs (Fig. 7a), although, in this case, there is no difference in the number of values that the expert has to define for the graph and for the table.

To make things easier for the expert, we took the graphical representation for each symbol type and defined some initial costs by comparing the area each symbol covered. These initial values were presented to the expert as a starting point.

The gapcosts used in (2) are plotted in the graphs shown in Fig. 7. It is clear from these graphs that there is a cost per symbol to which a cost per type associated with each symbol is added.

The normalization process is applied to the distances between each of the six components of the two sequences for comparison (these distances are denoted ED_X in Fig. 6, where x is the number of the component that has been compared). Then all the distances are defined in the interval [0, 1]. The normalization is based on dividing the obtained distance value by what would have been output in the worst case. In our domain, as all the sequences have six curvatures (two for each repetition), the worst case would be to have *substitute* operations for the curvatures and *substitute* operations for ascents or descent with the worst gapcost.

Once the normalized distances have been obtained for each component, their arithmetic mean is calculated. This outputs the symbolic isokinetics distance between the two compared sequences. This is useful for comparing symbolic time series with reference models to detect injuries or class a sportsperson in a give population group.

5 Using GGGP to Create Reference Models

As part of the research, a method, called SYRMO (SYmbolic Reference MOdels), has been defined to create symbolic reference models from symbolic temporal sequences. A symbolic reference model will be composed of a repetition, that is, an extension and flexion. To do this, we weighed up several alternatives.

The first possibility was to formulate a brute-force algorithm that generated all the possible symbol combinations to form a repetition that would be used as a trial model. Each trial model would be compared with what is currently the best model using the mean of all the symbolic distances of the model with all the existing repetitions. If the mean of the trial model is less than the mean of the existing model, then the model is updated. Evidently, this alternative is not feasible because, apart from producing a combinatorial explosion of symbols, there is no way of generating optimal trial solutions to form a trial model. Trial solutions are generated completely at random, and better and worse trial models are likely to appear in each algorithm run.

Then we weighed up the possibility of using the brute-force algorithm with an improved trial model generation mechanism. This mechanism used the SID algorithm to find the absolute alignments between the symbolic sequences from which the model was to be generated. Our research confirmed that this algorithm was hard to generalize for *n* symbolic sequences and did not assure the correct generation of trial models.

Finally, GGGP was applied to create reference models. GGGP is an extension of genetic programming. Genetic programming (GP) is a means of automatically generating computer programs by employing operations inspired by biological evolution [6]. First, the initial population is generated, and then genetic operators, such as

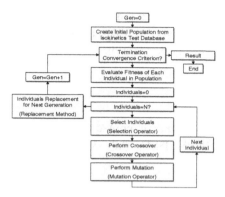

Fig. 8. Genetic Programming Algorithm

selection, crossover, mutation and replacement, are executed to breed a population of trial solutions that improves over time [7].

In this case, the initial population is composed of a selected set of symbolic sequences on which the set of genetic operators should be applied. Fig. 8 shows the genetic algorithm's steps. As the process advances, the individuals, in this case symbolic temporal sequences, are selected, crossed and mutated generation by generation to improve a fitness function, applying a generational replacement method. The algorithm ends when the convergence criteria are met. The result is an individual, the best individual in the last population, which is the model of the initial population.

In GP all algorithms start with the random generation of the initial population, which is composed of individuals that represent possible solutions to the search problem at hand. The main disadvantage of this process is that, being completely random, it can cause the generation of invalid individuals: sequences that are too large or that do not represent possible solutions to the problem. One way to overcome this drawback is to keep on generating individuals, discarding the invalid ones, until the initial population is complete. However, the computational cost of this approach is extremely high for problems requiring large population sizes [6]. The goal of grammar-guided genetic programming is to solve the closure problem [7]. This problem involves always generating valid individuals (points or possible solutions that belong to the search space), which directly affects the initial population-generating algorithm. To solve the closure problem, GGGP employs a context-free grammar (CFG) that establishes a formal definition of the syntactical restrictions of the problem to be solved and its possible solutions. Each of the individuals handled by GGGP is a derivation tree that generates and represents a sentence (solution) belonging to the language defined by the CFG [8].

In this case, the isokinetics language is defined by the IG (Isokinetics Grammar) CFG that uses the ISA alphabet, shown in (3).

Once the IG grammar had been defined, we went on to determine the fitness function according to which the individuals of the population were to evolve. After running several experiments, the chosen fitness function was the mean of an individual's symbolic distances from 80% of its nearest neighbours in the initial population. This individual is the reference model that is representative of the initial population.

Then we went on to select the best set of genetic operators and convergence criteria for the GGGP algorithm to converge as fast as possible. To do this, we had to run experiments on different genetic operators with different convergence criteria to find out how the algorithm converged and what percentage of fitness the fitness function achieved.

$IG = (\sum_N, \sum_T, \text{Exercise}, P)$

$\sum_N = $ {Exercise, Ext, Flex, Climb, Fall, preAsc, preDesc, prePeak, preTrough, postAsc, postDesc, postPeak, postTrough, Curvature, Asc, Desc, Peak, Trough}

$\sum_T = $ {Curv–sharp, Curv–flat, Curv–irregular, Asc–sharp, Asc–gentle, Desc–sharp, Desc–gentle, Peak–big, Peak–small, Trough–big, Trough–small, Transition}

$$P = \{\text{Exercise} \rightarrow \text{Ext Transition Flex}$$

Ext → Climb Curvature Fall

Flex → Climb Curvature Fall

Climb → presAsc Asc postAsc | preAsc Asc | Asc postAsc | Asc

Fall → preDesc Desc postDesc | preDesc Desc | Desc postDesc | Desc

preaAsc → preDesc Desc | prePeak Peak | preTrough Trough | Desc
          | Peak | Trough

preDesc → preAsc Asc | prePeak Peak | preTrough Trough | Asc | Peak
          | Trough   (3)

prePeak → preAsc Asc | preDesc Desc | preTrough Trough | Asc | Desc
          | Trough

preTrough → preAsc Asc | preDesc Desc | prePeak Peak | Asc | Desc
          | Peak

postAsc→ Desc postDesc | Peak postPeak | Trough postTrough | Desc
          | Peak | Trough

postDesc →Asc postAsc | Peak postPeak | Trough postTrough | Asc
          | Peak | Trough

postPeak → Asc postAsc | Desc postDesc | Trough postTrough | Asc
          | Desc | Trough

postTrough → Asc postAsc | Desc postDesc | Peak postPeak | Asc
          | Desc | Peak

Curvature → Curv–sharp | Curv–flat | Curv–irregular

Asc → Asc–sharp | Asc–gentle

Desc → Desc–sharp | Desc–gentle

Peak → Peak–big | Peak–small

Trough → Trough–big | Trough–small}

Table 2 shows the relationship of the genetic operators analysed to determine which ones will generate valid symbolic models. Any operator should output valid individuals, that is, individuals that comply with the grammar defined in (3). Therefore,

Table 2. Genetic Operators

Operator Type	Operator Name	Analysis
Selection	Tournament [9]	After analysing the results of the experiments, these selection operators were considered not to be applicable because they have to create a huge number of generations to achieve the required convergence of the population. The tournament selection operator is the least recommendable in this case, with 500 times more generations than generational selection operator, if it converges. The next least recommendable operator for this domain would be roulette selection, with 21 times more generations, followed by the scaling selection operator, which, although it comes fairly close to the generational selection operator, will still create 4% more generations.
	Roulette [10]	
	Scaling [11]	
	Generational [12]	In this case, instead of being formed at random, the first generation is composed of the population that is to produce the symbolic reference model. Therefore, the generational selection operator is the one that behaves best with respect to the number of generations produced, as the whole population is selected for crossover. This way, all the individuals have a chance of passing on their genetic load to future offspring.
Crossover	Koza [13]	These crossover operators were not applied, because, as they are not based on a CFG, they can generate invalid individuals, and therefore the resulting symbolic reference model would not be a correct reference model.
	SCPC [14]	
	Fair [15, 16]	
	Whigham [17]	Both crossover operators output valid individuals, as they comply with the CFG defined in the GGGP algorithm. The Whigham crossover operator is still now in use, because of its good performance [19, 20, 21]. However, in our experiments, the GBC operator has a 3% better convergence rate, as the GBC prevents the disproportionate growth of the size of the trees representing the individuals and takes advantage of the ambiguous grammar property. This property means that there is more than one different derivation tree for the same word, the key weakness of the Whigham operator [22].
	GBC [18]	
Mutation	Standard [23]	The Standard mutation operator is usually employed by GGGP. It substitutes the subtree whose root is the mutation node for another subtree whose symbol in the root node coincides with the one in the mutation node. This constraint on matching non-terminal symbols has a negative impact on the exploration capacity of the operator when an ambiguous CFG is used. GBM (Grammar Based Mutation) overcomes this weakness.
	GBM [18]	
Replacement	SSGA	De Jong applied the concept of generational replacement rate with the aim of implementing a controlled overlap between parents and offspring [12]. In this paper, a proportion t_{tg} of the population is selected for crossover. The resulting offspring will replace the worst-adapted members of the earlier population. These types of genetic algorithms, where only a few individuals are replaced, are known as SSGA (steady-state replacement genetic algorithms). This is used in the experiments run to replace the individuals of previous population.

the only operators that we were unable to apply because they could generate invalid individuals were: Koza, SCPC and Fair crossover operators.

Having eliminated these operators, we went on to experiment with the others to select the ones that led to the fastest convergence to the best possible model. As a result of the experiments described in Table 2, the best combination of operators was: the Generational selection operator, the GBC crossover operator, the GBM mutation operator and the SSGA replacement operator.

6   Results and Evaluation

SYRMO was evaluated in a two-stage experimentation process. In the first stage, 500 experimental reference models were created with the aim of tuning the algorithm parameters and determining what genetic operators were best, as discussed above.

The goal of the second stage of experimentation was to evaluate the results of SYRMO. To evaluate these results, the numerical method now in use was used to generate 20 reference models. The reference models were created from populations of football, basketball and handball players. Then SYRMO was used to create the symbolic reference models from the same populations as above. When the expert in isokinetics analysed both the symbolic and numerical references models, she found that the two were very alike. In most cases, there is a perfect match between the numerical and symbolic reference models, as illustrated in Fig. 9. This figure shows a numerical reference model, represented by the extension and flexion curve, and the symbolic reference model, shown as dashed lines, which is wholly equivalent to the numerical reference model. However, some differences, which were, from the viewpoint of the isokinetics expert, not significant, were found in 5% of the cases.

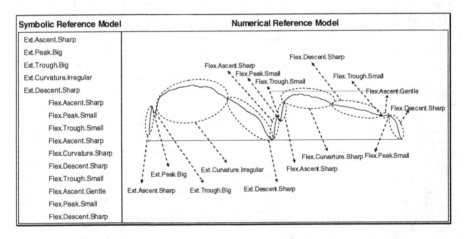

Fig. 9. Numerical Reference Model versus Symbolic Reference Model

7   Conclusions

This paper has presented ongoing work on the development of a comprehensive system to deal with isokinetics data, including symbolic data analysis.

Our earlier experience with numerical methods has been very positive, but experts did not have enough confidence in the system, because the information they received did not highlight the relevant aspects of the isokinetics series in a language they found easy to understand. This is the reason that led us to introduce symbolic methods, which use the same language as our experts.

This paper has focused on the I4 project's sKDD subsystem. sKDD transforms the original numerical temporal sequences into symbolic sequences, defines a symbolic isokinetics distance (SID) that can be used to compare symbolic isokinetics sequences, and provides a method, SYRMO, for creating symbolic isokinetics reference models using grammar-guided genetic programming.

The evaluation has shown that the numerical and symbolic reference models generated from isokinetics tests on top-competition sportsmen and women are, in the expert's opinion, similar. In view of these encouraging results, we are continuing our research in the field of symbolic data analysis to build new functionalities into I4 and add symbolic injury characterization to the sKDD subsystem.

References

1. Alonso, F., Martínez, L., Pérez, A., Santamaría, A., Valente, J.P.: Symbol Extraction Method and Symbolic Distance for Analysing Medical Time Series. In: Maglaveras, N., Chouvarda, I., Koutkias, V., Brause, R. (eds.) ISBMDA 2006. LNCS (LNBI), vol. 4345, pp. 311–322. Springer, Heidelberg (2006)
2. Alonso, F., Valente, J.P., Martínez, L., Montes, C.: Discovering Patterns and Reference Models in the Medical Domain of Isokinetics. In: Zurada, J.M. (ed.) New Generations of Data Mining Applications. IEEE Press/Wiley (2005)
3. Lin, J., Keogh, E., Lonardi, S., Patel, P.: Finding Motifs in Time Series. In: Proceedings of the 2nd Workshop on Temporal Data Mining. At the 8th ACM SIGKDD Int'l Conference on Knowledge Discovery and Data Mining, Edmonton, Alberta, Canada, pp. 53–68 (2002)
4. Lin, J., Keogh, E., Lonardi, S., Chiu, B.: A symbolic representation of time series, with implications for streaming algorithms. In: Proceedings of 8th ACM SIGMOD workshop on Research issues in data mining and knowledge discovery, San Diego, California, pp. 2–11 (2003)
5. Needleman, S.B., Wunsch, C.D.: A general method applicable to the search for similarities in the amino acid sequences of two proteins. J. Mol. Biol. 48, 443–453 (1970)
6. Luke, S.: Two Fast Tree-Creation Algorithms for Genetic Programming. IEEE Trans. on Evolutionary Computation 4(3), 274–283 (2000)
7. Koza, J.R., Keane, M.A., Streeter, M.J., et al.: Genetic Programming IV: Routine Human-Competitive Machine Intelligence. Kluwer Academic Publishers, Norwell (2005)
8. Langdon, W.B., Poli, R.: Foundations of Genetic Programming. Springer, London (2001)
9. Brindle, A.: Genetic Algorithms for Function Optimization. PhD Thesis, University of Alberta (1991)
10. Baker, J.E.: Reducing Bias and Inefficiency in the Selection Algorithm. In: Proceedings of the 1st International Conference on Genetic Algorithms, pp. 101–111 (1987)
11. Crow, J.F., Kimura, M.: An Introduction to Population Genetics Theory. Harper and Row, New York (1970)
12. De Jong, K.A.: Analysis of Behaviour of a class of Genetic Adaptive Systems. PhD Thesis, University of Michigan (1975)
13. Koza, J.R.: Genetic Programming: On the Programming of Computers by Means of Natural Selection. MIT Press, Cambridge (1992)

14. D'haesler, P.: Context Preserving Crossover in Genetic Programming. In: Proceedings of the 1994 IEEE World Congress on Computational Intelligence, Orlando, Florida, USA, vol. (1), pp. 379–407 (1994)
15. Langdon, W.B.: Size Fair and Homologous Tree Genetic Programming Crossovers. In: Proceedings of the Genetic and Evolutionary Computation Conference, GECCO 1999, pp. 1092–1097. Morgan Kaufmann, San Francisco (1999)
16. Crawford-Marks, R., Spector, L.: Size control via Size Fair Genetic Operators in the PushGP Genetic Programming System. In: Proceedings of the Genetic and Evolutionary Computation Conference, New York, USA, pp. 733–739 (2002)
17. Whigham, P.A.: Grammatically-Based Genetic Programming. In: Rosca, J.P. (ed.) Proceedings of the Workshop on Genetic Programming: From Theory to Real-World Applications, Tahoe City, California, USA, pp. 33–41 (1995)
18. Couchet, J., Manrique, D., Ríos, J., Rodríguez-Patón, A.: Crossover and mutation operators for grammar-guided genetic programming. Soft Computing - A Fusion of Foundations, Methodologies and Applications 11(10), 943–955 (2007)
19. Grosman, B., Lewin, D.R.: Adaptive genetic programming for steady-state process modelling. Comput. Chem. Eng. 28, 2779–2790 (2004)
20. Hussain, T.S.: Attribute grammar encoding of the structure and behavior of artificial neural networks. PhD Thesis, Queen's University, Kingston, Ontario (2003)
21. Rodrigues, E., Pozo, A.: Grammar-guided genetic programming and automatically defined functions. In: Proceedings of the 16th Brazilian symposium on artificial intelligence, Brazil, pp. 324–333 (2002)
22. Hoai, N.X., McKay, R.I.: Is ambiguity useful or problematic for grammar guided genetic programming? A case of study. In: Proceedings of the 4th Asia-Pacific conference on simulated evolution and learning, Singapore, pp. 449–453 (2002)
23. Wong, M.L., Leung, K.S.: Data mining using grammar based genetic programming and applications. Kluwer, Norwell (2002)

Data Mining with Neural Networks for Wheat Yield Prediction

Georg Ruß[1], Rudolf Kruse[1], Martin Schneider[2], and Peter Wagner[2]

[1] Otto-von-Guericke-University of Magdeburg
[2] Martin-Luther-University of Halle

Abstract. Precision agriculture (PA) and information technology (IT) are closely interwoven. The former usually refers to the application of nowadays' technology to agriculture. Due to the use of sensors and GPS technology, in today's agriculture many data are collected. Making use of those data via IT often leads to dramatic improvements in efficiency. For this purpose, the challenge is to change these raw data into useful information. In this paper we deal with neural networks and their usage in mining these data. Our particular focus is whether neural networks can be used for predicting wheat yield from cheaply-available in-season data. Once this prediction is possible, the industrial application is quite straightforward: use data mining with neural networks for, e.g., optimizing fertilizer usage, in economic or environmental terms.

Keywords: Precision Agriculture, Data Mining, Neural Networks, Prediction.

1 Introduction

Due to the rapidly advancing technology in the last few decades, more and more of our everyday life has been changed by information technology. Information access, once cumbersome and slow, has been turned into "information at your fingertips" at high speed. Technological breakthroughs have been made in industry and services as well as in agriculture. Mostly due to the increased use of modern GPS technology and advancing sensor technology in agriculture, the term *precision agriculture* has been coined. It can be seen as a major step from uniform, large-scale cultivation of soil towards small-field, precise planning of, e.g., fertilizer or pesticide usage. With the ever-increasing amount of sensors and information about their soil, farmers are not only harvesting, e.g., potatoes or grain, but also harvesting large amounts of data. These data should be used for optimization, i.e. to increase efficiency or the field's yield, in economic or environmental terms.

Until recently [13], farmers have mostly relied on their long-term experience on the particular acres. With the mentioned technology advances, cheaper sensors have eased data acquisition on such a scale that it makes them interesting for the data mining community. For carrying out an information-based field cultivation,

P. Perner (Ed.): ICDM 2008, LNAI 5077, pp. 47–56, 2008.

the data have to be transformed into utilizable information in terms of management recommendations as a first step. This can be done by decision rules, which incorporate the knowledge about the coherence between sensor data and yield potential. In addition, these rules should give (economically) optimized recommendations. Since the data consist of simple and often even complete records of sensor measurements, there are numerous approaches known from data mining that can be used to deal with these data. One of those approaches are artificial neural networks [4] that may be used to build a model of the available data and help to extract the existing pattern. They have been used before in this context, e.g. in [1], [7] or [12].

The connection between information technology and agriculture is and will become an even more interesting area of research in the near future. In this context, IT mostly covers the following three aspects: data collection, analysis and recommendation [6]. This work is based on a dissertation that deals with data mining and knowledge discovery in precision agriculture from an agrarian point of view [15]. This research led to economically optimized decision rules, but left out some of the details on the used techniques. Since we are dealing with the above-mentioned data records, the computer science perspective will be applied. The main research target is whether we can model and optimize the site-specific data by means of further computational intelligence techniques. We will therefore deal with data collection and analysis.

The paper is structured as follows: Section 2 will provide the reader with details on the acquisition of the data and some of the data's properties. Section 3 will give some background information on neural networks. In Section 4 we will describe the experimental layout and afterwards, we will evaluate the results that were obtained. The last section will give a brief conclusion.

2   Data Acquisition

The data available in this work have been obtained in the years 2003 and 2004 on a field near Köthen, north of Halle, Germany. All information available for this 65-hectare field was interpolated to a grid with 10 by 10 meters grid cell sizes. Each grid cell represents a record with all available information. During the growing season of 2004, the field was subdivided into different strips, where various fertilization strategies were carried out. For an example of various managing strategies, see e.g. [11], which also shows the economic potential of PA technologies quite clearly. The field grew winter wheat, where nitrogen fertilizer was distributed over three application times.

Overall, there are seven input attributes – accompanied by the yield in 2004 as the target attribute. Those attributes will be described in the following. In total, there are 5241 records, thereof none with missing values and none with outliers.

2.1   Nitrogen Fertilizer – N1, N2, N3

The amount of fertilizer applied to each subfield can be easily measured. It is applied at three points in time into the vegetation period. Since the site of

Table 1. Data overview

Attribute	min	max	mean	std	Description
N1	0	100	57.7	13.5	amount of nitrogen fertilizer applied at the first date
N2	0	100	39.9	16.4	amount of nitrogen fertilizer applied at the second date
N3	0	100	38.5	15.3	amount of nitrogen fertilizer applied at the third date
REIP32	721.1	727.2	725.7	0.64	red edge inflection point vegetation index
REIP49	722.4	729.6	728.1	0.65	red edge inflection point vegetation index
EM38	17.97	86.45	33.82	5.27	electrical conductivity of soil
Yield03	1.19	12.38	6.27	1.48	yield in 2003
Yield04	6.42	11.37	9.14	0.73	yield in 2004

application had also been designed as an experiment for data collection, the range of N1, N2, and N3 in the data is from 0 to 100 $\frac{kg}{ha}$, where it is normally at around $60\frac{kg}{ha}$.

2.2 Vegetation – REIP32, REIP49

The *red edge inflection point* (REIP) is a first derivative value calculated along the red edge region of the spectrum, which is situated from 680 to 750nm. Dedicated REIP sensors are used in-season to measure the plants' reflection in this spectral band. Since the plants' chlorophyll content is assumed to highly correlate with the nitrogen availability (see, e.g. [10]), the REIP value allows for deducing the plants' state of nutrition and thus, the previous crop growth. For further information on certain types of sensors and a more detailed introduction, see [15] or [8]. Plants that have less chlorophyll will show a lower REIP value as the red edge moves toward the blue part of the spectrum. On the other hand, plants with more chlorophyll will have higher REIP values as the red edge moves toward the higher wavelengths. For the range of REIP values encountered in the available data, see Table 1. The numbers in the REIP32 and REIP49 names refer to the growing stage of winter wheat.

2.3 Electric Conductivity – EM38

A non-invasive method to discover and map a field's heterogeneity is to measure the soil's conductivity. Commercial sensors such as the EM-38[1] are designed for agricultural use and can measure small-scale conductivity to a depth of about 1.5 metres. There is no possibility of interpreting these sensor data directly in terms of its meaningfulness as yield-influencing factor. But in connection with other site-specific data, as explained in the rest of this section, there could be coherences. For the range of EM values encountered in the available data, see Table 1.

---

[1] Trademark of Geonics Ltd, Ontario, Canada.

Table 2. Overview on available data sets for the three fertilization times (FT)

FT1	Yield03, EM38, N1
FT2	Yield03, EM38, N1, REIP32, N2
FT3	Yield03, EM38, N1, REIP32, N2, REIP49, N3

2.4 Yield 2003/2004

Here, yield is measured in $\frac{t}{ha}$. In 2003, the range for corn was from 1.19 to 12.38. In 2004, the range for wheat was from 6.42 to 11.37, with a higher mean and smaller standard deviation, see Table 1.

2.5 Data Overview

A brief summary of the available data attributes is given in Table 1.

2.6 Points of Interest

From the agricultural perspective, it is interesting to see how much the influencable factor "fertilization" really determines the yield in the current site-year. Furthermore, there may be additional factors that correlate directly or indirectly with yield and which can not be discovered using regression or correlation analysis techniques like PCA. To determine those factors we could establish a model of the data and try to isolate the impact of single factors. That is, once the current year's yield data can be predicted sufficiently well, we can evaluate single factors' impact on the yield.

From the data mining perspective, there are three points in time of fertilization, each with different available data on the field. What is to be expected is that, as more data is available, after each fertilization step the prediction of the current year's yield (Yield03) should be more precise. Since the data have been described in-depth in the preceding sections, Table 2 serves as a short overview on the three different data sets for the specific fertilization times.

For each data set, the Yield04 attribute is the target variable that is to be predicted. Once the prediction works sufficiently well and is reliable, the generation of, e.g., fertilization guidelines can be tackled. Therefore, the following section deals with an appropriate technique to model the data and ensure prediction quality.

3 Data Modeling

In the past, numerous techniques from the computational intelligence world have been tried on data from agriculture. Among those, neural networks have been quite effective in modeling yield of different crops ([12], [1]). In [14] and [15], artificial neural networks (ANNs) have been trained to predict wheat yield from fertilizer and additional sensor input. However, from a computer scientist's perspective, the presented work omits details about the ANN's internal settings, such as network topology and learning rates. In the following, an experimental layout will be given that aims to determine the optimal parameters for the ANN.

3.1   Neural Networks Basics

The network type which will be optimized here are multi-layer perceptrons (MLPs) with backpropagation learning. They are generally seen as a practical vehicle for performing a non-linear input-output mapping [4]. To counter the issue of overfitting, which leads to perfect performance on training data but poor performance on test or real data, cross-validation will be applied. As mentioned in e.g. [5], the data will be split randomly into a training set, a validation set and a test set. Essentially, the network will be trained on the training set with the specified parameters. Due to the backpropagation algorithm's properties, the error on the training set declines steadily during the training process. However, to maximize generalization capabilities of the network, the training should be stopped once the error on the validation set rises [2].

As explained in e.g. [3], advanced techniques like Bayesian regularization [9] may be used to optimize the network further. However, even with those advanced optimization techniques, it may be necessary to train the network starting from different initial conditions to ensure robust network performance. For a more detailed and formal description of neural networks, we refer to [3] or [4].

3.2   Variable Parameters

For each network there is a large variety of parameters that can be set. However, one of the most important parameters is the network topology. For the data set described in Section 2, the MLP structure should certainly have up to seven input neurons and one output neuron for the predicted wheat yield. Since we are dealing with more than 5000 records, the network will require a certain amount of network connections to be able to learn the input-output mapping sufficiently well. Furthermore, it is generally unclear and mostly determined experimentally how many layers and how many neurons in each layer should be used [2]. Therefore, this experiment will try to determine those network parameters empirically. Henceforth, it is assumed that two layers are sufficient to approximate the data set. This structure is generally assumed to be capable of approximating virtually any function of interest, provided that sufficiently many hidden connections are available [5]. To determine the exact number of neurons, a maximum size of 32 neurons in the first and second hidden layer has been chosen – this provides a maximum of 1024 connections in between the hidden layers, which should be sufficient. The range of the network layers' sizes will be varied systematically from 2 to 32. The lower bound of two neurons has been chosen since one neuron with a sigmoidal transfer function does not contribute much to the function approximation capabilities. The upper bound is generally problem-dependent; here, it was determined by preliminary experiments that showed that the generalization capabilities are reduced by using more than a certain number of neurons. Moreover, the maximum network size has also been chosen for reasons of computation time.

3.3 Fixed Parameters

In preliminary experiments which varied further network parameters systematically, a learning rate of 0.5 and a minimum gradient of 0.001 have been found to deliver good approximation results without overfitting the data. All of the network's neurons have been set to use the *tanh* transfer function, the initial network weights have been chosen randomly from an interval of $[-1, 1]$. Data have been normalized to an interval of $[0, 1]$.

3.4 Network Performance

The network performance with the different parameters will be determined by the mean of the squared errors on the test set since those test data will not be used for training. Overall, there are three data sets for which a network will be trained. The network topology is varied from 2 to 32 neurons per layer, leaving 961 networks to be trained and evaluated. The network's approximation quality can then be shown on a surface plot.

4   Results and Discussion

To visualize the network performance appropriately, a surface plot has been chosen. In each of the following figures, the x- and y-axes show the sizes of the first and second hidden layer, respectively. Figures 1(a), 1(b) and 2(a) show the mean squared error vs. the different network sizes, for the three fertilization times (FT), respectively. For the first FT, the mse on average is around 0.3, at the second FT around 0.25 and at the third FT around 0.2. It had been expected that the networks' prediction improves once more data (in terms of attributes) become available for training. There is, however, no clear tendency towards better prediction with larger network sizes. Nevertheless, a prediction accuracy of between 0.44 and 0.55 $\frac{t}{ha}$ (the figures only show the mean *squared* error) at an average yield of 9.14 $\frac{t}{ha}$ is a good basis for further developments with those data and the trained networks.

Furthermore, there are numerous networks with bad prediction capabilities in the region where the first hidden layer has much fewer neurons than the second hidden layer. Since we are using feedforward-backpropagation networks without feedback, this behaviour should also be as expected: the information that leaves the input layer is highly condensed in the first hidden layer if it has from two to five neurons – therefore, information is lost. The second hidden layer's size is then unable to contribute much to the network's generalization – the network error rises.

For the choice of network topology, there is no general answer to be given using any of the data sets from the different FTs. What can be seen is that the error surface is quite flat so that a layout with 16 neurons in both hidden layers should be an acceptable tradeoff between mean squared error and computational complexity.

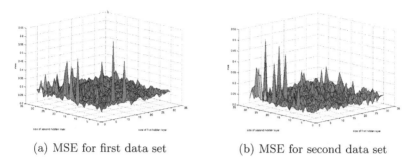

(a) MSE for first data set   (b) MSE for second data set

Fig. 1. MSE plots for first and second data set

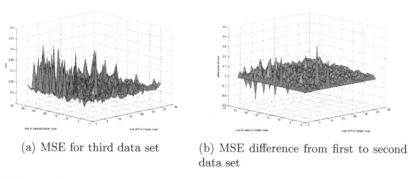

(a) MSE for third data set   (b) MSE difference from first to second
                               data set

Fig. 2. MSE plot for third data set, MSE difference plot for first data set

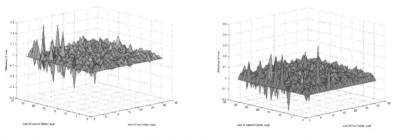

(a) MSE difference from second to third (b) MSE difference from first to third
data set   data set

Fig. 3. MSE difference plots for second and third data set

4.1 Difference Plots

Figures 2(b), 3(a) and 3(b) show the difference between the networks' mean squared errors vs. the different network sizes, respectively. Therefore, they illustrate the networks' performance quite clearly. In the majority of cases, the networks generated from later data sets, i.e. those with more information, can predict the target variable better than the networks from the earlier data sets.

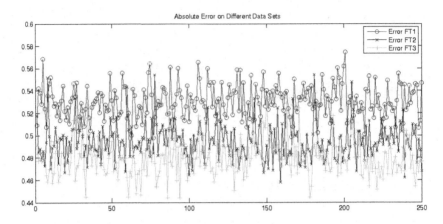

Fig. 4. Comparison of data sets, absolute error vs. trial index

4.2   Comparison of Data Sets FT1, FT2, FT3

In the preceding section we assumed that the networks trained on those data that were available later into the season perform better than the ones on less, earlier data. To substantiate this claim we fixed the network structure to the one that we established earlier: two hidden layers with 16 neurons each and fully connected. The three data sets FT1, FT2, and FT3 were divided randomly into training, validation and testing set at a ratio of $0.6/0.2/0.2$. The division and training steps were repeated 250 times and the absolute error was recorded. Figure 4 shows the error on the different data sets against the trial index. It can be seen quite clearly that our assumption could be substantiated: the average error on FT3 is considerably smaller than the one on FT1 or FT2. For FT1, the mean error is 0.53; for FT2, it is 0.49; and for FT3 it is 0.48. The error's standard deviation on all data sets is 0.015.

5   Conclusion

This paper contributes to finding and evaluating models of agricultural yield data. Starting from a detailed data description, we built three data sets that could be used for training. In earlier work, neural networks had been used to model the data. Certain parameters of the ANNs have been evaluated, most important of which is the network topology itself. We built and evaluated different networks and substantiated the assumption that the prediction accuracy of the networks rises once more data become available at later stages into the growing season.

5.1   Future Work

In subsequent work, we will compare ANNs with suitable further techniques (such as regression or SVMs) to find the best predictor. We will make use of those techniques to model site-year data from different years. It will be evaluated whether the data from one year are sufficient to predict subsequent years' yields. It will also be interesting to study to which extent one field's results can be carried over to modeling a different field. The impact of different parameters during cropping and fertilization on the yield will be evaluated. Finally, controllable parameters such as fertilizer input can be optimized, environmentally or economically.

Acknowledgements

Experiments have been conducted using Matlab 2007b and the corresponding Neural Network Toolbox 5.1. The field trial data came from the experimental farm Görzig of Martin-Luther-University Halle-Wittenberg, Germany. The matlab script that produced Figure 4 and can easily be tailored towards producing the remaining figures can be downloaded from http://tinyurl.com/2fmk2m or can otherwise be requested from the first author of this work.

References

1. Drummond, S., Joshi, A., Sudduth, K.A.: Application of neural networks: precision farming. In: The 1998 IEEE International Joint Conference on Neural Networks Proceedings, 1998. IEEE World Congress on Computational Intelligence, vol. 1, pp. 211–215 (1998)
2. Fausett, L.V.: Fundamentals of Neural Networks. Prentice Hall, Englewood Cliffs (1994)
3. Hagan, M.T.: Neural Network Design (Electrical Engineering). Thomson Learning (December 1995)
4. Haykin, S.: Neural Networks: A Comprehensive Foundation, 2nd edn. Prentice Hall, Englewood Cliffs (1998)
5. Hecht-Nielsen, R.: Neurocomputing. Addison-Wesley, Reading (1990)
6. Heimlich, R.: Precision agriculture: information technology for improved resource use. Agricultural Outlook, 19–23 (April 1998)
7. Kitchen, N.R., Drummond, S.T., Lund, E.D., Sudduth, K.A., Buchleiter, G.W.: Soil Electrical Conductivity and Topography Related to Yield for Three Contrasting Soil-Crop Systems. Agron J. 95(3), 483–495 (2003)
8. Liu, J., Miller, J.R., Haboudane, D., Pattey, E.: Exploring the relationship between red edge parameters and crop variables for precision agriculture. In: Proceedings of IEEE International Geoscience and Remote Sensing Symposium, 2004. IGARSS 2004, vol. 2, pp. 1276–1279 (2004)
9. MacKay, D.J.C.: Bayesian interpolation. Neural Computation 4(3), 415–447 (1992)
10. Middleton, E.M., Campbell, P.K.E., Mcmurtrey, J.E., Corp, L.A., Butcher, L.M., Chappelle, E.W.: "Red edge" optical properties of corn leaves from different nitrogen regimes. In: IEEE International Geoscience and Remote Sensing Symposium, 2002. IGARSS 2002, vol. 4, pp. 2208–2210 (2002)
11. Schneider, M., Wagner, P.: Prerequisites for the adoption of new technologies - the example of precision agriculture. In: Agricultural Engineering for a Better World, Düsseldorf. VDI Verlag GmbH (2006)

12. Serele, C.Z., Gwyn, Q.H.J., Boisvert, J.B., Pattey, E., Mclaughlin, N., Daoust, G.: Corn yield prediction with artificial neural network trained using airborne remote sensing and topographic data. In: Proceedings of IEEE International Geoscience and Remote Sensing Symposium, 2000. IGARSS 2000, vol. 1, pp. 384–386 (2000)
13. Sonka, S.T., Bauer, M.E., Cherry, E.T., John, Heimlich, R.E.: Precision Agriculture in the 21st Century: Geospatial and Information Technologies in Crop Management. National Academy Press, Washington (1997)
14. Wagner, P., Schneider, M.: Economic benefits of neural network-generated site-specific decision rules for nitrogen fertilization. In: Stafford, J.V. (ed.) Proceedings of the 6th European Conference on Precision Agriculture, pp. 775–782 (2007)
15. Weigert, G.: Data Mining und Wissensentdeckung im Precision Farming - Entwicklung von ökonomisch optimierten Entscheidungsregeln zur kleinräumigen Stickstoff-Ausbringung. PhD thesis, TU München (2006)

Experiences Using Clustering and Generalizations for Knowledge Discovery in Melanomas Domain

A. Fornells[1], E. Armengol[2], E. Golobardes[1], S. Puig[3], and J. Malvehy[3]

[1] Grup de Recerca en Sistemes Intel·ligents
Enginyeria i Arquitectura La Salle, Universitat Ramon Llull
Quatre Camins 2, 08022 Barcelona, Spain
{afornells,elisabet}@salle.url.edu
[2] IIIA - Artificial Intelligence Research Institute,
CSIC - Spanish Council for Scientific Research,
Campus UAB, 08193 Bellaterra, Catalonia, Spain
eva@iiia.csic.es
[3] Melanoma Unit, Dermatology Department
IDIBAPS, U726 CIBERER, ISCIII
Hospital Clinic i Provincial de Barcelona, Spain
{spuig,jmalvehy}@clinic.ub.es

Abstract. One of the main goals in prevention of cutaneous melanoma is early diagnosis and surgical excision. Dermatologists work in order to define the different skin lesion types based on dermatoscopic features to improve early detection. We propose a method called SOMEX with the aim of helping experts to improve the characterization of dermatoscopic melanoma types. SOMEX combines clustering and generalization to perform knowledge discovery. First, SOMEX uses Self-Organizing Maps to identify groups of similar melanoma. Second, SOMEX builds general descriptions of clusters applying the anti-unification concept. These descriptions can be interpreted as explanations of groups of melanomas. Experiments prove that explanations are very useful for experts to reconsider the characterization of melanoma classes.

Keywords: Melanoma, Skin Tumour, Dermoscopy, Medicine, Knowledge Discovery, Clustering, Self-Organizing Maps, Explanations.

1 Introduction

Early diagnosis and surgical excision are the main goals in the secondary prevention of cutaneous melanoma. Nowadays, the diagnosis of melanoma is based on the ABCD rule [9] which considers four clinical features commonly observed in this kind of tumour: asymmetry, border irregularity, colour variegation, and a diameter larger than 5 mm. Although most of melanomas are correctly diagnosed following this rule, a variable proportion of melanomas does not comply with these criteria. The current procedure when a suspicious skin lesion appears

P. Perner (Ed.): ICDM 2008, LNAI 5077, pp. 57–71, 2008.

is to excise and to analyse it by means of biopsy. Commonly, the result of the biopsy allows to determine the accurate malignity of the lesion.

Dermoscopy is a non-invasive technique for a more accurate evaluation of skin lesions introduced by dermatologists two decades ago. Dermoscopy provides the opportunity to avoid the excision of benign skin lesions. However, dermatologists need to achieve a good dermatoscopic classification of lesions previously to extraction [14]. Hofmann-Wellenhof et al [11] suggested a classification of benign melanocytic lesions. Recently, Argenziano et al [2] hypothesized that dermoscopic classification may be better than the classical clinico pathological classification of benign melenocytic lesions (nevi). Currently, there is no dermoscopic classification of melanoma located in trunk and extremitis. In the era of genetic profiling, molecular studies including microarrays suggest that there is more than one type of melanoma in these locations. The aim of the present work is to help dermatologists in the classification of early melanoma (*in situ melanoma*) based on dermoscopy characteristics. For this reason, dermatologists define several dermatoscopic classes of in situ melanoma based on their dermatoscopic features. Dermatopathologies also suggest another classification based on histological features.

The goal of this work is twofold: on one hand we want to confirm that the dermatoscopic classes are well defined and, on the other hand, we want to relate these classes to the histological classes of melanomas from the histopathological analysis of biopsies. The present paper describes a method called SOMEX to help dermatologists in their research. SOMEX is a combination of two machine learning approaches: clustering and generalization. In a first step, a Self-Organizing Map [12] clusters a set of skin lesions in patterns according to their similar characteristics. In a second step, a generalization method based on the notion of anti-unification [4] is used to explain clustering results. Results should help dermatologists to discover what fails in defining classes and why lesions that they consider belong to different classes have been clustered together.

The paper is organized as follows. The next section describes the combination of clustering and generalization in SOMEX. Section 3 explains briefly the melanoma domain and it also describes some particular results achieved with SOMEX application. Section 4 describes some related work. Finally, section 5 summarizes the article with conclusions and future work.

2   SOMEX

Let us suppose the following scenario: there is a set of objects belonging to several classes and we want to test whether or not these classes are correctly defined. The first idea is to apply some clustering technique in order to achieve natural groups of similar objects. By testing these groups taking into account the classes we can determine their commonalties. This is exactly what SOMEX achieves by means of generalization of the clusters defined by SOM. Next sections explain in detail how SOMEX works.

2.1   Self-organizing Maps

Self-Organizing Map (SOM) [12] is one of the major unsupervised learning paradigms in the family of artificial neural networks. It has many important properties which make it useful for clustering [10]: (1) It preserves the original topology; (2) It works well even though the original space has a high number of dimensions; (3) It incorporates the selection feature approach; (4) Although one class has few examples they are not lost; (5) It provides an easy way to show data; (6) It is organized in an autonomous way to be better adjusted to data. Moreover, SOM is a soft-computing technique that allows the management of uncertain, approximate, partial truth and complex knowledge. These capabilities are useful in order to manage real domains, which are often complex and uncertain.

Because SOM is a no supervised technique it has to discover by itself which commonalities, correlations and classes of the objects are. SOM projects the original space from an input layer of N neurons (many neurons as input data features) to an output layer of a new space with less dimensions (many neurons as maximum number of clusters expected) with the aim of identifying groups of similar elements. Figure 1 shows a typical 2-dimensional grid of $M \times M$ neurons, where each one is represented by a *director vector* of N dimensions (v_m). A director vector can be described as the expected value for each one of the N features. Moreover, each input neuron is connected to all the output neurons. The definition of clusters can be summarized in the next steps:

1. Director vectors of each neuron are randomly initialized.
2. Given a new input example e, the distance between e and each director vector is computed with the aim of identifying the most suitable neuron where the e should be mapped. For example, the winner neuron is the one with the value most similar to 1 if the normalized Euclidean distance (see Eq. 1).

$$similarity(e, m) = |d(\boldsymbol{e}, \boldsymbol{v_m})| = \left| \sqrt{\frac{\sum_{n:1..N}(e(n) - v_m(n))^2}{N}} \right| \qquad (1)$$

3. Directors vectors are adjusted. The director vector of the winner neuron is adjusted for improving the match with new objects similar to the current one. In contrast, the rest of directors vectors are modified to weakly represent the current example.

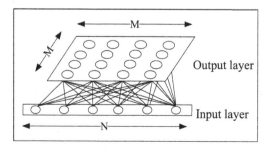

Fig. 1. SOM groups similar elements according to their data features

4. Steps 2 and 3 are repeated for all training examples until director vectors are representative enough. Usually their representativeness is determined by establishing a minimal error value computed as the global sum of distance between the set of cases of each cluster and its respectively director vector. Nevertheless, other common criteria is to establish a maximum number of algorithm iterations.

5. When the training process ends, step 2 is the procedure used to map the new input example in the most suitable clusters.

The main drawback of the method is the definition of the training parameters. First aspect is to determine the *map size*, which is related to the final number of clusters. Thus, a big size of maps will produce a high number of clusters, where each cluster will contain few objects. Conversely, small maps will produce few clusters containing a lot of objects and, consequently director vectors of clusters will be overgeneralised. A second aspect to take into account is *neighbourhood factor*, which is the influence of each cluster over others. The third aspect is the learning factor, which determines the convergence of algorithm. High values of this factor could produce a random behaviour of learning procedure and low values could produce slow ratio of convergence. Finally, the last aspect is the distance measure used to make comparisons.

To conclude, SOM is a smart technique to identify hidden and complex relationships between elements and also to identify the most relevant features thanks to its knowledge discovery and soft-computing capabilities. This is exactly what experts need: to discover relationships between elements to improve the precision of the classes proposed by them.

2.2   How to Explain a Cluster

Director vectors can be described as the expected values of each attribute for belonging to a cluster. However, from the user's point of view these tuples do not give an easy intuition of why some objects have been clustered together. Because of this in [5] we propose to build symbolic explanations of the clusters with the purpose of justifying why a set of cases have been clustered together (this is a concept similar to *charaterization* used in data mining terminology [16]). Experts found symbolic explanations more understandable than director vectors since the former are constructed using the same representation language than they used to describe the domain objects.

Thus, we propose to explain a cluster using a symbolic description that is a generalization of all objects contained in the cluster. This generalization is based on the anti-unification concept [4] although with some differences. The anti-unification (AU) of a set of objects is a description defined as their most specific generalization. The AU contains attributes shared by the set of objects and where each attribute takes as value the most specific of all the values holding in the original set. In this paper we only work with the idea of shared attributes among a set of objects.

Let M_i be a cluster and let $c_1, ..., c_n$ be the set of objects that belong to that cluster after the application of SOM to a set of objects. Each object c_j is described by a set of attributes $\mathcal{A}$. The explanation D_i of why a subset of objects have been clustered in M_i is built in the following way:

- D_i contains attributes which are common to all the objects in M_i. Attributes with *unknown* value in some object $c_j \in M_i$ are not in D_i.
- Let a_k be an attribute common to all objects in M_i such that a_k takes symbolic values on a set $\mathcal{V}_k$. The attribute a_k will not be in D_i when the union of the values that a_k takes in M_i is exactly $\mathcal{V}_k$.
- An attribute a_i takes in D_i the union of all values that a_i holds in the objects in M_i.

Let us illustrate with an example how to build explanations for a cluster. Let M_5 be the cluster formed by the three cases (see Fig. 2). Let D_5 be the explanation of why these cases are clustered (see Fig. 3). An attributes such as C_Max-Diam is not in D_5 because it is not common to all cases (i.e., C_Max-Diam is not present in obj-61). In contrast, attributes such as C_Sex, D_Pseudopigment-Network or D_Peppering are not in D_5 because they take all possible values. For instance, the feature C_Sex takes the value M in objects obj-68 and obj-9 and the value F in obj-61, this means that the value of this attribute is irrelevant to describe M_5.

Summarizing, explanations provide a symbolic description that contains the commonalties among all objects of a cluster. We chosen to show for each attribute the union of possible values (instead of the average or the mode as is

(Define (Object :Id Obj-68)	(Define (Object :Id Obj-61)	(Define (Object :Id Obj-9)
(C_Sex M)	(C_Sex F)	(C_Sex M)
(C_Age 69)	(C_Age 36)	(C_Age 64)
(C_Max_Diam 10)	(C_Site Forearm)	(C_Max_Diam 20)
(C_Site Back)	(D_Pattern Reticular)	(C_Site Upper_Extr)
(D_Pattern Multicomponent)	(D_Pigment_Network 1)	(D_Pattern Reticular)
(D_Pigment_Network 1)	(D_Atypical_Pn 1)	(D_Pigment_Network 1)
(D_Atypical_Pn 1)	(D_Pseudopigment_Network 0)	(D_Atypical_Pn 1)
(D_Pseudopigment_Network 0)	(D_Dots_And_Globules 0)	(D_Pseudopigment_Network 0)
(D_Dots_And_Globules 0)	(D_Atypical_D_And_G 0)	(D_Dots_And_Globules 1)
(D_Atypical_D_And_G 0)	(D_Streaks 1)	(D_Atypical_D_And_G 0)
(D_Streaks 1)	(D_Irregular_Streaks 1)	(D_Streaks 1)
(D_Irregular_Streaks 1)	(D-Regression_Structures 1)	(D_Irregular_Streaks 1)
(D_Regression_Structures 1)	(D_Peppering 1)	(D_Regression_Structures 1)
(D_Peppering 1)	(D_White_Areas 0)	(D_Peppering 0)
(D_White_Areas 1)	(D_Bw_Veil 0)	(D_White_Areas 1)
(D_Bw_Veil 0)	(D_Bloches 0)	(D_Bw_Veil 0)
(D_Bloches 0)	(D_Irregular_Bloches 0)	(D_Bloches 0)
(D_Irregular_Bloches 0)	(D_Vessels 0)	(D_Irregular_Bloches 0)
(D_Vessels 0)	(D_Dotted_Vessels 0)	(D_Vessels 0)
(D_Dotted_Vessels 0)	(D_Atypical_Vessels 0)	(D_Dotted_Vessels 0)
(D_Atypical_Vessels 0)	(D_Millia_Like_Cyst 0)	(D_Atypical_Vessels 0)
(D_Millia_Like_Cyst 0)	(H_Diagnosis P))	(D_Millia_Like_Cyst 0)
(H_Diagnosis Mnevus))		(H_Diagnosis Nonc))

Fig. 2. Description of three classes included in a cluster, say M_5

```
(C_Age (36 64 69))
(C_Site (Back Forearm Upper_Extr))
(D_Pattern (Multicomponent Reticular))
(D_Pigment_Network (1))
(D_Atypical_Pn (1))
(D_Atypical_D_And_G (0))
(D_Streaks (1))
(D_Irregular_Streaks (1))
(D_Regression_Structures (1))
(D_Bw_Veil (0))
(D_Bloches (0))
(D_Irregular_Bloches (0))
(D_Vessels (0))
(D_Dotted_Vessels (0))
(D_Atypical_Vessels (0))
(D_Millia_Like_Cyst (0))
(H_Diagnosis (Mnevus P Nonc))
```

Fig. 3. Explanation of why the objects included in cluster M_5 (Fig. 2) have been clustered together

the usual approach) because the expert finds more useful knowing all possible values. This is the explicit information that an expert extracts from SOMEX; but, our question would be if there is some implicit information from the explanations. The answer would be affirmative. Two aspects of the explanations are specially relevant from the point of view of the knowledge discovery: one is the number of attributes composing explanations and the other is the number of values that take these attributes. Both aspects give an idea of how similar the objects contained in a cluster are.

Concerning the number of attributes, explanations with a high number of attributes represent very similar objects whereas explanations with few attributes mean that these objects have few aspects in common. Nevertheless, the number of values holding the attributes of an explanation also plays a crucial role. Thus, the more values an attribute holds the more irrelevant this attribute is. Notice that the explanation is built using common attributes and taking as values for these attributes the union of all values hold by the objects of a cluster. Thus, a common attribute that takes several values, means that has a high variability and this attribute is probably not too relevant. Conversely, attributes holding only one value represent aspects of the objects that could be taken as candidates to characterize a cluster.

In short, clusters explained by means of descriptions composed of a high number of attributes where each attribute holds one value, can be interpreted as good clusters in the sense that all the objects included in them are very similar. On the other hand, if the object class is known two situations can happen: 1) all objects of the cluster belong to the same class, or 2) objects belong to several classes. This second situation is the interesting one from the point of view of knowledge discovery since it means that objects that, in principle, belong to different classes according to the dermatological point of view are highly similar.

This should be a starting point from the expert to reconsider the definition of classes (for instance, by merging classes to which objects belong).

Similarly, clusters explained by means of descriptions with a lot of attributes holding almost all possible values, can be interpreted as imprecise clusters in the sense that objects in the cluster have not many similarities. From the knowledge discovery point of view, this situation is interesting when all objects of such clusters belong to the same class, since it means that although they have been classified as belonging to the same class, these objects are not actually similar.

These situations will be illustrated in more detail in the next section where SOMEX is applied to support dermatologists in the definition and validation of some classes of malignant skin lesions.

3 Using SOMEX for Knowledge Discovery

Dermatologists take into account dermoscopic aspects of skin lesions with the aim of determining whether or not it will become a melanoma (malignant skin lesions) prior to lesion excision. That is why the aim of this work is to support them to extract melanoma patterns through SOMEX application. First, SOM clusters together objects (descriptions of skin lesions) that are similar independently of the class. So, symbolic explanations show dermatologists the common features of objects clustered together, allowing them to consider some modifications in the class definition. This section describes briefly the melanomas domain and results achieved by SOMEX support.

3.1   Testbed: The Melanomas Domain

A skin lesion can be described from two different aspects: *dermoscopic* and *histologic*. Dermoscopic aspects are those obtained using a technique called dermoscopy. This technique combines an image magnification process for making bigger the image (i.e. x30) and a system to decrease the reflex ion and the refraction of the light through polarized light and polarization filters. Thus, dermoscopy allows to identify global patterns (*D_Pattern*) and local features (attributes inside the rectangle in the right part of Fig. 4), which are used for experts to suggest a hypothetical diagnosis (*D_Diagnosis*). In contrast, histological aspects (attributes inside the rectangle in the middle part of Fig. 4) are obtained from the analysis of a excised and biopsies of suspicious skin lesion. Moreover, this clinical practice allows experts to confirm the real diagnosis (*H_Diagnosis*). Both kind of aspects are summarized in Fig. 4. In addition to these attributes, the description of a lesion is completed with the clinical profile of the patient such as for example age and sex among others.

Although the experimental dataset used contains only 75 melanomas, the small set of examples is considered as a representative sample of the domain since there is a consensus among several experts around their characterization.

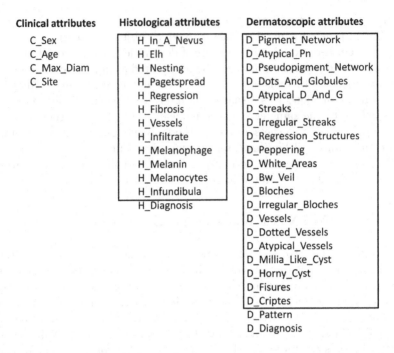

Fig. 4. Clinical, histological and dermoscopic attributes used to describe a melanoma

With our experiments we want to support dermatologists in finding 1) how to dermoscopically describe histologic classes, and 2) to test whether or not histologic classes have been correctly defined. The next section describes the conditions under which experiments have been performed and also some interesting results obtained from SOMEX application.

3.2   Experiments

Since our purpose was to support dermatologists in determining the dermatoscopic features that describe the histologic classes, we only focused on the clinical and dermatoscopic attributes (see Fig. 4). We also included the histological class represented by the H_Diagnosis attribute. the attribute which is the histologic class. Dermatologists defined the following histologic classes: *LTG_M, LMM, nonc, PL_M, P, P_LTG, Mnevus,* and *SKlMM.*

Bearing in mind this information, we performed several SOM configurations in order to find out interesting results. SOM was tested using several map sizes of 2-dimensions (3×3, 4×4 and 5×5 to analyse several data dispersions), two different distance measures (normalized Euclidean distance and normalized Hamming distance) and 10 random seeds (to minimize the random effects of initialization). The learning factor, the neighbour factor and maximum iterations were set to typical values: from 0.6 to 0.01, from M to 1 and 500 iterations by neuron.

3.3   Discussion of the Results

Independently of the map size and of the distance measure used to build the clusters, SOMEX results showed that the definition of histological classes should be adjusted. The reason is that most clusters include objects belonging to several histologic classes and explanations show that these objects have a lot of common aspects. This is reflected in the fact that most of explanations are very specific, i.e. they have a lot of common attributes holding a unique value. Notice that the more numerous the common features with only one value, the more similar objects are. Conversely, explanations with features holding more than one value mean that, although objects are described by similar features, they have a lot of variability and they are not so similar.

The use of clustering techniques allowed a natural group of similar objects. Then, as a result of generalizing SOMEX explains why a subset of objects have been clustered together. Results show that some melanomas that dermatologists considered as belonging to different classes actually are not so different since they belong to the same cluster. Moreover, the explanation supports the user in discovering the common aspects and also characteristics that are different among objects of a same cluster. In fact, this provides them a clue to reconsider the definition of histological classes. Prior to SOMEX experiments, dermatologists had the hypothesis that the *pattern* (feature D_Pattern) of a skin lesion could be an important aspect to determine the classification of a lesion. As we will detail later, from SOMEX experiments we point out that the *pattern*, at least taken it isolated from other characteristics, is not enough to classify.

A conclusion from the experiments is that criteria used by dermatologists when defining histologic classes (*H_Diagnosis*) do not take into account all aspects describing a melanoma. In fact, clusters almost always contain melanomas of several histological classes. thus, from the predictivity point of view, clusters are not appropriate. However, when experts analyze the explanations of clusters they find them interesting despite their entropy. Experts noted that attributes shared by melanomas into a cluster usually are those considered as important for experts (for instance, D_dots_and_globules or D_Pigment_network). For this reason we prefer to show the analysis that experts performed of the SOMEX explanations instead of giving predictivity measures. Moreover, experts used explanations the reconsider the initial descriptions of histological classes.

Experiments produced three types of clusters: 1) clusters with a reasonable number of objects belonging to different classes, 2) clusters with few objects belonging all of them to the same class, and 3) clusters with few objects of several classes. SOMEX results show that there are not clusters with a high number of objects belonging all of them to the same class nor clusters with few objects with a general explanation. Let us see some results obtained by SOMEX.

Example 1. Let us suppose the cluster M_{15} containing 10 objects. The explanation of this cluster can be seen in Fig. 5. Concerning the number of attributes of the explanation, we see that there is a subset of 15 attributes (from the 28 composing a complete description of an object) shared by all the objects of the

Cluster 15

The cluster is composed of the objects :
(<Obj-69> <Obj-70> <Obj-28> <Obj-31> <Obj-56>
<Obj-15> <Obj-10> <Obj-11> <Obj-37> <Obj-44>)

The explanation is the following
((C_Site (Leg Arm Lower_Extr Upper_Extr Trunk
  Back))
(C_Max_Diam (5 6 7 8 9 18))
(C_Age (28 43 49 50 54 65 66 68)))
(D_Millia_Like_Cyst (0))
(D_Atypical_Vessels (0))
(D_Irregular_Bloches (0))
(D_Bloches (0))
(D_Peppering (0))
(D_Regression_Structures (0))
(D_Irregular_Streaks (0))
(D_Streaks (0))
(D_Atypical_D_And_G (1))
(D_Dots_And_Globules (1))
(D_Pattern (Globular Reticular Unspecific
  Multicomponent))
(H_Diagnosis (Mnevus P_Ltg Pl_M Nonc Ltg_M))

Cluster 24

The cluster is composed of the objects :
(<Obj-12> <Obj-13> <Obj-16>)

The explanation is the following
((C_Sex (F))
(C_Age (38 40 62))
(C_Max_Diam (4 6 28))
(C_Site (Trunk Lower_Extr))
(D_Pattern (Unspecific
  Reticular))
(D_Atypical_Pn (0))
(D_Dots_And_Globules (0))
(D_Atypical_D_And_G (0))
(D_Streaks (0))
(D_Irregular_Streaks (0))
(D_Bw_Veil (0))
(D_Bloches (0))
(D_Irregular_Bloches (0))
(D_Vessels (1))
(D_Atypical_Vessels (1))
(D_Millia_Like_Cyst (0)))
(H_Diagnosis (Pl_M))

Fig. 5. Explanations justifying the clusters M_{15} and M_{24}

cluster. Focusing on values of these common attributes, we seen that most of them have an unique value, meaning that the explanation is specific enough.

In the explanation of the cluster M_{15} there are five attributes with more than one value: C_Age, C_Max-Diam, C_Site, D_Pattern and H_Diagnosis. Two of these attributes, C_age and C_Max-Diam are numerical and currently we cannot extract any conclusion from them. This is because explanations are not able to handle with continuous attributes. Dermatologists plan to establish some kind of discretization to establish ranges of equivalent values for these attributes. Concerning the values of attributes C_Site and D_Pattern, SOMEX shown that they hold a lot of values (almost all the possible values in the case of D_Pattern). In particular, the role of a lesion *pattern* as potential relevant aspect of a melanoma seems to be compromised according to this cluster explanation.

Finally, an interesting analysis can be carried out from values of H_Diagnosis. This is, in fact, the classification proposed by dermatopathologists; therefore, according to their criterion objects of M_{15} belong to five different classes (*LTG_M*, *nonc*, *PL_M*, *P_LTG* and *Mnevus*). However SOMEX show that these objects have a high similarity and the explanation suggests to dermatologists a possible analysis of the relevance of object commonalties so as that they should reconsider the criteria used to classify objects in different histological classes. An analysis of the differences among the objects in M_{15} could also clarify the class definition.

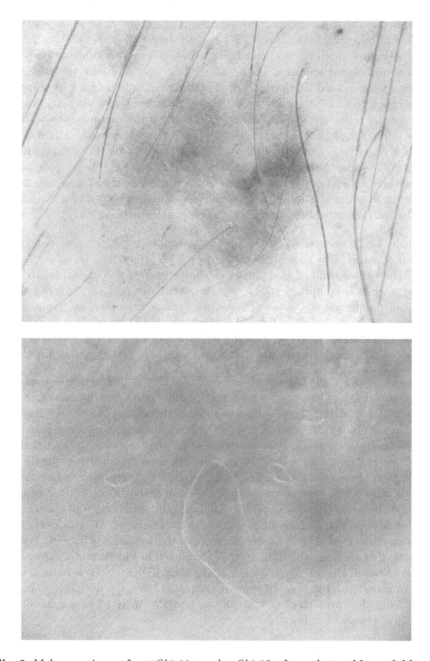

Fig. 6. Melanoma image for $<Obj$-$11>$ and $<Obj$-$13>$ from cluster M_{15} and M_{24} respectively. $<Obj$-$11>$ presents under dermoscopy dots and globules (that are atypical) characteristic of this cluster of lesions, in the absence of all negative features (values 0 in the definition) and also the presence of typical pigment network and typical vessels (may be present in this cluster). $<Obj$-$13>$ presents under dermoscopy only vessels that are atypical, all other criteria are negative.

Example 2. The explanation of cluster M_{24} is composed of three objects with the same histological class. There are 4 multi-valued attributes: C_Age, C_Max-Diam (both numerical), C_Site, and D_Pattern. Globally this explanation seems a good partial characterization for the class PL_M since a further analysis of the numerical values could produce a more specific explanation. An important aspect to take into account is that the attribute D_Pattern has two possible values, *unspecific* and *reticular*. The importance of this fact is that according to SOMEX results, dermatologists should consider the possibility to reject D_Pattern as relevant for classifying a melanoma, since this feature holds different values in objects of the same histological class. In our current experiments we do not consider neither the relationship among attributes nor the weight of some attributes in order to bias the clustering. A possibility is that the pattern of a melanoma could be relevant in relation to the value of any other attribute.

Example 3. The cluster M_5 shown in Fig. 2 is an example of a small one with elements of several histological classes (in fact, each object belongs to a different class). The explanation of this cluster is shown in Fig. 3. This explanation is specific since it is composed by 17 common attributes and all of them except 4 hold an unique value. Notice that as in previous examples, attributes with multiple values are C_Age, C_Site, D_Pattern and H_Diagnosis. Once again the conclusion should be to reconsider the definition of histological classes and to analyse the relevance of attributes that dermatopathologists used to define them.

An important point from the application of SOMEX is that symbolic explanations obtained from clusters give to dermatopathologists descriptions of groups of melanomas that they commonly recognize as different. For instance, the explanation for cluster M_{15} (see Fig. 6) describes lesions that under dermoscopy presents both dots and globules and typical pigment network (notice that all other features have as value 0, meaning *absence*). This description is clearly recognized from the dermatological point of view since they provided us the picture shown in Fig. 5 left, that corresponds to a lesion belonging to cluster M_{15} (in particular, it is the object <*Obj-11*>. Similarly, the explanation of cluster M_{24} describes lesions, completely different that those of cluster M_{15}. In particular, lesions in cluster M_{24} have as unique feature the presence of typical vessels, dermatologists recognized lesions such as the shown in Fig. 5 right (corresponds to <*Obj-13*>) of cluster M_{24}). Summarizing, SOMEX provides a natural clustering of objects and the use of symbolic explanations supports dermatologists in analysing the correctness of the clusters and also in redefining some of the histological classes they propose.

4   Related Work

Clustering techniques are a smart way to extract relationships from huge data amounts. Consequently, this useful property has been widely used in medical domains such as the one in which this work contextualizes.

The types of work mainly depend on the data topology and the usage of extracted relations from analysis. There are melanoma studies focused in the identification of relationships between malignant melanoma and familiar or hereditary tumours (i.e. breast cancer, ovarian cancer, colon cancer, pancreatic cancer) such as in [15]. On the other hand, others works analyse thousands of genes with the aim of extracting the 'guilty' genes [6, 7] related to the cancer. Anyway, both approaches help experts to be aware and detect melanoma formation in early stages. The main difference between our work and others is that we use SOMEX to help experts to improve their melanoma definition and classification. Thus, their precision in the diagnosis can be improved.

The idea of using symbolic descriptions for characterizing clusters can be interpreted as a memory organization. In this sense, our approach is similar to Perner's [13] and Abidi's [1] works. Perner proposes to organize the cases following a hierarchy similar to a decision tree where each node c_i is described by a symbolic description (prototype). Each symbolic description subsumes descriptions of all nodes included in the subtree rooted by c_i until reach the leaves that contain the individual cases. Somehow, nodes of that hierarchy could be interpreted as explanations, i.e. why a subset of domain objects (cases) have been grouped under a node. This work relies on the context of case-based reasoning where the main aim is to classify a new problem, therefore prototypes are used to select a subset of cases to solve a new problem. In previous works such as [8], we also proposed the use of explanations during the retrieval phase of the case-based reasoning, nevertheless in SOMEX the use of explanations is different. SOMEX does not take into account the class of cases, since we assume that these classes could no be accurately defined.

The procedure proposed by Abidi et al [1] is similar to SOMEX because they produce rules that describe objects included in a cluster without using the class information. Firstly, domain objects are clustered according to their similarity, secondly continuous values are discretized, and finally they use rough sets to generate symbolic rules for each cluster. In fact, the explanation generated by SOMEX could also be interpreted as a domain rule (as we suggested in [3]).

The basic difference among SOMEX and the works above is the use of explanations. Perner uses symbolic descriptions to organize the memory of cases with the purpose of achieving a more efficient retrieval. Abidi et al. propose a procedure to obtain symbolic rules from clusters. In SOMEX, explanations are used as a basis for knowledge discovery since they support experts in comparing the classification of cases they proposed with explanations of clusters, where cases have been grouped without taking into account the class information. The analysis of this informations gives to experts some clues for redefining the classes.

5   Conclusions and Further Work

On this paper we propose the use of SOMEX, a combination of clustering and generalizations, to support dermatologists in discovering knowledge about melanomas. The purpose of dermatologists was to define several classes of melanomas and finding dermatoscopic features characterizing these classes. SOMEX supported dermatologists in focusing on groups of similar objects and commonalties among them. In particular, they can analyse the entropy of clusters, i.e. why melanomas that they consider as belonging to different histological classes are actually so similar. Dermatologists can also analyse the relevance of attributes for classification. A particular example is the melanoma *pattern*, considered as a relevant aspect prior to SOMEX application and that results proved that taken in isolation is not a good classifier.

As future work we plan to modify some parameters of the clustering in two ways. Firstly we want to confirm the relevance of *pattern* and we plan to weight some of these features in order to highlight the relationship of this feature with others. A second kind of experiments could be focused on enforcing the number of clusters and experimentally determining the best group of melanomas in order to empirically define their histological classes. Finally, from the point of view of the explanations, we could analyse relations between them.

Acknowledgments

We would like to thank the Spanish Government for the support in MID-CBR project under grant TIN2006-15140-C03 and the *Generalitat de Catalunya* for the support under grants 2005-SGR-00093, 2005-SGR-302 and 2007FIC-0976. We would like to thank *Enginyeria i Arquitectura La Salle* of Ramon Llull University for the support to our research group as well.

On the other hand, we also would like to thank the clinicians involved in the confection of the dataset: Dr Paolo Carli (Dermatologist), Dr Vinzenzo di Giorgi (dermatologist), Dr Daniela Massi (dermatopathologist) from the University of Firenze; Dr Josep Malvehy (dermatologist and co-author), Dr Susana Puig (dermatologist and co-author) and Dr Josep Palou (dermatopathologist) from Melanoma Unit in *Hospital Clinic i Provincial de Barcelona*. Part of the work performed by S. Puig and J. Malvehy is partially supported by: *Fondo de Investigaciones Sanitarias* (FIS), grant 0019/03 and 06/0265; Network of Excellence, 018702 GenoMel from the CE.

References

[1] Abidi, S.S.R., Hoe, K.M., Goh, A.: Analyzing data clusters: A rough sets approach to extract cluster-defining symbolic rules. In: Hoffmann, F., Adams, N., Fisher, D., Guimarães, G., Hand, D.J. (eds.) IDA 2001. LNCS, vol. 2189, pp. 248–257. Springer, Heidelberg (2001)
[2] Argenziano, G., Zalaudek, I., Ferrara, G., Hofmann-Wellenhof, R., Soyer, H.: Proposal of a new classification system for melanocytic naevi. Br. J. Dermatol. 157(2), 217–227 (2007)
[3] Armengol, E.: Usages of generalization in cbr. In: Weber, R.O., Richter, M.M. (eds.) ICCBR 2007. LNCS (LNAI), vol. 4626, pp. 31–45. Springer, Heidelberg (2007)

[4] Armengol, E., Plaza, E.: Bottom-up induction of feature terms. Machine Learning 41(1), 259–294 (2000)

[5] Corral, G., Armengol, E., Fornells, A., Golobardes, E.: Data security analysis using unsupervised learning and explanations. In: Corchado, E., Corchado, J.M., Abraham, A. (eds.) Innovations in Hybrid Intelligent Systems, vol. 44, pp. 112–119. Springer, Heidelberg (2007)

[6] Dang, H., Segaran, T., Le, T., Levy, J.: Integrating database information in microarray expression analyses: Application to melanoma cell lines profiled in the nci60 data set. Journal of Biomolecular Techniques 13, 199–204 (2002)

[7] Eisen, M.B., Spellman, P.T., Brown, P.O., Botstein, D.: Cluster analysis and display of genome-wide expression patterns. National Academy Scientific 25(95), 14863–14868 (1998)

[8] Fornells, A., Armengol, E., Golobardes, E.: Explanation of a clustered case memory organization. In: Artificial Intelligence Research and Development, vol. 160, pp. 153–160. IOS Press, Amsterdam (2007)

[9] Friedman, R.J., Rigel, D.S., Kopf, A.W.: Early detection of malignant melanoma: The role of physician examination and self-examination of the skin. Ca-A Cancer J. Clinicians 35, 130–151 (1985)

[10] Haykin, S.: Neural Networks: A Comprehensive Foundation, 2nd edn. Prentice Hall, Englewood Cliffs (1999)

[11] Hofmann-Wellenhof, R., Blum, A., Wolf, I., Zalaudek, I., Piccolo, D., Kerl, H., Garbe, C., Soyer, H.: Dermoscopic classification of clark's nevi (atypical melanocytic nevi). Clin. Dermatol. 20(3), 255–258 (2002)

[12] Kohonen, T.: Self-Organizing Maps, 3rd edn. Springer, Heidelberg (2000)

[13] Perner, P.: Case-base maintenance by conceptual clustering graphs. Engineering Applications of Artificial Intelligence 9, 381–393 (2006)

[14] Puig, S., Argenziano, G., Zalaudek, I., Ferrara, G., Palou, J., Massi, D., Hofmann-Wellenhof, R., Soyer, H., Malvehy, J.: Melanomas that failed dermoscopic detection: a combined clinicodermoscopic approach for not missing melanoma. Dermatol. Surg. 33(10), 1262–1273 (2007)

[15] Stefano, P., Fabbrocini, G., Scalvenzi, M., Emanuela, B., Pensabene, M.: Malignant melanoma clustering with some familiar/hereditary tumors. Annals of Oncology 3, 100 (2002)

[16] Witten, I.H., Frank, E.: Data Mining: Practical Machine Learning Tools and Techniques, 2nd edn. Morgan Kaufmann Series in Data Management Systems. Morgan Kaufmann, San Francisco (2005)

Noisy Image Segmentation by a Robust Clustering Algorithm Based on DC Programming and DCA

Le Thi Hoai An[1], Le Hoai Minh[1], Nguyen Trong Phuc[1], and Pham Dinh Tao[2]

[1] Laboratory of Theoretical and Applied Computer Science (LITA EA 3097) UFR MIM, University of Paul Verlaine - Metz, Ile du Saulcy, 57045 Metz, France
[2] Laboratory of Modelling, Optimization & Operations Research, National Institute for Applied Sciences - Rouen, BP 08, Place Emile Blondel F 76131 Mont Saint Aignan Cedex, France

Abstract. We present a fast and robust algorithm for image segmentation problems via Fuzzy C-Means (FCM) clustering model. Our approach is based on DC (Difference of Convex functions) programming and DCA (DC Algorithms) that have been successfully applied in a lot of various fields of Applied Sciences, including Machine Learning. In an elegant way, the FCM model is reformulated as a DC program for which a very simple DCA scheme is investigated. For accelerating the DCA, an alternative FCM-DCA procedure is developed. Moreover, in the case of noisy images, we propose a new model that incorporates spatial information into the membership function for clustering. Experimental results on noisy images have illustrated the effectiveness of the proposed algorithm and its superiority with respect to the standard FCM algorithm in both running-time and quality of solutions.

Keywords: Image Segmentation, Fuzzy C-Means, DC programming, DCA.

1   Introduction

Image segmentation plays an important role in a variety of application such as robot vision, object recognition and medical imaging [3,17]. Segmentation of the medical images is a primary step in most applications of computer vision to image analysis. Classically, image segmentation is defined as the partitioning of an image into non-overlapped, consistent regions which are homogeneous with respect to some characteristics such as gray value or texture.

There are a lot of algorithms developed for image segmentation problem which can be classified as four categories [20]: the classical methods such as thresholding, region growing, edge based technique; the statistical methods such as the maximum-likelihood-classifier (MLC); the neural networks methods; the fuzzy clustering methods.

P. Perner (Ed.): ICDM 2008, LNAI 5077, pp. 72–86, 2008.

This work is included in the last category: it concerns with a novel fuzzy clustering method for the segmentation of images with noise.

Fuzzy C-Means (FCM) clustering is undoubtedly a most widely used fuzzy clustering method. It was originally introduced by Bezdek in 1981 [2] as an extension to Dunn's algorithm [5]. FCM algorithm is one of the most powerful methods for image segmentation, and its success chiefly attributes to the introduction of fuzziness for the belongingness of each image pixels [2,6]. An image can be represented in various feature spaces, and the FCM algorithm classifies the image by grouping similar data points in the feature space into clusters. This clustering is achieved by iteratively minimizing a cost function that is dependent on the distance of the pixels to the cluster centers in the feature domain.

Let $X := \{x_1, x_2, .., x_n\}$ denotes an image with n pixels to be partitioned into c $(2 \leq c \leq n)$ homogeneous clusters $C_1, C_2, .., C_c$ where $x_k \in \mathbb{R}^d$ $(k = 1, .., n)$ represents multispectral (features) data. In an MRI image $x_k \in \mathbb{R}$ $(d = 1)$ corresponds to the intensity of value of the pixel while in an colour image each pixel can be represented by, for example, the values of tree colours: red, green and blue $(d = 3)$. Consider the matrix $U = (u_{i,k})_{c \times n}$ called the *fuzzy partition matrix* in which each element $u_{i,k}$ indicates the membership degree of each pixel x_k in the cluster C_i (the probability that a pixel x_k belongs to the cluster C_i). The FCM technique is based on optimizing the objective function

$$J_m(U, V) = \sum_{k=1}^{n} \sum_{i=1}^{c} u_{i,k}^m ||x_k - v_i||^2, \tag{1}$$

where $||.||$ is, in this whole paper, the Euclidean norm in the corresponding space, and V is the $(c \times d)$ - matrix whose the ith row is $v_i \in \mathbb{R}^d$, the centre of cluster C_i. The parameter $m \geq 1$ is called the fuzziness index of membership of each datum. The mathematical model of FCM is given by

$$\begin{cases} \min J_m(U, V) := \sum_{k=1}^{n} \sum_{i=1}^{c} u_{i,k}^m ||x_k - v_i||^2 \\ s.t \quad u_{i,k} \in [0, 1] \text{ for } i = 1, .., c \quad k = 1, .., n \\ \sum_{i=1}^{c} u_{i,k} = 1, \ k = 1, .., n \end{cases} \tag{2}$$

One disadvantage of the standard FCM is that it does not consider any spatial information in image segmentation, this makes the algorithm very sensitive to noise and other imaging artifacts. In fact, the spatial relationship of neighboring pixels is an important characteristic that can be of great aid in imaging segmentation, because the pixels on an image are highly correlated, i.e. the pixels in the immediate neighborhood possess nearly the same feature data and, by the way, the probability that they belong to the same partition is great. Recently, many researchers have incorporated the local spatial information into the conventional FCM algorithm to improve the performance of image segmentation (see e.g. [1,4,16,17,25]).

The purpose of our work is twofold. Firstly we develop a fast and robust clustering algorithm via FCM model for image segmentation. The clustering model in image segmentation is, in general, a very large dimension problem for which the research of efficient methods is still current. Secondly, for the segmentation of noisy images, we consider an adaptive FCM model (called Spatial FCM) that incorporates the spatial information into the membership function for clustering.

Our optimization approach is based on DC (Difference of Convex functions) programming and DCA (DC Algorithms) that were introduced by Pham Dinh Tao in their preliminary form in 1985. They have been extensively developed since 1994 by Le Thi Hoai An and Pham Dinh Tao (see [8,9,10,18,19] and references therein) and become now classic and more and more popular (see e.g. [13,14,15,21,22,24]). DCA has been successfully applied to many large-scale (smooth or nonsmooth) nonconvex programs in various domains of applied sciences [9,10], in particular in Machine Learning [11,12,13,15,21,24]. In this work, we reformulate, in an elegant way the nonconvex Spatial FCM model in the form of a *nice* DC program for which the resulting DCA is explicitly described via a very simple formula. For computing a good initial point and accelerating the convergence of DCA we propose an alternative FCM-DCA procedure that combines the DCA with the classical FCM algorithm. Experimental results on several images with noise have illustrated the effectiveness of the proposed algorithm and its superiority with respect to the standard FCM algorithm in both running-time and quality of solutions. Moreover, with the model Spatial FCM, our algorithm greatly reduce the effect of noise and biases the algorithm toward homogeneous clustering.

The rest of the paper is organized as follows. In Section 2, we present the adaptive Spatial FCM model for noisy image segmentation. The new fuzzy clustering method based on DC programming and DCA is developed in Section 3 while the acceleration of DCA by an alternative FCM-DCA procedure is presented in Section 4. Finally, numerical experiments on the noisy images are reported in Section 5.

2   Spatial FCM Model

In the image segmentation by the standard FCM model, each pixel $x_k \in \mathbb{R}^d$ represents the multispectral (features) data. However, as mentioned before, one of the important characteristics of an image is that neighboring pixels possess similar feature values, therefore the spatial relationship is interesting for image segmentation by clustering. The spatial information is the relation between the pixel and its neighborhoods. There are different ways to incorporate the neighboring information. In this work, we consider the spatial information of x_k as the average value of its neighborhoods 3×3, and each data point x_k in (2) has now two groups of values: the values of the pixel and the average values of its neighborhoods 3×3.

Let N_k be the neighborhoods 3×3 of the pixel x_k, the data input x_k in our spatial FCM model are $x_k = (x_{k1}, x_{k2})$ where x_{k1} represents the values of the

k^{th} pixel in the image and $x_{k2} = (x_{k1} + \sum_{i \in N_k} x_{i1})/9$. In this case, the number of variable U is not changed and V becomes a $c \times 2d$ matrix whose the ith row is $v_i \in \mathbb{R}^{2d}$, the centre of cluster C_i.

The Spatial FCM model in our approach is nothing else but the classical FCM (2) in which $x_k \in \mathbb{R}^d$ is replaced by $x_k = (x_{k1}, x_{k2}) \in \mathbb{R}^{2d}$. Hence any algorithm for the classical FCM (2) can be applied to the Spatial FCM model. From numerical points of view the Spatial FCM problem is more difficult, because the number of variables with respected to V is doubled.

3   The Fuzzy C-Means Algorithm Based on DC Programming and DCA

In the sequence we consider the general model of the form (2) in which $x_k \in \mathbb{R}^p$. The Spatial FCM model (resp. classical FCM) corresponds to the case of $p = 2d$ (resp. $p = d$). We propose instead to optimize nonconvex problems (2) using the DC optimization Algorithm DCA.

3.1   DC Programming and DCA

DC Programming and DCA constitute the backbone of smooth/nonsmooth non-convex programming and global optimization. They address the problem of minimizing a function f which is difference of convex functions on the whole space $\mathbb{R}^p$ or on a convex set $C \subset \mathbb{R}^p$. Generally speaking, a DC program takes the form

$$\alpha = \inf\{f(x) := g(x) - h(x) : x \in \mathbb{R}^p\} \quad (P_{dc}) \tag{3}$$

where g, h are lower semicontinuous proper convex functions on $\mathbb{R}^p$. Such a function f is called DC function, and $g - h$, DC decomposition of f while g and h are DC components of f. The convex constraint $x \in C$ can be incorporated in the objective function of (P_{dc}) by using the indicator function on C denoted χ_C which is defined by $\chi_C(x) = 0$ if $x \in C$, $+\infty$ otherwise. Let

$$g^*(y) := \sup\{\langle x, y \rangle - g(x) : x \in \mathbb{R}^p\}$$

be the conjugate function of g where $\langle . \rangle$ denotes a scalar product. Then, the following program is called the dual program of (P_{dc}):

$$\alpha_D = \inf\{h^*(y) - g^*(y) : y \in \mathbb{R}^p\}. \quad (D_{dc}) \tag{4}$$

One can prove that $\alpha = \alpha_D$, and there is the perfect symmetry between primal and dual DC programs: the dual to (D_{dc}) is exactly (P_{dc}).

For a convex function θ, the subdifferential of θ at $x_0 \in \text{dom } \theta := \{x \in \mathbb{R}^p : \theta(x) < +\infty\}$, denoted $\partial\theta(x_0)$, is defined by

$$\partial\theta(x_0) := \{y \in \mathbb{R}^n : \theta(x) \geq \theta(x_0) + \langle x - x_0, y \rangle, \forall x \in \mathbb{R}^p\}. \tag{5}$$

The subdifferential $\partial\theta(x_0)$ generalizes the derivative in the sense that θ is differentiable at x_0 if and only if $\partial\theta(x_0) \equiv \{\theta'(x_0)\}$.

The idea of DCA is quite simple: each iteration of DCA approximates the concave part $-h$ by one of its affine majorization defined by $y^k \in \partial h(x^k)$ and minimizes the resulting convex function (i.e., computing $x^{k+1} \in \partial g^*(y^k)$).

DCA Scheme

INPUT

– Let $x^0 \in \mathbb{R}^p$ be a best guest, $0 \leftarrow k$.

REPEAT

– Calculate $y^k \in \partial h(x^k)$.
– Calculate

$$x^{k+1} \in \arg\min \left\{ g(x) - h(x^k) - \langle x - x^k, y^k \rangle \quad s.t. x \in \mathbb{R}^p \right\}. \qquad (P_k)$$

– $k + 1 \leftarrow k$.

UNTIL{convergence of x^k.}

Convergence properties of DCA and its theoretical basis can be found in [8,10,18,19]. It is important to mention that

– DCA is a descent method (the sequences $\{g(x^k) - h(x^k)\}$ and $\{h^*(y^k) - g^*(y^k)\}$ are decreasing) *without linesearch*;
– If the optimal value α of problem (P_{dc}) is finite and the infinite sequences $\{x^k\}$ and $\{y^k\}$ are bounded then every limit point x^* (resp. $\tilde{y}$) of the sequence $\{x^k\}$ (resp. $\{y^k\}$) is a critical point of $g - h$ (resp. $h^* - g^*$), i.e. $\partial h(x^*) \cap \partial g(x^*) \neq \emptyset$ (resp. $\partial h^*(y^*) \cap \partial g^*(y^*) \neq \emptyset$).
– DCA has a *linear convergence* for general DC programs.

It is interesting to note that [8,10,18,19] DCA works with the convex DC components g and h but not the DC function f itself. Moreover, a DC function f *has infinitely many DC decompositions which have crucial impacts on the qualities* (speed of convergence, robustness, efficiency, globality of computed solutions,...) of DCA.

We remark that the convex concave procedure (CCCP) for constructing discrete time dynamical systems mentioned in [23] is nothing else but a special case of DCA. In the last five years DCA has been successfully applied in several works in Machine Learning for SVM-based Feature Selection [15], for improving boosting algorithms [7], for implementing-learning [13,14,22], for Transductive SVMs [21] and for unsupervised clustering [11,12].

3.2   A Nice DC Reformulation of FCM Model

In the problem (2) the variable U is a priori bounded. One can also bound the variable V. Indeed, the first order necessary optimality conditions in (U, V) imply that $\nabla_V J_m(U, V) = 0$, i.e.,

$$\partial_{v_i} J_m(U, V) = \sum_{k=1}^{n} u_{i,k}^m 2(v_i - x_k), \quad \forall i = 1, .., c, \forall k = 1, .., n$$

or $v_i \sum_{k=1}^{n} u_{i,k}^m = \sum_{k=1}^{n} u_{i,k}^m x_k$.

On the other hand, the nonemptiness of all clusters imposes that $\sum_{k=1}^{n} u_{i,k}^m > 0$, for $i = 1, .., c$. Hence

$$\|v_i\|^2 \leq \frac{(\sum_{k=1}^{n} u_{i,k}^m \|x_k\|)^2}{(\sum_{k=1}^{n} u_{i,k}^m)^2} \leq \sum_{k=1}^{n} \|x_k\|^2 := r^2.$$

We introduce below a *nice* DC reformulation of the problem (2) for which the DCA resulting is explicitly determined via a very simple formula.

Let us consider now the new variables $t_{i,k}$ such that $u_{i,k} = t_{i,k}^2$. The constraint $\sum_{i=1}^{c} u_{i,k} = 1$ becomes

$$\sum_{i=1}^{c} t_{i,k}^2 = 1 \text{ or } \|t_k\|^2 = 1 \text{ with } t_k \in \mathbb{R}^c.$$

Let S_k (resp. R_i) be the Euclidean sphere (resp. ball) centered at the origin and of radius 1 (resp. r) in $\mathbb{R}^c$ (resp. $\mathbb{R}^p$). We can reformulate the problem (2) as:

$$\begin{cases} \min J_{2m}(T, V) := \sum_{k=1}^{n} \sum_{i=1}^{c} t_{i,k}^{2m} \|x_k - v_i\|^2 \\ s.t \quad T \in \mathcal{S} := \Pi_{k=1}^{n} S_k, \ V \in \mathcal{C} := \Pi_{i=1}^{c} R_i \end{cases} \tag{6}$$

For finding a DC decomposition of the objective function of (6) we express it in the form

$$J_{2m}(T, V) = \frac{\rho}{2}(\|T\|^2 + \|V\|^2) - \left[\frac{\rho}{2}\|(T, V)\|^2 - J_{2m}(T, V)\right]$$

and then for all $(T, V) \in \mathcal{S} \times \mathcal{C}$ we have

$$J_{2m}(T, V) = \frac{\rho}{2}n + \frac{\rho}{2}\|V\|^2 - H(T, V)$$

with

$$H(T, V) := \frac{\rho}{2}\|(T, V)\|^2 - J_{2m}(T, V). \tag{7}$$

In the Proposition below we will give the conditions which ensure the convexity of the function H.

Proposition 1. *Let $\mathcal{B} := \Pi_{k=1}^{n} B_k$, where B_k is the ball of centre 0 and radius 1 in $\mathbb{R}^c$. The function $H(T, V)$ is convex on $\mathcal{B} \times \mathcal{C}$ for all values of ρ such that*

$$\rho \geq \frac{m}{n}(2m-1)\alpha^2 + 1 + \sqrt{\left[\frac{m}{n}(2m-1)\alpha^2 + 1\right]^2 + \frac{16}{n}m^2\alpha^2}, \qquad (8)$$

where

$$\alpha = r + \max_{1 \leq k \leq n} \|x_k\|. \qquad (9)$$

Proof. first, we note that $\rho > 0$ because $m \geq 1$.

Since

$$H(T, V) = \sum_{k=1}^{n} \sum_{i=1}^{c} \left[\frac{\rho}{2}t_{i,k}^2 + \frac{\rho}{2n}\|v_i\|^2 - t_{i,k}^{2m}\|x_k - v_i\|^2\right],$$

H is convex when all the functions

$$h_{i,k}(t_{i,k}, v_i) := \frac{\rho}{2}t_{i,k}^2 + \frac{\rho}{2n}\|v_i\|^2 - t_{i,k}^{2m}\|x_k - v_i\|^2$$

are convex for $i = 1, .., c$, $k = 1, .., n$.
Consider the next function :

$$\begin{array}{l} f : \quad \mathbb{R} \times \mathbb{R} \to \mathbb{R} \\ f(x, y) = \frac{\rho}{2}x^2 + \frac{\rho}{2n}y^2 - x^{2m}y^2 \cdot \end{array} \qquad (10)$$

The Hessian of f is given by:

$$J_f(x, y) = \begin{pmatrix} \rho - 2m(2m-1)y^2 x^{2m-2} & -4mx^{2m-1}y \\ -4mx^{2m-1}y & \frac{\rho}{n} - 2x^{2m} \end{pmatrix}. \qquad (11)$$

For all $(x, y) : 0 \leq x \leq 1; \|y\| \leq \alpha$, we have: ($|J_f(x, y)|$ is the determinant of $J_f(x, y)$)

$$|J_f(x, y)| = \left(\rho - 2m(2m-1)y^2 x^{2m-2}\right)\left(\frac{\rho}{n} - 2x^{2m}\right) - 16m^2 x^{4m-2}y^2$$

$$\geq \frac{1}{n}\rho^2 - \left[2\frac{m}{n}(2m-1)y^2 x^{2m-2} + 2x^{2m}\right]\rho - 16m^2 x^{4m-2}y^2$$

$$\geq \frac{1}{n}\rho^2 - 2\left(\frac{m}{n}(2m-1)\alpha^2 + 1\right)\rho - 16m^2\alpha^2.$$

Consequently, with ρ and α defined in (8) and (9) respectively, we have $|J_f(x, y)| \geq 0$, for all $(x, y) \in \mathbb{R}^2$ such that $0 \leq x \leq 1, |y| \leq \alpha$.

Hence, the function f is convex on $[0, 1] \times [-\alpha, \alpha]$. Therefore, the functions

$$\theta_{i,k}(t_{i,k}, v_i) := \frac{\rho}{2}t_{i,k}^2 + \frac{\rho}{2n}\|x_k - v_i\|^2 - t_{i,k}^{2m}\|x_k - v_i\|^2$$

are convex on $\{0 \leq t_{i,k} \leq 1, \|v_i\| \leq r\}$.
Likewise, the function $h_{i,k}$ is convex, because

$$h_{i,k}(t_{i,k}, v_i) = \theta_{i,k}(t_{i,k}, v_i) + \frac{\rho}{n}\langle x_k, v_i \rangle - \frac{\rho}{2n}\|x_k\|^2.$$

Finally the function $H(T, V)$ is convex on $\mathcal{B} \times \mathcal{C}$ with ρ defined in (8) and α given in (9).

In the sequel we will work with these values of ρ and α.

It is clear that for all $T \in \mathcal{B}$ and a given matrix $V \in \mathcal{C}$, the function $J_{2m}(T, V)$ is concave in variable T (since $H(T, V)$ is convex). Hence $\mathcal{S}$ contains minimizers of $J_{2m}(T, V)$ on $\mathcal{B}$, i.e.,

$$\min \left\{ \frac{\rho}{2} \|V\|^2 - H(T, V) : (T, V) \in \mathcal{B} \times \mathcal{C} \right\}$$

$$= \min \left\{ \frac{\rho}{2} \|V\|^2 - H(T, V) : (T, V) \in \mathcal{S} \times \mathcal{C} \right\}.$$

The problem (6) can be now reformulated as

$$\min \left\{ \frac{\rho}{2} \|V\|^2 - H(T, V) : (T, V) \in \mathcal{B} \times \mathcal{C} \right\},$$

or again

$$\min \begin{cases} \chi_{\mathcal{B} \times \mathcal{C}}(T, V) + \frac{\rho}{2} \|V\|^2 - H(T, V) \\ s.t. \ (T, V) \in \mathbb{R}^{c \times n} \times \mathbb{R}^{c \times p}. \end{cases} \tag{12}$$

This is a DC program with the following DC decomposition:

$$\chi_{\mathcal{B} \times \mathcal{C}}(T, V) + \frac{\rho}{2} \|V\|^2 - H(T, V) := G(T, V) - H(T, V),$$

where $G(T, V) := \chi_{\mathcal{B} \times \mathcal{C}}(T, V) + \frac{\rho}{2} \|V\|^2$ is evidently convex function, due to the convexity of $\mathcal{B}$ and $\mathcal{C}$.

3.3   Solving (12) by DCA

According to the above description of DCA, solving FCM via the DC formulation (12) by DCA consists in the construction of two sequences $(Y^l, Z^l) \in \partial H(T^l, V^l)$ and

$$(T^{l+1}, V^{l+1}) \in \arg\min \begin{cases} \frac{\rho}{2} \|V\|^2 - \langle (T, V), (Y^l, Z^l) \rangle \\ s.t. \ (T, V) \in \mathcal{B} \times \mathcal{C}. \end{cases} \tag{13}$$

The function H is differentiable and its gradient at the point (T^l, V^l) is given by:

$$\nabla H(T^l, V^l) = \rho(T^l, V^l) -$$
$$(2m t_{i,k}^{2m-1} \|x_k - v_i\|^2, 2 \sum_{k=1}^{n} (v_i - x_k) t_{i,k}^{2m}). \tag{14}$$

The solution of Problem (13) is explicitly computed as (Proj denotes the projection)

$$T^{l+1} = \mathrm{Proj}_{\mathcal{B}}(Y^l), \quad V^{l+1} = \mathrm{Proj}_{\mathcal{C}}\left(\frac{1}{\rho} Z^l\right)$$

More precisely:

$$V_{i,.}^{l+1} = \begin{cases} \frac{(Z^l)_{i,.}}{\rho} & \text{if } \|(Z^l)_{i,.}\| \le \rho r \\ \frac{(Z^l)_{i,.} r}{\|(Z^l)_{i,.}\|} & \text{otherwise} \end{cases}, \quad i = 1, .., c, \tag{15}$$

$$T^{l+1}_{.,k} = \begin{cases} Y^l_{.,k} & \text{if } \|Y^l_{.,k}\| \leq 1 \\ \frac{(Y^l)_{.,k}}{\|(Y^l)_{.,k}\|} & \text{otherwise} \end{cases} \quad , k = 1, .., n. \tag{16}$$

where, for a matrix A, $A_{i,.}$ and $A_{.,k}$ denote its ith row and k^{th} column respectively.

Algorithm 1. DCA applied to (12)

INPUT

- $T^0 \in \mathbb{R}^{c \times n}$ and $V^0 \in \mathbb{R}^{c \times p}$.
- $l = 0$. Let $\epsilon > 0$ be sufficiently small number.

REPEAT

- Calculate $(Y^l, Z^l) = \nabla H(T^l, V^l)$ via (14);
- Calculate (T^{l+1}, V^{l+1}) via (15) and (16);
- $l + 1 \leftarrow l$.

UNTIL$\{\|(T^{l+1}, V^{l+1}) - (T^l, V^l)\| \leq \epsilon(\|(T^{l+1}, V^{l+1})\|)\}$

Segmentation: Let (T^*, V^*) be the solution calculated by DCA and $u_{i,k} = t^{*2}_{i,k}$. The point x_k belongs to the class C_i if $u_{ik} = \max_{j=1..c} u_{j,k}$.

4   Accelerating DCA by an Alternative FCM-DCA Procedure

Finding a good starting point plays a crucial role in the solution of DC programs by DCA. The research of such a point depends on the structure of the problem being considered and can be done by, for example, a heuristic procedure. Generally speaking a good starting point for DCA must not be a *local minimizer*, because DCA is stationary from such a point. Moreover, we observe that from any non local minimizer, the objective function is decreasing rapidly during some first iterations of DCA. We have the same remark for the classical FCM algorithm. That is why we propose an alternative FCM-DCA procedure for Problem (12) called Algorithm 2.

Algorithm 2. Combined FCM-DCA algorithm

INPUT

- Let U^0 and V^0 be the membership and the cluster centers randomly generated.
- Set $l = 0$. Let $\epsilon > 0$ be sufficiently small number.

REPEAT
i. One iteration of FCM:

- Compute the cluster centers V^l via

$$v_i = \sum_{k=1}^{n} u_{ik}^m x_k \Big/ \sum_{k=1}^{n} u_{ik}^m \quad \forall i = 1, .., c. \tag{17}$$

- Compute the membership U^l via

$$u_{ik} = \left[\sum_{j=1}^{c} \frac{\|x_k - v_i\|^{2/(m-1)}}{\|x_k - v_j\|^{2/(m-1)}} \right]^{-1}. \tag{18}$$

- Set $t_{ik} = \sqrt{u_{ik}}$, $\forall i = 1, .., c$ and $\forall k = 1, .., n$.

ii. One iteration of DCA:

- Calculate $(Y^l, Z^l) = \nabla H(T^l, V^l)$ via (14);
- Calculate (T^{l+1}, V^{l+1}) via (15) and (16);
- $l + 1 \leftarrow l$

UNTIL$\{\|(T^{l+1}, V^{l+1}) - (T^l, V^l)\| \leq \epsilon(\|(T^{l+1}, V^{l+1})\|)\}$.

If we use the combined FCM-DCA Algorithm until its convergence, the efficiency of DCA may not be well exploited. An efficient algorithm based on DCA may be a two phase DCA algorithm in which the Phase 1 deals with some iterations of the combined FCM-DCA Algorithm and the phase 2 consists of applying DCA from the point given by Phase 1. The algorithm can be summarized as **Algorithm 3**.

Algorithm 3. The two phase algorithm

INPUT

- Let U^0 and V^0 be the membership and the cluster centers randomly generated.
- Set $l = 0$. Let $\epsilon > 0$ be sufficiently small number.

PHASE 1:

- Perform q iterations of **Algorithm 2** for obtaining (T^{q+1}, V^{q+1}).
- Update $(T^0, V^0) \leftarrow (T^{q+1}, V^{q+1})$

PHASE 2:

- Apply **Algorithm 1** from the initial point (T^0, V^0) until the convergence.

---

5   Implementations and Results

The algorithms are coded in C++ and run on PC Pentium[R] 4 CPU 3.00GHz 1.00Go RAM. For testing the efficiency of the proposed method, we first use an original image of multiple regions where the gray level inside each region varies within certain limits. We add then, in the same image, Gaussian noise with different signals to noise ratio and obtain the noisy image. We have tested the four algorithms: the classical FCM algorithm, and our **Algorithms 2, 3** on four images among them two MRI images. The computational results ares reported in the figures below. In these figures

- (a) (resp. (b)) corresponds to the original image without (resp. with) noise;
- (c) (resp. (d)) represents the resulting image given by FCM Algorithm without (resp. with) spatial information.
- (e) represents the resulting image given by **Algorithm 2** without spatial information
- (f) (resp. (g)) represents the resulting image given by **Algorithm 3** without (resp. with) spatial information.

In Table 1 we report the comparative computational results of FCM Algorithm and our **Algorithms 2 and 3** tested on 10 images. We use the following notations: $Size$: size of each image; c: the number of clusters of image; N^oF: the number iteration of FCM Algorithm; N^oI: the number iteration of **Algorithm 2**; q: the number of iterations of FCM-DCA in phase 1 in **Algorithm 3**; N^oD: the number of iterations of DCA in phase 2 of **Algorithm 3**; $Time$: the CPU of each algorithm in seconds.

Table 1. Computation time of **FCM Algorithm** and **Algorithm 2, 3**

Data			FCM		Algorithm 2		Algorithm 3		
N^o	Size	c	N^oF	Time	N^oI	Time	q	N^oD	Time
1	128^2	2	24	1.453	16	1.312	12	10	1.219
2	128^2	2	17	1.003	12	0.985	10	2	0.765
3	256^2	3	36	15.340	24	13.297	20	2	10.176
4	256^2	3	75	31.281	57	30.843	30	12	26.915
5	256^2	3	39	15.750	27	14.687	20	14	13.125
6	256^2	5	91	84.969	75	86.969	40	78	61.500
7	256^2	3	73	31.094	62	34.286	15	21	24.188
8	256^2	3	78	34.512	52	32.162	20	13	29.182
9	512^2	3	49	92.076	41	102.589	30	46	74.586
10	512^2	5	246	915.095	196	897.043	120	86	691.854

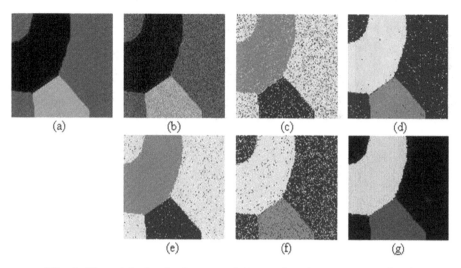

Fig. 1. The original noisy image and the results of segmentation ($c=3$)

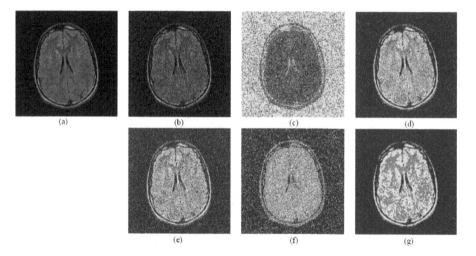

Fig. 2. The medical noisy image and the results of segmentation ($c=3$)

Comments. From the experimental results we see that:

- Unlike FCM algorithm, our **Algorithm 2, 3** without spatial information can overcome the noisy image segmentation in some cases.
- **Algorithm 3** with spatial information works well on all noisy images - it can segment the target from the noisy background more effectively.
- **Algorithm 3** is faster than FCM Algorithm, because the computation of the projections of points on Euclidean balls are explicit.

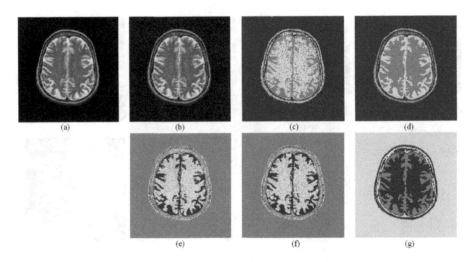

Fig. 3. The medical noisy image and the results of segmentation (c=3)

Fig. 4. The **Blume** noisy image and the results of segmentation (c=5)

Conclusion. We have proposed an efficient approach for the image segmentation based on Fuzzy C-Means clustering model via DCA. The FCM model is beforehand reformulated, in an elegant way, as a DC program so that the resulting DCA is very simple and fast. The alternative FCM-DCA procedure is efficient for finding a good initial point of DCA and for accelerating its convergence. The two phase DCA algorithm can then handle large scale clustering problems. On the other hand, the use of the spatial information for noisy image segmentation seems to be efficient. Preliminary numerical simulations show that the proposed algorithms are promising for the noisy image segmentation.

References

1. Ahmed, M.N., Yamany, S.M., Mohamed, N., Farag, A.A., Moriarty, T.: A modified fuzzy C-means algorithm for bias field estimation and segmentation of MRI data. IEEE Trans. on Medical Imaging 21, 193–199 (2002)
2. Bezdek, J.C.: Pattern Recognition with Fuzzy Objective Function Algorithm. Plenum Press, New York (1981)
3. Bezdek, J.C., Hall, L.O., Clake, L.P.: Review of MR image segmentation techniques using pattern recognition. Medical Physics 20, 1033–1048 (1993)
4. Chuang, K.S., Tzeng, H.L., Chen, S., Wu, J., Chen, T.J.: Fuzzy c-means clustering with spatial information for image segmentation. Computerized Medical Imaging and Graphics 30, 9–15 (2006)
5. Dunn, J.C.: A Fuzzy Relative of the ISODATA Process and Its Use in Detecting Compact Well-Separated Clusters. Journal of Cybernetics 3, 32–57 (1973)
6. Hung, W.L., Yang, M.S., Chen, D.H.: Parameter selection for suppressed fuzzy c-means with an application to MRI segmentation. Pattern Recognition Letters 27, 424–438 (2006)
7. Krause, N., Singer, Y.: Leveraging the margin more carefully. In: International Conference on Machine Learning ICML (2004)
8. Le Thi, H.A.: Contribution á l'optimisation non convexe et l'optimisation globale: Théorie, Algorithmes et Applications. Habilitation á Diriger des Recherches, Université de Rouen (1997)
9. Le Thi, H.A., Pham Dinh, T.: Large scale molecular optimization from distance matrices by a D.C. optimization approach. SIAM Journal on Optimization 14, 77–116 (2003)
10. Le Thi, H.A., Pham Dinh, T.: The DC (difference of convex functions) Programming and DCA revisited with DC models of real world nonconvex optimization problems. Annals of Operations Research 133, 23–46 (2005)
11. Le Thi, H.A., Belghiti, T., Pham Dinh, T.: A new efficient algorithm based on DC programming and DCA for Clustering. Journal of Global Optimization (July 2006) (in press)
12. Le Thi, H.A., Le Hoai, M., Pham Dinh, T.: Optimization based DC programming and DCA for Hierarchical Clustering. European Journal of Operational Research (June 2006) (in press)
13. Liu, Y., Shen, X., Doss, H.: Multicategory ψ-Learning and Support Vector Machine: Computational Tools. Journal of Computational and Graphical Statistics 14, 219–236 (2005)
14. Liu, Y., Shen, X.: Multicategory ψ-Learning. Journal of the American Statistical Association 101, 500–509 (2006)
15. Neumann, J., Schnörr, C., Steidl, G.: SVM-based Feature Selection by Direct Objective Minimisation. In: Pattern Recognition, Proc. of 26th DAGM Symposium, pp. 212–219 (2004)
16. Pham, D.L.: Image segmentation using probabilistic fuzzy c-means clustering. Image Processing 1, 722–725 (2001)
17. Pham, D.L.: Fuzzy Clustering with spatial constraints. In: Proc.IEEE Intern. Conf. on Image Processing, New Yord, USA (August 2002)
18. Pham Dinh, T., Le Thi, H.A.: Convex analysis approach to DC programming: Theory, Algorithms and Applications. Acta Mathematica Vietnamica 22, 289–355 (1997)

19. Pham Dinh, T., Le Thi, H.A.: DC optimization algorithms for solving the trust region subproblem. SIAM J. Optimization 8, 476–505 (1998)
20. Rajapakse, J.C., Giedd, J.N., Rapoport, J.L.: Statistical Approach to Segmentation of Singke-Chanel Cerebral MR Images. IEEE Trans. On Medical Imaging 16 (April 2004)
21. Ronan, C., Fabian, S., Jason, W., Léon, B.: Trading Convexity for Scalability. In: International Conference on Machine Learning ICML (2006)
22. Shen, X., Tseng, G.C., Zhang, X., Wong, W.H.: On ψ-Learning. Journal of American Statistical Association 98, 724–734 (2003)
23. Yuille, A.L., Rangarajan, A.: The Convex Concave Procedure (CCCP). In: Advances in Neural Information Processing System, vol. 14, MIT Press, Cambridge (2002)
24. Weber, S., Schüle, T., Schnörr, C.: Prior Learning and Convex-Concave Regularization of Binary Tomography. Electr. Notes in Discr. Math. 20, 313–327 (2005)
25. Zhang, D.Q., Chen, S.C.: A novel kernelized fuzzy C-means algorithm with application in medical image segmentation. Artificial Intelligence in Medicine 32, 37–50 (2004)

An Application for Electroencephalogram Mining for Epileptic Seizure Prediction

Bruno Direito[1], António Dourado[1], Francisco Sales[2], and Marco Vieira[1]

[1] Centro de Informática e Sistemas da Universidade de Coimbra
Department of Informatics Engineering
Polo II Universidade 3030-290 Coimbra, Portugal
{brunodireito,dourado,mvieira}@dei.uc.pt
[2] Hospitais da Universidade de Coimbra,
Praceta Mota Pinto 3000-075 Coimbra, Portugal
franciscosales@huc.min-saude.pt

Abstract. A computational framework to support seizure predictions in epileptic patients is presented. It is based on mining and knowledge discovery in Electroencephalogram (EEG) signal. A set of features is extracted and classification techniques are then used to eventually derive an alarm signal predicting a coming seizure. The epileptic patient may then take steps in order to prevent accidents and social exposure.

1 Introduction

EEG signals have a high content informative potential for brain working conditions. Software developments for biomedical signal processing tools offer more and more the possibility to focus mainly on features relevant segments for the intended study. The free availability of more general and easy-to-use signal processing software for EEG data might encourage the wider adoption of more inclusive approaches in order to a better extraction of the EEG relevant information. For example the EEGLab [1] or Biosig [2] software toolboxes for Matlab are some of these advances, allowing the processing of data through a collection of various techniques such as Independent Component Analysis (ICA) and spectral analysis as well as data averaging techniques.

Among the relevant information extracted from EEG analysis one can find several abnormalities which can indicate the existence of structural brain anomalies, head injury, haemorrhage or seizure disorders (such as epilepsy).

Epilepsy is among the most common neurological disorders, and represents temporary and reversible electric activity in the brain. Epilepsy is characterized by occasional, excessive and disorderly discharging of neurons, which can be detected by clinical manifestations, the seizures. This disturbed activity, can cause strange sensations, emotions, and behaviors or sometimes convulsions, muscle spasms, and loss of consciousness. Several researchers consider that the period preceding seizures, the pre-ictal period, may present identifiable electrical variations. Patient-specific algorithms based on several mathematical and computational methods have shown a high potential for successful development of seizure prediction algorithms.

P. Perner (Ed.): ICDM 2008, LNAI 5077, pp. 87–101, 2008.
© Springer-Verlag Berlin Heidelberg 2008

This area of research generally includes the analysis of nonlinear dynamics, wavelet transform, signal quantification, signal synchronization among others [11] [12] [13] [14] [15].

This work presents an application under development integrating several methodologies and tools in a common platform, in Matlab environment, for seizure prediction studies. The goal of this application is to apply computational approaches such as time-frequency analysis and spectral analysis, multidimensional scaling, computational intelligence techniques, to EEG processing and epileptic seizure prediction. The development of such application is the result of the observation of the need to provide the research and medical communities with an easy to use general signal processing and data mining application, capable of reading the most common data formats and with a friendly user interface.

The basis of the software under development is the integration of free software in a common framework in Matlab environment [1] [3] [4][16] and the programming of several algorithms for classification based on energy concepts, wavelets and nonlinear theory.

2 Overview of the Application

The source code was developed in Matlab, and is the result of integration of several toolboxes (such as Matlab wavelet toolbox, EEGlab, TSTool, VISRED) with some specific algorithms developed for seizure prediction (Fig.1).

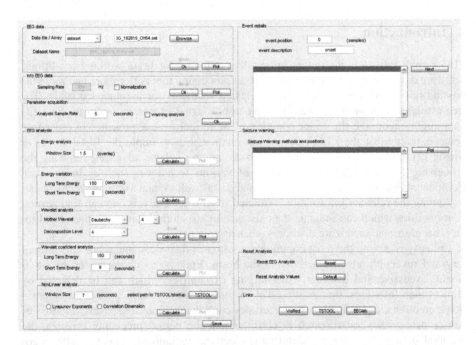

Fig. 1. Application overview, developed algorithms and data selection on the left, and warnings and event panels on the right

2.1 Data Structure

Since EEGlab accepts most of data formats used in clinics, two possibilities are presently available: .set file (EEGlab dataset) or an ASCII file representing a single channel (if the selection is ASCII an EEGlab dataset is created, and the file is saved according to the user selection).

Data information – to process the signal, the user has to define two fields: sampling rate (Hz) and, if required, normalization around zero, in the interval [-100, 100].

Parameter acquisition – Analysis interval represents the time interval between two sets of features computations (default value is 5 seconds). In other words, this value represents the window shift. The segment length is determined for each method.

2.2 EEG Signal Analysis- Basic Functions

Accumulated energy – One of the concepts usually used in the processing of EEG signals is the concept of signal energy. Several authors connect energy variations to epileptic seizure prediction [5] [6].

The algorithm presented consists in the average value of the signal energy during the analysis interval. Consists in the determination of average energy of short segments; the average energy values are determined through a sliding window, overlapping with the previous one to obtain a better resolution.

Energy variation represents a two windowed algorithm, with the purpose of identifying variations between EEG signal long-term energy and short-term energy [5].

Wavelet analysis – presents an algorithm based on the wavelet transform of the EEG signal. The signal processing is made using Matlab wavelet toolbox. The user has to define the mother wavelet and the decomposition level. The down sample associated with each decomposition is surpassed with simple upsampling of the resulting feature.

Wavelet coefficients energy - the coefficients obtained through wavelet decomposition are processed in two separate windows with different lengths. The average energy of each window is then determined and compared. This is the former energy analysis applied to the wavelet coefficients of each decomposition level.

Nonlinear analysis – calculates two features based on nonlinear concepts: Lyapunov exponents and correlation dimension. These features are determined through nearest neighbors theory implemented in the toolbox TSTool[3][7].

2.3 Multidimensional Scaling

The user interface allows calling VISRED [16] for reduction of the information space dimension in order to visualize in a two or three dimensional space. Clustering techniques may then be applied to classify data in the reduced space.

2.4 Event Details

The developed toolbox allows the user to compare graphically the time course of the selected features. Vertical lines representing events, whose inputs consist in two

Event details

event position   55555   (samples)

event description   end

Event: description "onset" and position 0 (samples);
Event: description "end" and position 55555 (samples);

Next

Fig. 2. Event details panel – position and description of events

variables, event position (sample number where the event occurs or begins) and the event description (brief description of the event, Fig.2).

2.5 Seizure Warning

Throughout the algorithms, when the values of the features are above a threshold established, the time stamps are added to vectors. These are represented in the listbox in the seizure warning panel. The values can be observed in a figure (Fig 3).

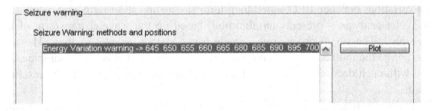

Seizure warning

Seizure Warning: methods and positions

Energy Variation warning -> 645 650 655 660 665 680 685 690 695 700

Plot

Fig. 3. Seizure warning panel – the user can visually identify where the values of the algorithm are above the threshold established

2.6 Saving Data

The processed data can be saved in two formats, Matlab .mat and in excel .xls file. If the user intends to save in a .mat file, a Matlab structure is saved with all the main variables processed (data and acquisition times associated with each feature).

2.7 Data Visualization

A plot option is associated with all features in analysis. Events can be added to these plots in the event panel (Fig. 4).

In the algorithms, thresholds have been settled to identify abnormal values. When the feature values exceed these thresholds, the time stamp is added to vectors that can be represented in the panel seizure warning (Fig.5).

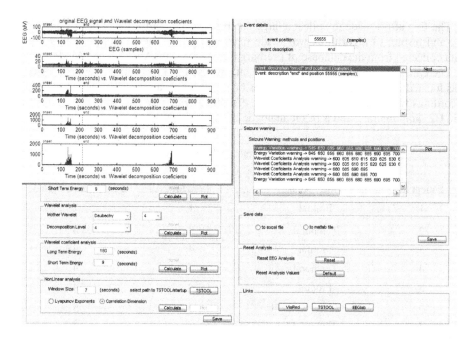

Fig. 4. Interface example. Blue dotted vertical lines represent events in the plots and on the right, event panel (position, description and listbox.)

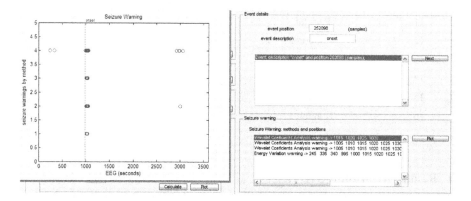

Fig. 5. Seizure warning plot example – the red circles represent the warning obtained by the thresholds settled in each method

3 Case Study, Freiburg Database

EEG Signal processing and seizure prediction requires the existence of large publicly available datasets from a variety of patients. One of the aspects that are clearly limiting the results in the seizure prediction research is the inability to access epileptic EEG signal.

The first effort made to collect a large database, sufficient for the appropriate training and testing of algorithms was made in Freiburg [8]. The results presented used data collected from the database of Freiburg Center for Data Analysis and Modeling.

The epileptic focus in the Freiburg patients, whose EEG recordings were processed, was located in neocortical brain structures or in the hippocampus [8]. The intracranial recordings were acquired using Neurofile NT digital video system with 128 channels, 256 Hz sampling rate, and a 16 bit analogue-to-digital converter (table 1).

Table 1. The Freiburg database [8]

Patient Id	Onset Area	# seizure	Total data proc-
1	Frontal	1	2
2	Temporal	3	4
3	Frontal	4	4
4	Temporal	3	4
5	Frontal	4	4
6	Temporo/Occipital	3	4
7	Temporal	2	3
8	Frontal	2	3
9	Temporo/Occipital	3	4
10	Temporal	2	3
11	Parietal	3	4
12	Temporal	2	3
13	Temporo/Occipital	2	3
14	Fronto/Temporal	4	5
15	Temporal	3	3
16	Temporal	4	4
19	Frontal	3	5

3.1 Results, Algorithms Overview

In this study, various techniques for EEG processing were developed to confirm the presence of quantifiable variations before the onset of epileptic seizures. Methods based on various mathematical concepts (energy, wavelet transform and nonlinear dynamics), were the foundation of the development of this Matlab toolbox.

The individual analysis of the extracted features did not present the expected results. After the analysis of previous investigations and respective results, our conclusions, obtained by approaches based on those investigations, produced inferior results.

Nevertheless, in several events, undeniable variations occur before seizures, suggesting that these variations could be related to pre-seizure activity.

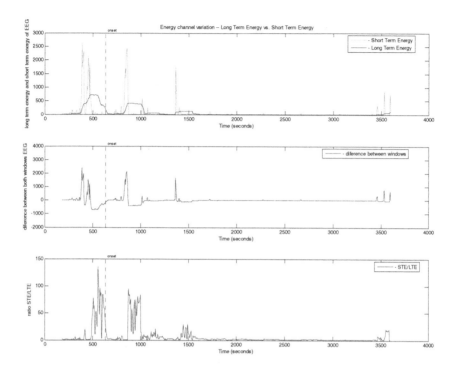

Fig. 6. The software allows to plot the obtained data by the analysis of the energy variation algorithm, the plot clearly demonstrate an energy increase before the seizure onset

3.2 Energy Analysis

The windowed average power and accumulated energy were analyzed and plotted for the entire data sets. Generally, the computation of windowed energy on intracranial EEG recordings did not reveal consistent increases or changes prior to seizures. In certain cases and for certain seizures, a distinctive increase of the STE/LTE factor occurred some time before the seizure onset.

These energy events have already been discussed by several authors [9]. However, the most significant increases occurred at or after the onset of seizures.

3.3 Wavelets Transform Coefficients Analysis

In general, some frequency bands present variations before seizure onsets; unfortunately, various electrographic events induce the existence of similar variations in inter-ictal periods. The existence of these pre-ictal variations, have already been described by several authors [10] [11].

Hence, the computation of coefficients did not present consistent variations prior to seizures; the existence of specific pre-ictal patterns was not proved by the analysis. In specific cases and for certain seizures, a distinctive increase of the STE value occurred some time before the seizure onset.

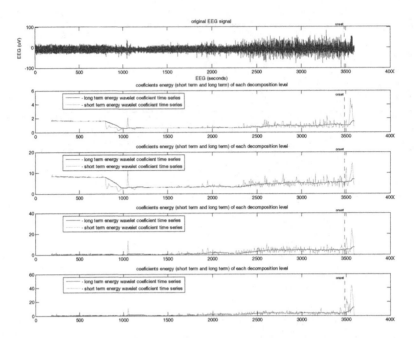

Fig. 7. The application under development allows the user to identify the signal segments presenting severe frequency variations through the analysis of the energy of the wavelet coefficients. The plot presents increasing variations before the seizure onset.

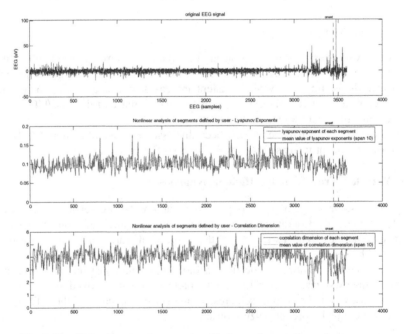

Fig. 8. Identifying the signal segments with decreasing nonlinear features

The developed tool allows the users to seek frequency variations throughout the signal, allowing the correspondence of these variations with pre-selected events and warning alarms when these variations exceed a previously settled threshold.

These variations may be involved in pre-seizure mechanisms; further investigations are necessary to recognize consistent pattern variations.

3.4 Nonlinear Dynamics Analysis

Several authors report evidences for a characteristic pre-seizure state transition and seizure predictability identifiable through nonlinear quantification. Two nonlinear parameters, Lyapunov exponents and correlation dimension, were considered and implemented in the application.

Some of the processed recordings present a decrease in the complexity of attractors of pre-seizure signal segments.

4   Results and Discussion

One of the main concerns of this research was to understand the importance the specificity of each method parameters. The uniqueness of each patient brain activity led us to investigate the importance of training each method with signal of each patient, and onset area. The evaluation of the results would represent the first key step to guide our research.

Using a univariate analysis approach to process intracranial EEG, for each method used (energy variation, wavelet decomposition coefficient energy analysis, Lyapunov exponents) patient-specific parameters were calculated. These values were grouped by onset area, and the parameters to each group of patient were calculated by the mean of each individual in the groups. Finally, the mean values of the parameters processed for each patient were calculated to compare the sensitivity with the two previous parameters sets.

The results presented were obtained using a seizure prediction horizon of 15 minutes. The following tables show the result of these efforts.

4.1 Energy Analysis

As can be observed in Table 2, the best results are obtained using patient-specific parameters.

Considerable differences both in sensitivity and false positive rate occur when the threshold value is determined by non specific parameters. A decrease in the number of correct predictions and an increased number of false positive result from a decrease in the specificity of the parameters associated to the method threshold. The results confirm the assumption that each patient has specific properties and the methods should take these properties into account.

Table 2. Energy analysis, prediction statistics comparison between parameteres determined according to patient-specific, onset area and total. CP- Correct predictions, FP- False Positives.

On-set Area	Patient	N° seizures	Patient-specific		Onset area		total	
			CP	FP	CP	FP	CP	FP
F	pac1	1	1	0	1	2	0	0
T	pac2	3	2	1	0	0	0	0
F	pac3	4	2	1	0	0	0	0
T/O	pac6	3	2	1	2	5	2	6
T	pac7	1	0	1	0	1	0	2
F	pac8	2		2	0	2	0	2
T/O	pac9	3	3	2	0	1	0	1
T	pac10	3	2	2	1	4	1	4
P	pac11	3	2	1	2	0	0	0
T	pac12	2	2	1	2	1	2	1
T/O	pac13	2	1	1	1	2	1	2
F/T	pac14	3	2	1	1	2	1	3
T	pac15	3	2	2	1	0	1	1
T	pac16	3	1	2	1	4	1	5
F	pac19	2	2	1	2	1	0	1
total		38	24	19	14	25	9	28

4.2 Wavelet Decomposition Coefficients Energy Analysis

Similar results were obtained on the method based in the wavelet decomposition coefficients. Significant variations occur, including the decrease of sensitivity and increase in false positives. Table 3 shows the results.

Table 3. Wavelet decomposition coefficients energy analysis, prediction statistics comparison between parameteres determined according to patient-specific, onset area and total. CP- Correct predictions, FP- False Positives.

Onset Area	Patient	Number seizures	Patient-specific		Onset area		total	
			CP	FP	CP	FP	CP	FP
F	pac1	1	1	0	0	0	0	0
T	pac2	3	1	2	0	0	0	0
F	pac3	4	2	2	2	3	0	0

Table 3. (*continued*)

T/O	pac6	3	1	2	1	4	1	5
T	pac7	1	0		0		0	
F	pac8	2	1	1	1	1	0	1
T/O	pac9	3	3	1	2	1	3	2
T	pac10	3	2	3	2	2	2	2
P	pac11	3	0		0	0	0	0
T	pac12	2	1	1	1	1	1	1
T/O	pac13	2	1	1	1	1	1	4
F/T	pac14	3	2	1	1	3	1	3
T	pac15	3	2	2	1	7	1	8
T	pac16	3	3	5	0	0	0	1
F	pac19	2	0		1	2	0	1
total		38	20	21	13	25	10	28

Table 4. Nonlinear features analysis, prediction statistics comparison between parameteres determined according to patient-specific, onset area and total. (CP, FP same as before)

Onset Area	Patient	Number seizures	Patient-specific	
			CP	FP
F	pac1	1	1	1
T	pac2	3	3	1
F	pac3	4	0	0
T/O	pac6	3	1	2
T	pac7	1	0	0
F	pac8	2	2	1
T/O	pac9	3	1	1
T	pac10	3	0	0
P	pac11	3	0	0
T	pac12	2	0	0
T/O	pac13	2	1	1
F/T	pac14	3	3	2
T	pac15	3	2	1
T	pac16	3	2	3
F	pac19	2	0	0
total		38	16	13

4.3 Nonlinear Analysis, Maximum Lyapunov Exponent

The method based in nonlinear measures presents different results. The results presented in the document only refer patient specific parameters because there are no significant variations between patients' parameters (Table 4).

4.4 Methods Overview

The analysis presented through several methodologies, suggest that valuable elements can be extracted through the comparison of the results obtained by these methods; several events were preceded by significant variations in these features. Furthermore, the results presented led us to believe that each patient has a better-suited method. Considering correct prediction calculated by each method, each patient has more satisfactory results through a determined method.

Table 5. Comparison between the sensitivity of each method according to the determined parameteres

Onset Area	Patient	Energy (predictons)	Wavelet (predictions)	Lyapunov (predictions)	Number Seizures
F	pac1	1	1	1	1
T	pac2	2	1	3	3
F	pac3	2	2	0	4
T/O	pac6	2	1	1	3
T	pac7	0	0	0	1
F	pac8		1	2	2
T/O	pac9	3	3	1	3
T	pac10	2	2	0	3
P	pac11	2	0	0	3
T	pac12	2	1	0	2
T/O	pac13	1	1	1	2
F/T	pac14	2	2	3	3
T	pac15	2	2	2	3
T	pac16	1	3	2	3
F	pac19	2	0	0	2
total		24	20	16	38

5 Conclusions and Further Work

The idea of continuously tracking the changes in several features for seizure prediction is based on clinical and experimental observations that the transition between states may not be abrupt but rather gradual [9].

Techniques usually used in seizure prediction include methods based on computational methods, statistical methods and mathematical approaches to EEG signal.

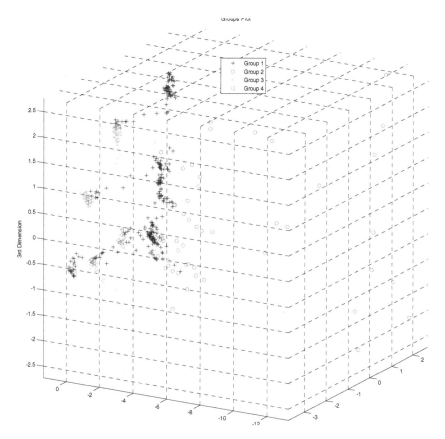

Fig. 9. VISRED analysis of the 14 features extracted by the developed application, Group 1 – inter ictal, Group 2 – pre-ictal, Group 3 – ictal, Group 4 pos ictal

The goal of the application under development is to provide the research and medical community with an easy to use signal processing application, capable of reading the most common formats.

The first results obtained through the methods implemented in this application allowed several conclusions.

For any prediction method to be clinically viable it must be formulated in a manner that allows it to operate fully prospectively and using only informations that would be reasonably available to the system at the time of predictions are made. This makes it necessary that the prediction algorithm operates in real-time. The study suggests that the prediction method and its evaluation scheme, optimized values of the prediction horizons, and preferred brain structures for EEG recording have to be determined for each patient and prediction method individually.

The ability to choose better-suited features for individual patients will be a critical point in the development of these applications, since the results suggest that some features describe more accurately certain patterns associated with each patient. The comparison of the results obtained by each method in different patients support these ideas.

The development of tools for the processing of large datasets of EEG signal is absolutely essential to confirm the indications presented in the literature which reveal a diversity of approaches. With the improvement of the various methodologies already explored, a reliable classification of states is possible.

Computational intelligence techniques to classify the data sets composed by the features extracted in this study will be faced to improve the application. Dynamic neural networks, Support Vector Machines, and neuro-fuzzy systems have a great potential and will be tested.

The implemented techniques use only one EEG channel. Multichannel techniques such as the phenomena of synchronization and nonlinear dynamics evolution in different regions fn the brain, might provide new insights on the seizure generation and on seizure prediction.

Another planned step is the introduction of computational tools for state classification. Initial introductory studies with space reduction, neural networks and neurofuzzy systems have already been made and show a good potential for the patient specific oriented approach, as shown in Fig. 9.

Only with these developments and with the understanding of the mechanisms underlying the seizures generation, the methods will produce knowledge which ultimately will result in development of functional alarming devices and safe therapies.

Acknowledgments

This research is supported by European Union FP7 Framework Program, EPILEPSIAE Project Grant 211713, to which this work is a contribution, an by CISUC (FCTUnit 326).

References

1. Delorme, A., Makeig, S.: EEGLAB: an open source toolbox for analysis of single-trial EEG dynamics. Journal of Neuroscience Methods 134, 9–21 (2004)
2. http://biosig.sf.net
3. http://www.dpi.physik.uni-goettingen.de/tstool/indexde.html
4. The MathWorks, Inc.
5. Esteller, R., Echauz, J., D'Alessandro, M., Worrell, G., et al.: Continuous energy variation during the seizure cycle: towards an on-line accumulated energy. Clinical Neurophysiology 116, 517–526 (2005)
6. Litt, B., Esteller, R., Echauz, J., D'Alessandro, M., Shor, R., Henry, T., et al.: Epileptic seizures begin hours in advance of clinical onset: a report of five patients. Neuron 30, 51–64 (2001)
7. Wichard, J., Parlitz, U.: Applications of nearest neighbours statistics. In: International Symposium on Nonlinear Theory and Its Applications (NOLTA 1998) (1998)
8. Freiburger Zentrum fur Datenanalyse und mollbildung, http://www.fdm.uni-freiburg.de/groups/timeseries/epi/EEGData/download/infos.txt
9. Litt, B., Esteller, R., Echauz, J., D'Alessandro, M., Shor, R., Henry, T., et al.: Epileptic seizures begin hours in advance of clinical onset: a report of five patients. Neuron 30, 51–64 (2001)

10. Gigola, S., Ortiz, F., D'Atellis, C., Silva, W., Kochen, S.: Prediction of epileptic seizures using accumulated energy in a multiresolution framework. Journal of Neuroscience Methods 138, 107–111 (2004)
11. Jirsch, J.D., Urrestarazu, E., LeVan, P., Olivier, A., Dubeau, F., Gotman, J.: High-frequency oscillations during human focal seizures. Brain129 (Pt 6), 593–608 (June 2006)
12. Winterhalder, M., Schelter, B., Maiwald, T., Brandt, A., Schad, A., Schulze-Bonhage, A., Timmer, J.: Spatio-temporal patient–individual assessment of synchronization changes for epileptic seizure prediction. Clinical Neurophysiology 117, 2399–2413 (2006)
13. Le van Quyen, M., Martinerie, J., Navarro, V., Boon, P., D'Have, M., Adam, C., et al.: Anticipation of epileptic seizures from standard EEG recordings. Lancet 357, 183–188 (2001b)
14. Lehnertz, K., Andrzejak, R., Arnhold, J., Kreuz, T., Mormann, F., Rieke, C., et al.: Nonlinear EEG analysis in epilepsy: Its possible use for interictal focus localization, seizure anticipation, and prevention. J. Clin. Neurophysiol. 18, 209–222 (2001)
15. Mormann, F., Kreuz, T., Rieke, C., Lehnertz, K., et al.: On the predictability of epileptic seizures. Clinical neurophysiology 116, 569–587 (2005)
16. Dourado, A., Ferreira, E., Barbeiro, P.: VISRED - Numerical Data Mining with Linear and Nonlinear Techniques. In: Perner, P. (ed.) ICDM 2007. LNCS (LNAI), vol. 4597, pp. 92–106. Springer, Heidelberg (2007), http://eden.dei.uc.pt/~dourado

An Infrastructure for Mining Medical Multimedia Data

Sara Colantonio, Ovidio Salvetti, and Marco Tampucci

Institute of Information Science and Technologies, Italian national Research Council,
Via G. Moruzzi 1, 56124 Pisa, Italy
{Name.Surname}@isti.cnr.it

Abstract. Biomedical research processes related to disease diagnosis, prognosis and monitoring would great benefit from advanced tools able not exclusively to store and manage multimodal data but also to process and extract significant relations and then novel knowledge from them. Indeed, making a prediction on a disease outcome usually requires considering heterogeneous pieces of information obtained from several sources which should be compared and related. Mining medical multimedia objects is aimed at discovering and making available the hidden useful knowledge embedded in collections of data and is, then, of key importance for supporting clinical decision-making. In this paper, we report current results of a medical warehouse we are developing in an integrated environment for mining clinical data acquired by different media. In particular, focus is herein given to the infrastructure of the warehouse and its current functionalities not limited to storage and management but including intelligent representation and annotation of multimedia objects.

Keywords: Multimedia Mining, Features Extraction, Metadata Standards.

1 Introduction

The success of clinical healthcare organization and delivery is more and more relying on an effective integration of information technologies within human-based clinical decision-making workflows. Indeed, healthcare practitioners have to continually face a wide range of challenges, trying to make difficult diagnoses, avoid errors, ensure highest quality, maximize efficacy, and save money all at the same time [12]. In particular, clinicians usually have to make decisions by considering and relating, as fast as possible, a large amount of diagnostic test results, medications and past treatment responses. However, the limit of human capability to process information is well known, i.e. just recall the so called *'bounded rationality'* [21]. Nevertheless, there are increasing amounts of medical knowledge, of new information about disease causes, of diagnostic and therapeutic options, and mountains of patient data from ever-expanding numbers of diagnostic tests and imaging studies. Medical data are becoming always largely multimodal; just for making an example, investigations for the same patient can produce text and numeric data (e.g., demographic and blood analysis data), images (one or more of the many medical imaging modalities, e.g., Magnetic Resonance Imaging, Computerized Tomography), signals (e.g., Electrocardiography) and videos (e.g., ultrasound examinations).

P. Perner (Ed.): ICDM 2008, LNAI 5077, pp. 102–113, 2008.
© Springer-Verlag Berlin Heidelberg 2008

These conditions have offered a fertile environment for the growing of computerized systems able to tackle this "information explosion" and aid clinicians in their decision-making processes. Initially, focus was given to the fundamental problem of storing data and allowing their consistent retrieval. Maturity in this field is testified by the growing spread of Electronic Health Records (EHR) for collecting information about care delivery [19], and of Picture Archiving and Communication Systems (PACS) for the acquisition, storage, transmission, processing and display of digital medical images [9]. Also, several standards, such as IHE (Integrating the Health Enterprise) [14] on its turn based on other standards like HL7 (Health Level 7) [13] and DICOM (Digital Imaging and Communications in Medicine) [8], have been introduced for facing the demand for data sharing and communication among healthcare information resources.

Recently, the awareness of the great value inherent in clinical data has fostered the development of medical data warehouses devoted also to exploring these large amounts of data for discovering novel, implicit knowledge, such as, for instance, the relevance of a biomarker for the prognosis of heart failure. Research in this field can exploit assessed disciplines such as Data Mining and Knowledge Discovery in Databases, but systems able to cope with really multimodal and multimedia objects are still far to be completely developed and used in practice. Few medical image mining systems have been presented [17], especially within histology and cytology [3], [4], [11]. However the development of most of them has been mainly focused on either the data mining problem, i.e., on the extraction of knowledge from suitably extracted features without any care of advanced image storage and retrieval, or to image retrieval by content, thus missing the real interesting challenge of extracting new knowledge.

The previous considerations motivated our activity in the field that are aimed at developing a *Multimedia Mining Medical Warehouse* (MMMW) by exploiting experiences and expertise matured within several international projects, i.e. the EU Network of Excellence MUSCLE (Multimedia Understanding through Semantics, Understanding and Learning), the STREP EU project HEARTFAID (A knowledge based platform of services for supporting medical-clinical management of the heart failure within the elderly population) and the Italian-Russian bilateral project [6]. Actually, within MUSCLE, focus has been given to general purpose multimedia data retrieval by content, in particular, to metadata standards for advanced data storage and retrieval [1], and to ontological models for media analysis [2]. Within HEARTFAID, research activities are concentrated on a specific medical problem (i.e., heart failure) and aim at developing an integrated platform able to support decision making, diagnostic signal and image processing, knowledge discovering in patients' data repository [5]. Within the bilateral project, novel methods for medical image categorization have been developed by exploiting ontological models for image analysis (namely, for cytological images analysis) [7]. Results obtained in the above mentioned projects are being combined for implementing the MMMW, which, though currently in the early stage of its development, has been designed for supplying all the multimedia mining functionalities, ranging from multimedia storage to novel knowledge discovery. In particular, it is being realized by integrating and extending the infrastructure we have developed within the MUSCLE initiative for managing multimedia metadata (the so-called 4M infrastructure) [1].

In this paper, main results of the design activity are presented since interesting aspects have come forth. The main components of MMMW are also introduced and future activities discussed.

2 Medical Multimedia Data Mining

The Multimedia Mining Medical Warehouse has been conceived for aiding the investigation processes of medical problems, by providing local and remote access to known cases, the facility for retrieving multimedia objects by content, and the possibility of mining multimedia patterns relevant to medical decision making, e.g., the processes of diagnosis and prognosis.

An illustrative scenario can be useful for better explaining the motivations at the basis of MMMW. Consider the problem of the prognostic evaluation of heart failure patients: a very difficult task due to the lack of large and randomized studies and to many comorbidities that can affect the prognosis. Currently, a set of parameters are considered relevant as suggested by clinical guidelines [22]; they belong to five categories:

- *demographic data*: age, comorbidities such as diabetes and so on;
- *clinical data*: heart rate, blood pressure, weight loss, blood and electrolytic parameters, ...;
- *functional data*: cardiopulmonary stress responses, such as the maximum or the relative oxygen uptake and so on;
- *electrophysiological data*: heart rate variability, ventricular rhythms, ...;
- *haemodynamic data*: left ventricle *ejection fraction*, left ventricle volume, and so forth.

Such pieces of information come from different investigations, this means that can be obtained and represented by different media: Table 1 shows the modalities for acquiring the listed information.

Currently, the relevant markers are considered separately; but we can depict an ideal situation in which all the information could be collected from different sources and stored in the same warehouse (Fig. 1), which could also be distributed. This

Table 1. The different modalities for acquiring data relevant to the prognosis evaluation of heart failure patients

Relevant Data	Acquisition Modalities
Demographic	Patient's interview
Clinical	Patient's interview, blood analysis, sphygmomanometer, ...
Functional	Exercise testing: cardiopulmonary stress testing, 6-minute walking test, ...
Electrophysiological	Electrocardiography
Hemodynamic	Echocardiography, Cardiac Magnetic Resonance Imaging, ...

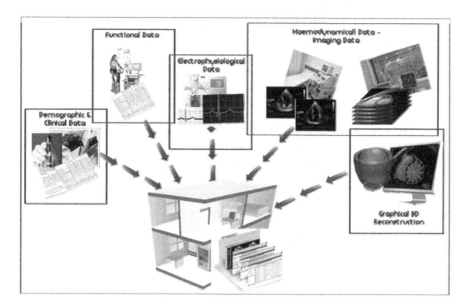

Fig. 1. Multimedia data relevant for the prognostic evaluation of heart failure

should be a multimedia warehouse that contains textual, logical and numerical data about the patients (i.e., demographic, clinical and functional); signal data (i.e., electrocardiogram recordings and correspondent parameters); imagery data (i.e., 2D Transthoracic Echocardiography or Magnetic Resonance Imaging); video data (i.e., M-mode Echocardiography). Graphical 3D models could also be considered as part of the relevant clinical data, think for instance to the 3D reconstruction of the left ventricle used for assessing the hemodynamic dysfunctions and impairment of the heart.

This large collection of multimedia objects could then be used for two main purposes:

- supporting case-based reasoning of clinicians, i.e. presenting them stored cases of patients having similar features to the one at hand, and then aiding the prognostic evaluation on the basis of previous responses;
- extracting novel knowledge, i.e., applying intelligent techniques for discovering the relations among the different parameters and individuating the relevant markers for the prognosis. This way, also new parameters could be computed from signal, imagery and video data so that new patterns not explicitly stored could be discovered.

MMMW has been devised for addressing the kind of needs illustrated within the example scenario. Four main functionalities have been then identified as illustrated in Fig. 2:

- multimedia data storage and management inside a repository;
- multimedia indexing and retrieval, for finding multimedia objects ranked in accordance to some requirements on their content. The retrieval functionality

will allow for explicit text query or query by content with a reference multimedia object, e.g., a diagnostic image showing specific features;

- multimedia semantic annotation, for associating lists of semantic keywords or meaningful sentences (e.g., diagnostic evaluations) to multimedia objects. Such pieces of information can then be used for data retrieval, in particular for text queries. A structured terminology will be supplied for aiding annotations by the users;

- multimedia mining, for extracting valid, novel and understandable knowledge about the diagnostic, prognostic and monitoring processes. Advanced data mining methods will be available for application to patterns built by correlating features extracted from data and domain concepts.

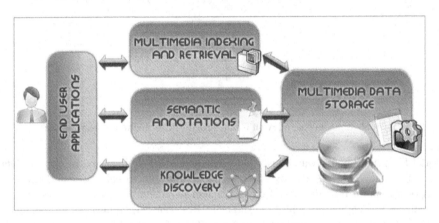

Fig. 2. The main MMMW functionalities

For supplying such functionalities, a fundamental challenge regards determining how low-level representation contained in a multimedia object can be processed to identify high-level information and relationships among data. This is strictly related to the process of representing the informative content of multimedia objects by extracting a set of meaningful *features*. In particular, two general categories of features can be identified: *description-based* and *content-based*. The former uses metadata, such as keywords, caption, size and time of creation; the latter is based on the content of the multimedia object itself. As to data representation within MMMW, we can then envisage two paths multimedia data will move along (Fig. 3).

For storing and retrieving purposes, data can be processed for extracting a number of general description features, corresponding to a medical standard such as DICOM, and of content features, corresponding to a multimedia standard such as MPEG-7 [16]. Semantic annotations can be considered as part of this process.

For knowledge discovery purposes, additional features could be extracted by analyzing the multimedia objects. This can requires several steps that consist in preprocessing, for improving the multimedia object quality; segmentation, for extracting relevant structures contained in the object; quantitative analysis, for extracting the interesting and significant features. Once adequately transformed and

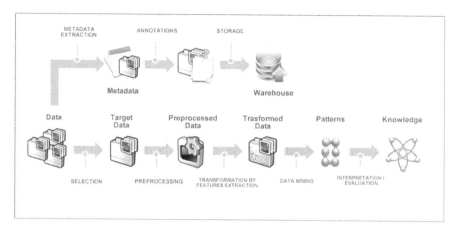

Fig. 3. The two paths followed by multimedia objects within MMMW

represented the multimedia object, data mining algorithms can be applied so that resulting patterns can be transformed in novel knowledge.

Such considerations resulted in the design of a MMMW constituted by the following components:

- a repository for storing, accessing and retrieving images and information extracted at different levels from them;
- the facility for applying a collection of image analysis algorithms for data processing and analysis (to improve data quality and specificity, and extract features from them);
- a library of data mining and pattern recognition algorithms for discovering and interpreting novel pattern extracted from data;
- a suite of ontologies including specific domain ontologies related to medical problems and a general image understanding ontology, for suggesting algorithms to be applied to different types of data and for adding terminologically controlled annotations;
- a user interface for accessing, uploading, browsing and annotating images.

Currently, some of the components are already developed inside the MultiMedia Metadata Management - 4M - infrastructure which has been realized within the MUSCLE initiative. Precisely, 4M includes a repository, algorithms for extracting a set of features belonging to metadata standards and a user interface as described in the next section.

2.1 The Infrastructure for Multimedia Data Management

The 4M infrastructure has been developed in order to guarantee and promote multimedia data and metadata storage, retrieval, maintenance and exchange. It has been devised by using (i) standardized Semantic Web technologies promoted by the W3C office, since they can facilitate the overall vision of distributed, machine readable metadata; (ii) multimedia metadata standards, in particular the MPEG-7 standard; and (iii) open-source software.

The system is able to provide:

- a complete user management;
- tools for the features extraction from different media sources;
- tools to perform semantic annotation;
- an efficient search and retrieval query system.

The main components are sketched in Fig. 4 and can be described as follows:

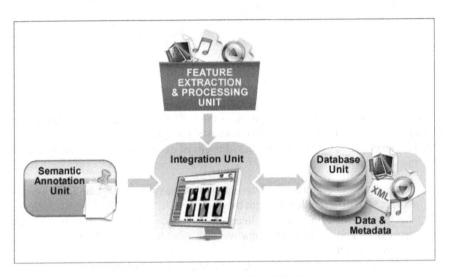

Fig. 4. The main components of the 4M infrastructure

Features extraction and processing unit: Devoted to the extraction of a set of description and content metadata based on the MPEG-7 standard. Extracted features are inserted into an XML [23] document opportunely stored which is divided into two parts: the *Multimedia Description Scheme* (MDS) and the values of the extracted MPEG-7 features. The MDS fields contain all the information regarding the media document (e.g., insertion date; XML document creator role, name and surname; media type; media extension; media size; height and width for the image and length for the audio) and annotations inserted by the user for describing the data content. All the MPEG-7 features can be extracted from text, signal, image and video data and are used for describing multimedia object content. Some examples of the features are: for image *color descriptors* (*layout, structure, dominant, scalable*), texture descriptor (*homogeneous texture, edge histogram,...*), and *shape descriptors* (*region and contour descriptors*); for videos the same descriptors used for still images plus *motion descriptors* (*camera motion, motion trajectory, motion activity descriptors*); for signals and audios basic descriptors (*audio waveform* and *audio power*), *signal descriptors* (*Audio Fundamental Frequency, Audio Harmonicity*).

Database unit: Developed by using an XML native database eXist [10], which was selected since provides efficient, index-based processing, extensions for full-text search, and a Java interface. The database is composed of multimedia objects, and collections, which can contain other collections and/or objects. Exploiting this peculiarity, we structured the database in order to increase the search and retrieval operator efficiency and to facilitate the access right. Besides, the database unit allows for comparison among the stored multimedia data. Such a functionality is at the basis of the *Similarity Query System*: exploiting the support to XQuery [24] provided by eXist, the System retrieves objects having feature values similar to the one supplied as reference for the search. A textual search is also available and performed via a *Google*-like interface. By using a *basic search* option, the user can specify a sentence or a list of keywords to be searched for by compiling a dedicated text area and, optionally, restricts the search by selecting the media type to be scanned. The search will retrieve all the objects which have at least an annotation containing the searched text. The *advanced search* option allows the user to perform textual queries about the content of media objects parts (e.g., image regions) and other descriptive metadata.

Semantic annotation unit: Currently devoted to add annotations only to imagery data, it provides functionalities to select image regions and to perform annotations on them, to manage the performed annotations and to view them. Annotations relative to the same image are stored into a RDF [20] document stored in the database. Such a document is created for the first annotations and then extended, by means of a XUpdate operation, for each new annotation added subsequently. Semi-automatic annotations are also possible, since, once selected an image region, MPEG-7 features can be extracted from it and add to an opportune XML file associated to the image. Fig. 5 shows an example of annotation to a mammographic image.

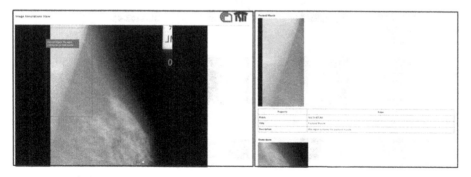

Fig. 5. Manual annotation of a mammogram: on the left the textual annotation reported onto the image; on the right the annotated region as extracted and characterized by the same annotation

Integration unit: Devoted to information storage and retrieval through suitable tools that manage the interaction of and access to all the other units. The main purpose of this unit is to provide interfaces and controllers between the individual units and the user and to act as a query mediator between other units. It is constituted by a java *servlet* [15] and a web-based user interface. The java servlet supplies to the task of unit coordination by recognizing the operation requested by the user and invoking the correspondent units. Once operation is performed, the servlet updates the user interface to notify a successful conclusion of the operation and return the result when requested.

3   Current Results and Future Activities

Current results of the MMMW consist in an infrastructure which allows for:

(i)   data upload;
(ii) data analysis algorithm for extracting a set of description and content metadata features;
(iii) data storage along with the computed metadata features;
(iv) data retrieval by content.

We are populating the warehouse with images belonging to cytopathology, histology, cardiology and oncology fields. Fig. 5 shows a sample of the mammography dataset. Also, cardiological signals obtained from public available datasets (e.g., Electrocardiograms from the Physionet dataset [18]) have been stored.

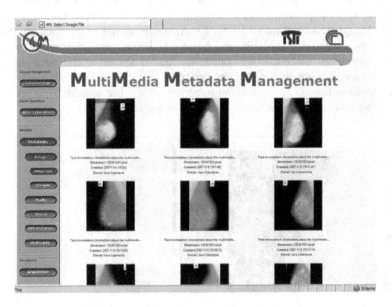

Fig. 6. A sample of the mammography dataset

Currently, when uploading an object, information regarding the data acquisition procedure (e.g., medical devise characteristics) is specified in the MDS; while patient's data are inserted as annotations. The content-based retrieval is performed by similarity on extracted features. Fig. 6 shows the results of a query performed with a reference image. Images are ranked according to the similarity value.

Obviously, the features currently extracted are not satisfactory for medical research purposes: MPEG-7 features can have a general value for broad-use multimedia

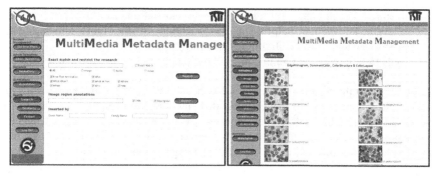

Fig. 7. On the left, the advanced textual search, on the right a result of a similarity query on a cytological image

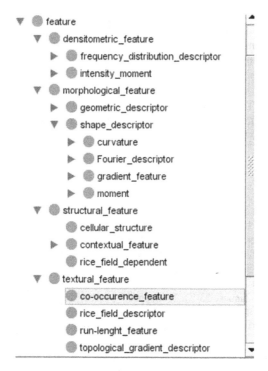

Fig. 8. Fragment of the feature extraction ontology

objects, but lack of specificity and relevance when dealing with the important content of medical multimedia data. In this context, future activities will consist in the extension of the 4M by, first of all, adding the extraction of typical medical metadata in accordance to the up-to-date standards (e.g., DICOM for images). Other components will be introduced, namely:

- tools for data processing devoted to their improvement and segmentation;
- tools for data analysis aimed at extracting a larger set of meaningful features;

- a suite of ontologies for aiding textual annotations and representing the relevant knowledge helpful in the selection of the most suitable data processing and analysis algorithms according to the media type;
- tools for applying data mining algorithms and extract novel knowledge.

Currently, a suite of ontologies, which has been developed for aiding the analysis of cytological images [7], is being integrated into the infrastructure. It contains a conceptualization of the cell biology, microscopy and microscopic images domains, also establishing relations among them. Moreover, an image analysis ontology has been developed for structuring the domain of image feature extraction. So far it is focused on the cytology domain, i.e., contains the features mostly used for extracting quantitative measurements for cytological images, but has a very general valence. Figure 8 shows an excerpt of the class hierarchy of the ontology. It has been equipped with a library of algorithms, this way, when integrated into MMMW, it will allow the user to select a subset of features, compute them on a pile of images, store results and launch data mining algorithms for discovering relations among them. For the latter service, an algorithm package is under development for supplying a library of machine learning methods usually employed for data mining, such as Decision Trees, Random Forest, Support Vector Machines and so on.

Acknowledgments. This work was partially supported by EU Network of Excellence MUSCLE - FP6- 507752, and by the EU STREP project HEARTFAID (FP6-IST-2004-027107, 07-07- 13545).

References

1. Asirelli, P., Little, S., Martinelli, M., Salvetti, O.: MultiMedia Metadata Management: a Proposal for an Infrastructure. In: SWAP 2006, Semantic Web Technologies and Applications, December 18-20, Pisa, Italy (2006)
2. Asirelli, P., Colantonio, S., Little, S., Martinellli, M., Salvetti, O.: Media Analysis and the Algorithm Ontology. In: Gurevich, I., Niemann, H., Salvetti, O. (eds.) IMTA 2008, the First International Workshop on Image Mining: Theory and Applications, Funchal, Madeira, Portugal, January 22-23, 2008, pp. 37–50. INSTICC Press (2008)
3. Cebron, N., Berthold, M.R.: Mining of Cell Assay Images Using Active Semi-Supervised Clustering. In: ICDM 2005 Workshop on Computational Intelligence in Data Mining, pp. 63–69 (2005)
4. Chen, W., Meer, P., Georgescu, B., He, W., Goodell, L.A., Foran, D.J.: Image Mining for Investigative Pathology Using Optimized Feature Extraction and Data Fusion. Computer Methods and Programs in Biomedicine 79, 59–72 (2005)
5. Chiarugi, F., Tsiknakis, M., Salvetti, O., et al.: Support for the Medical-Clinical Management of Heart Failure within Elderly Population: the HEARTFAID Platform. In: ITAB 2006, Int. Special Topic Conference on Information Technology in Biomedicine, Ioannina, Greece, October 26-28, Greece (2006)
6. Colantonio, S., Gurevich, I., Salvetti, O., Trusova, Y.O.: An Image Mining Medical Warehouse. In: Gurevich, I., Niemann, H., Salvetti, O. (eds.) IMTA 2008, the First Int. Workshop on Image Mining: Theory and Applications, Funchal, Madeira, Portugal, January 22-23, pp. 83–92. INSTICC Press (2008)

7. Colantonio, S., Martinelli, M., Salvetti, O., Gurevich, I.B., Trusova, Y.O.: Cell Image Analysis Ontology. Int. Journal of Pattern Recognition and Image Analysis 18(2), 332–341 (2008)

8. DICOM – Digital Imaging and Communications in Medicine (2008), http://medical.nema.org/

9. Dwyer, S.J.: A personalized view of the history of PACS in the USA. In: James Blaine, G., Siegel, E.L. (eds.) Proceedings of the SPIE, Medical Imaging 2000: PACS Design and Evaluation: Engineering and Clinical Issues, vol. 3980, pp. 2–9 (2000)

10. eXist (2007), http://exist.sourceforge.net/

11. Gholap, A., Naik, G., Joshi, A., Rao, C.V.K.: Content-Based Tissue Image Mining. In: CSBW 2005, IEEE Computational Systems Bioinformatics Conference Workshops (2005)

12. Greenes, R.A.: Clinical Decision Support: The Road Ahead. Elsevier, London (2007)

13. HL7 – Health Level 7 (2008), http://www.hl7.org/

14. IHE – Integrating the Healthcare Enterprise. IHE Radiology: Mammography User's Handbook. ACC/HIMSS/RSNA (2008), http://www.ihe.net/Resources/upload/ IHE_Mammo_Handbook_rev1.pdf

15. Java Servlet (2008), http://java.sun.com/products/servlet/

16. MPEG-7 Overview - International Organization for Standardization, ISO/IEC JTC 1/SC 29/WG 11, Coding of Moving Pictures and Audio - N5525, Pattaya (March 2003), http://www.chiariglione.org/mpeg/standards/mpeg-7/mpeg-7.htm

17. Perner, P.: Mining Knowledge in Medical Databases. In: Data Mining and Knowledge Discovery: Theory, Tools and Technology, SPIE, vol. 4057, pp. 359–369 (2000)

18. Physionet (2008), http://www.physionet.org/physiobank/database/chfdb/

19. Poissant, L., Pereira, J., Tamblyn, R., Kawasumi, Y.: The impact of electronic health records on time efficiency of physicians and nurses: a systematic review. J. Am. Med. Inform. Assoc. 12(5), 505–516 (2005)

20. RDF(2008), http://www.w3.org/RDF/

21. Simon, H.: A mechanism for social selection and successful altruism. Science 250(4988), 1665–1668 (1990)

22. Swedberg, K., et al.: Guidelines for the diagnosis and treatment of Chronic Heart Failure: full text (update 2005). European Heart J. 45 pages (2005)

23. W3C, eXtensible Markup Language (XML) 1.1 (2008), http://www.w3.org/XML/

24. XQuery language (2007), http://www.w3.org/TR/xquery/

Realizing Modularized Knowledge Models for Heterogeneous Application Domains

Klaus-Dieter Althoff, Kerstin Bach, and Meike Reichle

Intelligent Information Systems Lab
University of Hildesheim, Hildesheim, Germany
{althoff,bach,reichle}@iis.uni-hildesheim.de

Abstract. This paper addresses the realization of modularized knowledge models within a heterogeneous application domain using an existing knowledge management tool. The application domain we deal with is travel medicine, which combines medical aspects with geography, climate, holiday activities and associated traveling conditions. In this paper we present the application's requirements and show how knowledge models can be developed using an industrial strength application. Furthermore we present the challenges of a knowledge model based on multi-case bases whereas each case base represents its own area of expertise. Hence, we introduce our knowledge model for the travel medicine application and exemplify the implementation of typical data types and similarity measures.

1 Introduction

Knowledge-based systems provide a technology for covering a topic in a very comprehensive and scalable way. Realizing such systems requires high quality data sources, knowledge models and maintenance techniques. To achieve this, knowledge has to be acquired, analyzed, stored, and retrieved, which challenges a knowledge-based system and is crucial for its continuous existence over a longer period of time.

The Collaborative Multi-Expert-Systems (CoMES) approach [1] describes the concept of knowledge-lines[1] and how they can be applied when building collaborative Multi-Expert-Systems and communicate via software agents. In this paper we follow the SEASALT[2] architecture [2], which is based on the CoMES approach, to develop an intelligent information system for travel medicine in which different aspects of the domain have to be combined. Each knowledge

---

[1] Knowledge Lines use the principle of product lines as it is known from software engineering and apply it to the knowledge in a knowledge-based system, thus splitting rather complex knowledge in smaller, reusable units.

[2] SEASALT – Sharing Experience using an Agent-based System Architecture LayouT. The SEASALT architecture proposes an intelligent information system with several cross-linked case bases that are used to store information on different aspects of a complex knowledge domain and are filled with information mined from the online communication of a community of experts.

P. Perner (Ed.): ICDM 2008, LNAI 5077, pp. 114–128, 2008.

base for an aspect or topic is based on a case-based reasoning (CBR) system and, to some degree, provides its expertise like a human expert would do. The combination and collaboration between the individual CBR systems is realized via software agents on top of each CBR system. The application of software agents in CBR systems has been introduced by Plaza [3], even though Plaza focused on case bases with the same case format. Our case bases contain information on different topics and we decided to model different topic-based case bases, each with an own, individual case format. CBR systems with more than one case base were also used by Leake [4], but in their systems the case bases contained similar data and in comparison to our approach the retrieval results of the case bases had to be compared, refined and finally the best result was chosen. Instead, we take each result and use software agents to combine them to a valid solution.

In this paper we introduce the knowledge models of our heterogeneous application domain which are required to cover the field of travel medicine and create prevention information for travelers. Like the most medical information systems (i.e. Isabel[3] or GIDEON[4]) we also focus on one area of medicine, although travel medicine combines numerous fields of medicine. Furthermore we add non-medical but consultation relevant information to the given answer, which again requires different kinds of case bases. This additional information with covers the areas of geography, time and seasons, activities, and medicaments as well as their pharmaceutical ingredients. Each knowledge base can be updated individually, because after an initial model and case base has been implemented, we will integrate our information system in a travel medicine community. The results of discussions and information given in a community will be collected and after its evaluation it will be inserted in the appropriate case base.

Another advantage of using modular case bases to cover a heterogeneous application domain is their maintenance which is much easier than in monolithic case bases [5]. Also, modular case bases make it easier to include small "factoids" (for instance "There was a case of yellow fever in Indonesia") that do not represent a full case but are still worth being included.

In the course of this paper we will describe the realization of our knowledge models and illustrate in which way the individual case bases are linked with each other. First of all, in chapter 2 we will introduce the application domain travel medicine including the case bases we have determined. In chapter 3.3 we describe the requirements of the travel medicine domain, especially under the aspect of modularized knowledge bases. We will describe a typical question to the system and its respective answer to explain our concept of shared data types. Furthermore this chapter shows examples of our similarity measures. Chapter 4 contains the description of the implementation of our approach and the software

---

[3] Isabel is a clinical decision support system designed to enhance the quality of diagnosis decision making. home page: http://isabelhealthcare.com/

[4] GIDEON is an online application to diagnose infectious diseases and give information on epidemiology, treatment and microbiology.
home page: http://www.gideononline.com/

used. Chapter 4 picks up on the requirements of chapter 3.3 and shows the realization of our modular, cross-linked knowledge model according to those requirements. The paper closes in chapter 5 with a brief examination of our application project in the context of SEASALT [2] and CoMES [1] and finally gives a conclusion and outlook in chapter 6.

2   Travel Medicine as an Application Domain

Travel medicine is an interdisciplinary speciality concerned with the prevention, management and research of health problems associated with travel, and covers all medical aspects a traveler has to take care of before, during and after a journey. For that reason it covers many medical areas and combines them with further information about the destination, the activities planned and additional conditions which also have to be considered when giving medical advise to a traveler. Travel medicine starts when a person moves from one place to another by any mode of transportation and stops after returning home without diseases or infections. In case a traveler gets sick after a journey a travel medicine consultation might also be required.

The research project presented in this paper is supported by mediScon worldwide, a Germany based company with a team of physicians specialized on travel medicine and TEMOS[5], a tele-medical project of the Institute of Aerospace Medicine at the German Aerospace Center DLR[6]. Together we will develop docQuery, an intelligent information system on travel medicine which provides relevant information for each traveler about their individual journey. First of all we will focus on prevention work, followed by information provision during a journey and information for diseased returnees.

Since there are currently no sources on medical information on the World Wide Web, that are authorized by physicians and/or experts, we aim at filling this gap by providing trustworthy travel medical information for everybody.

Based on the SEASALT architecture [2], we propose building a web community which provides information to travelers and physicians (non-experts in the field of travel medicine) from experts on travel medicine. docQuery will provide an opportunity to share information and ensure a high information quality because it is maintained by experts. Travelers and experts can visit the website to get the detailed information they need for their journey. A traveler will give docQuery the key data on their journey (like travel period, destination, age(s) of traveler(s), activities, etc.) and docQuery will prepare an information leaflet the traveler can take to his or her general practitioner to discuss the planned journey. The leaflet will contain all necessary information to be prepared and provide detailed information if it is required. In the event that docQuery cannot answer the traveler's question, the request will be sent to experts who will answer it.

---

[5] TElemedicine for a MObile Society, see http://www.temos-network.org

[6] http://www.dlr.de/me/

The computation of the answer follows the steps a physician would take during a consultation. Therefore several knowledge bases have to be queried and their answers have to be combined according to the given constraints. We will realize the combination of information using software agents, where each agent represents one domain and provides information of its domain according to the query.

2.1   The docQuery Case Bases

The travel medicine domain covers heterogeneous aspects and according to the knowledge containers introduced in [6], each area requires a case representation vocabulary, similarity measure, case base and adaptation methods. Our implementation of the following case bases shows, that there are several data types which can be used in more than one case base.

The *destination* case base contains country characteristics which are important for a visitor, containing vaccination requirements and vaccination-preventable infectious diseases, pre-travel information on different kinds of diseases which might occur in a certain country or region, as well as hygiene and prevention advise. The destination case base differentiates not only by country, we can also handle regions, continents, etc. Diseases can be assigned to each degree of generalization.

The *disease* case base contains infectious diseases, the major area of travel medicine, and holds information about the disease itself as well as its transmission, the symptoms, the disease pattern, precaution advise, vaccination, prophylaxis advise and/or the treatment. Diseases provided in this case base are derived from the travel related branch of several traditional medical disciplines like tropical medicine, infectious diseases, gastroenterology, internal medicine, paediatrics, geriatrics, obstetrics-gynaecology, dermatology, etc. as described in [7].

The *medicament* case base holds information about medicaments such as their ingredients, their field of application, the therapeutic field, the immunogenicity, the efficacy, the dose, possible reverse reactions and its management, contraindications and interdependencies as well as advantages and disadvantages. Medicaments in this case base are related to diseases in the disease and chronic illness case bases ensuring that the docQuery application will provide information for each medicament it suggests.

The *activity* case base contains advise for travelers who plan to engage in activities that cause particular physiological stress, such as diving, do mountain climbing or sojourns at a high altitude, or who prepare for a long term stay like backpackers or expatriates, as well as miscellaneous travel activities like visiting caves, desert tourism or extreme sports [8]. Each of those activities requires individual information regarding the traveler's physical fitness or the conditions at the destination.

In our model we distinguish between *chronic illness* and disease because of the fact that chronic illnesses is a part of the travelers' anamnesis and on the other hand the chronic illnesses' descriptions vary in the the attributes and concepts we have used for the modeling. This case base contains general information on the illness and characteristics as well the restrictions it imposes on certain activities, medicaments or travel regions.

For each destination we provide information about *hospitals*. For each hospital we will be able to give information about the kind of hospital (primary, secondary, tertiary care of specialized hospital) including their facilities. In order to be able to tell a traveler aboard which hospital he or she should consult for his or her needs.

docQuery only provides information directly associated with the travelers healthiness. The *information resource* case base contains further sources provided by other travelers how to get more advise, for example about actual news, weather conditions or customs duty.

In travel medicine not only the prevention of diseases have to be mentioned, we also has to consider the travelers themselves (children, elderly people, pregnant women, disabled people, etc.), the reason for traveling or the way of traveling (aviation, navy cruise, road traffic, driving, etc.). The *associated condition* case base provides information on how people do travel and what advise they have to be given.

3   Requirements of the Travel Medicine Domain

3.1   Requests and Responses in docQuery

Typical Request. A traveler will be guided through the request and will be asked several questions by the system (see Table 1 for the questions and the range of answers) in order to receive the information necessary for the "consultation". The order of the questions is almost the same a physician would ask and can be divided into two parts: The first three questions only concern the traveler and his or her physical condition, ensuring the patient is able to travel and to identify the first constraints we have to pay attention to when creating an individual sequence of queries. Questions four to seven ask for the traveler's destination and the duration of his or her stay. Information on the destination will be required in order to provide the traveler with information about the country, local conditions, and potential dangers. The date of the journey is required to deduce more constraints and define more precisely what kind of infectious diseases have to be considered regarding the season abroad. The duration is needed to find an appropriate way of vaccination and medication, because travelers who only stay for a very short period of time need a different preparation as those who stay for a longer amount of time. Based on the given information a sequence of questions is created and the case bases are queried.

Typical Answer. Based on the given answers in the first step a request strategy is chosen. We use rules to determine the order in which the case bases are queried and create those queries. A typical way to start is to query the destination case base which returns information on vaccinations, stand-by-medications, behaviour advise, and local diseases. Regarding the given example in Table 1 our case base would return the country information of Indonesia containing standard, obligatory, and risk vaccination information. Listed infectious diseases will be used to query the diseases case base for more information. Each case base

Table 1. Typical Questions of the docQuery System

No.	Question	Range of Answer	Example
1.	Type of the traveller(s)	[child|teenager| young person| pregnant woman| elderly person]	young person
2.	Gender of the traveller(s)	[male|female]	male
3.	Current/Chronic illness(es) of the traveller(s)	[list of diseases]	none
4.	Destination of the journey	<destination>	Bali
5.	Time/Date of Travel	date <from> - <to>	March 5th, 2008 until March 19th, 2008
6.	Duration	[no of days/weeks/months]	2 weeks
7.	Planned Activities during the Journey	[list of activities]	road trip, beach

introduced in Section 2.1 is involved in creating the answer. For the communication and elaboration of the individual results we use software agents (topic agents according to the SEASALT architecture [2]) which perform the request and prepare an answer for each topic. Refining steps are also done by the topic agents until a valid answer can be given.

Table 2 shows the simplified structure of an answer (column Answer) with possible content (column Example) docQuery gives. The example does not contain the explanation, instead it shows the keywords. Rows 1 and 2 are extracted out of the destination case base. The descriptions of diseases and the constraints for medications are retrieved from the disease case base . In case chronic illnesses have to be considered, the according case base is queried. Both answers are used as constraints when querying the medicament and activities database. For each given information we provide explanations based on the given conditions in both case bases information resource and associated conditions.

3.2   Realization of the Modular, Cross-Linked Knowledge Model

As described in [5], the maintenance of modularized case bases is more effective than the maintenance of monolithic case bases, a further motivation for multiple case bases is the increase of solution quality, a more selective case addition or the usage of individual cases and their local competence as discussed in [4,9].

Each case base requires a vocabulary of its own containing, for example in the destination case base, geographical terms like countries, regions, capitals, etc. as well as medical terms to describe diseases. Both, geographical terms and diseases, can be modeled in taxonomies to define similarities and our knowledge model allows sharing each attribute ensuring a conjoint vocabulary is used. The importance of a shared vocabulary is explained in [10]. Combined similarity

Table 2. Typical Answers of the docQuery System

No.	Answer	Example
1.	Advise for the journey	Destination: Indonesia
2.	Infectious diseases	**Vaccination Standard:** Diphtheria, Hepatitis A, Measles, Poliomyelitis, Tetanus, Typhoid Fever **Vaccination Obligatory:** Yellow Fever **Vaccination Risk:** Cholera, Hepatitis B, Japanese Encephalitis, Rabies
4.	Information about infectious diseases abroad	AIDS, Malaria
5.	Stand-by medication	not necessary
6.	Behaviour	Explanation how to avoid Malaria infections Explanation how to avoid mosquito bites Explanation how to travelers diarrhoea (concerning food, water, etc.) Explanation how to avoid infections like HIV, Hepatitis
7.	Hospital nearby	List of addresses of tertiary hospitals in Bali
8.	Information about possible injuries abroad	bites, stings, wounds
9.	Information about medicaments if required	Explanation on Malaria Prophylaxis: Atovaquon&Proguanil (Malarone), Doxyciclin Manahydrat, Mefloquin (Lariam)
10.	Activities	Explanation how to avoid sun burned skin

measures in a medical domain have been presented in [11] as well as [12]. [11] combines similarity measures for image segmentation and similarity measures for textual information in order to retrieve cases containing non-image and image information. Our approach deals with textual information that has a different semantic. The approach presented in [12] uses individual similarity measures for two different groups of parameters (group one describes the clinical stage, gleason score, etc. and group two contains the degree of radiation). In comparison to this approach we switch the similarity measure according to the user's input.

3.3 Requirements with Respect to Similarities

Combined Similarity Measures. The complexity of the domain also requires some rather advanced similarity measures. For instance in the case of geographic information when planning a journey a *similar* destination would be one nearby, or a region that is maybe farther away but with comparable features with respect to climate, hygienic conditions or other associated risks. In order to accommodate this, similarities have to be defined with respect to different criteria and they are combined into a combined similarity measure that takes all possible similarities into account (see section 4.3 for an example).

Contextual Similarities. Another case is the modeling of diseases. They can be modeled in taxonomies but again with regard to different aspects such as incidents (where has the disease occurred), vectors (animals that carry the disease such as mice, rats or mosquitoes), ways of infection (airborne infection, contact infection, smear infection or infection through exchange of body fluids) or pathogen (virus, virusoid, bacteria, prions, worms, fungi etc.). Which aspect really determines a disease's similarity in the end depends on the query context. If the traveler is looking for information on prevention, aspects such as incidents, vectors or possible ways of transmission are relevant. If a traveler has already arrived at the destination and is looking for ways to treat an existing or supposed disease the disease's pathogen is the determining factor. This makes it necessary to not only be able to combine similarity measures but also to switch between them depending on the query context.

Circular Similarities. Another speciality is the modeling of the concept of time, which requires a circular similarity measure. When dealing with travel medicine we usually consider the month of a journey. The months cannot be modeled as a simple list where each month has a maximal similarity with its neighbors though, since in reality they form a circle. January should be as similar to December as it is to February.

3.4   Requirements with Respect to CBR Tools

Earlier we already introduced our concept of multiple case bases. A tool or system that is used to model the travel medicine domain according to our approach needs to be able to hold several case bases that are isolated from each other and can be queried individually, but also needs to be able to share concepts or data types (for instance the data types for regions is needed in the destination case base as well as in the diseases case base, for indicating occurrences of a disease).

Since we plan to offer docQuery as a freely accessible online source of information, we will have to deal with a high amount of system load which requires high availability, high processing speed as well as scalability. Also, being freely accessible, it may be exposed to possible attempts of misuse which require the system to also be as secure and failsafe as possible in order to keep up the service and protect the system's data from tampering or destruction.

We do not expect to find a tool that perfectly fits all of our needs, but instead expect it to be necessary to add self-written components integrated into the system, for example the subsequent querying of the individual case bases and the combination of the respective results. In order to be able to freely interact with the tool some kind of programming interface or API is necessary.

4   The Empolis Information Access Suite

The empolis Information Access Suite (e:IAS) [13] is being developed by empolis[7] a Bertelsmann company. It consists of a server-client based knowledge provision component, the Knowledge Server, and a knowledge modeling tool, the Creator. e:IAS uses case-based reasoning (CBR) as its main underlying methodology. The Knowledge Server is used to query the system's case base(s) and the Creator is used to model the individual case format(s), for configuring the processing and possible enrichment of imported data, as well as the actual case retrieval. In research projects empolis allows us to use the e:IAS under research and education license.

4.1   Basic Concepts

The Creator is based on the free eclipse integrated development environment[8] adopting its workbench concepts and extending it with different components of its own. Different tools within the Creator, so called Managers, are used to create a model of a respective domain (ModelManager), configure the import of existing data into the model (DataManager) and define the workflow of their processing (PipeManager). When modeling a new knowledge domain the domain is represented by an aggregate that includes freely definable attributes of different types. The data types can also be freely defined, by extending predefined basic data types such as text, integers, and floating point numbers. Further on attributes can be defined as a set of instances of a particular data type, or as a compound type, including atomic data types.

Data types can be constrained or specialized for instance by defining ranges or a fixed set of possible values. In the case of text based data types these values are called concepts. When defining new data types it is also necessary to define their similarity measures, which are crucial for case-based reasoning. These similarity measures can be defined as simple mathematical functions, in the case of numerical data types, or, in case of text based data types, as simple string distance measures. In case of concepts more elaborate similarity measures are also possible. Concepts can be arranged in lists or orders, forming taxonomies or similarity matrices. More complex similarities can be modeled using combined similarity measures where different measures are defined and the highest similarity is taken into account. Such a combined similarity measure can for example be used to create a model in which two leaves of a given taxonomy are more similar than the others, like for instance a pair of twins in a set of siblings. In this example a given family taxonomy would be combined with a similarity matrix in which the two twins are marked with a higher similarity than the similarity given by the taxonomy for its respective child leafs.

Once a model has been created it can be filled with data from a multitude of sources such as simple text or xml files, data bases, data collected by a web crawler or provided by an MS Exchange or Tamino server. Additional tools such

---

[7] http://www.empolis.com
[8] http://www.eclipse.org/

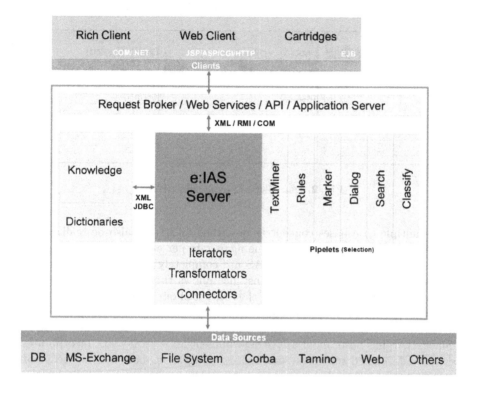

Fig. 1. e:IAS Architecture, source:[13]

as the TextMiner and different rules pipelets (components of the PipeManager) can be used for processing the data at import time.

e:IAS itself uses a custom XML dialect to store its models, similarities, pipe/ import definitions etc. The index that is generated from the imported data can be queried using Java, JSP or web services.

4.2 Requirement Compatibility

e:IAS also meets most of the requirements described in section 3.3 that arise from the actual knowledge domain. It offers the possibility to define custom similarity measures for data types and combine any number of them into a combined similarity measure, which is implemented in such a way that all included measures, are consulted and the highest value is taken into account. Switching between individual similarity measures during the system's runtime is realized using adaptation rules that are a part of e:IAS.

The aforementioned circular similarity measure which is needed in order to model temporal attributes has been realized by modelling the individual months as textual concepts and using a similarity matrix with manually defined similarities.

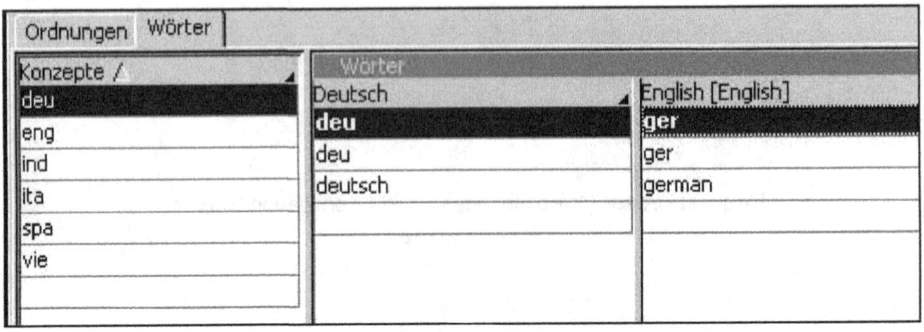

Fig. 2. Data Type Language in e:IAS

The multiple case bases concept as described in 2.1 can also be realized using e:IAS. We did this by creating one aggregate per case base within a single instance of e:IAS. Aggregates in e:IAS are completely independent from each other, but having them inside one instance allows them to share data types, thus creating a shared vocabulary. In the context of our application domain this is particularly useful since we do have shared data types such as region or disease that would otherwise have to be copied and synchronized manually.

Since e:IAS has been under active development for several years, incorporates previously developed successful knowledge management tools such as orenge or SmartFinder and has also been successfully deployed in numerous industrial applications like the SIMATIC Knowledge Manager [14] we are confident that its overall performance and security will meet our requirements. The required open API is also provided in the form of a Java API that can be used in Java programs or from within JSP.

4.3   Shared Data Types and Similarities in e:IAS

The shared data types within our knowledge model are languages, diseases, geographic regions, medication, transmission ways, symptoms or parts of disease patterns. In this section we will show some examples how we modeled different kinds of data types and their according similarities.

The data type *language* contains a list of languages in which data is stored. Each case base contains this attribute, so we have to define it once and can use it in all case bases. Figure 2 shows the definition of this data type in e:IAS. On the left hand side of the figure the concepts are displayed and for each concept we are able to define synonyms ensuring an appropriate mapping.

Another data type which is used by more than one case base is region, because it describes the region of the destination, the regions in which a disease might occur, as well as the region a hospital is situated in. Figure 3 shows how concepts can be modeled in a taxonomy including their similarity definitions. The numbers underneath each concept in the taxonomy in figure 3 are the similarities if *Jersey* is requested.

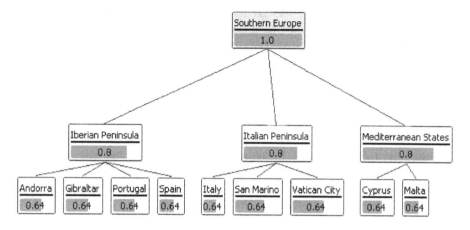

Fig. 3. Taxonomy Excerpt of the Data Type Region in e:IAS

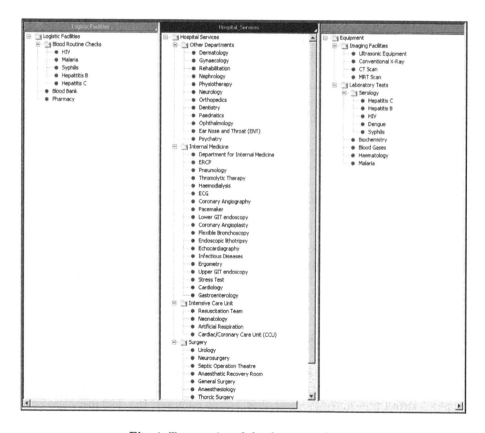

Fig. 4. Taxonomies of the data type disease

This taxonomy represents the geographical classification and furthermore e:IAS allows to combine more types of classification, for instance, using climatic aspects. The taxonomy represents only one view to define the similarities. For more context dependable similarities we can define a matrix containing all countries and their subsumptions and define another similarity measure. During the retrieval process both measures are considered and the highest similarity will be included in the calculation.

Contextual similarities can be defined by creating concepts with more than one taxonomy which enables the system to represent information in one attribute using multiple taxonomies. Each taxonomy represents its respective similarity model. Figure 4 shows the data type facilities containing different kinds of facilities and multiple orders. Each order can assign its own similarity measure and during indexing process all orders are considered.

5   docQuery in the Context of SEASALT and CoMES

The realization of the modular, cross-linked knowledge model with e:IAS is only a part of the general development of the SEASALT architecture [2], which is based on the CoMES approach, our underlying research vision [1]. docQuery as a real-life application gives us the opportunity to evaluate the concepts and techniques we developed in CoMES and SEASALT by implementing them on an existing knowledge domain, using a well-established industrial strength tool and thus testing their adequacy and feasibility. Of course not all parts of the SEASALT architecture can be implemented using e:IAS, which is why this paper focuses on docQuery's knowledge provision part. The knowledge acquisition part which is mostly concerned with natural language processing and classification will be implemented separately.

The implementation of an architecture such as docQuery's is not a typical application of e:IAS, since it uses several cross-linked case bases and demands features like the switching of case bases or similarity measures during runtime, which is not yet implemented. We plan further extensions, using semantic descriptions of knowledge sources for instance based on [15], with respect to improving the hybrid and decentralized/distributed reasoning capabilities as well as the problem solving capabilities in configuration-like tasks.

6   Conclusion and Outlook

In this paper we presented the realization of a modular, cross-linked knowledge model with e:IAS, an industrial knowledge management and CBR tool. We introduced travel medicine as our application domain and the docQuery project which aims to provide complete and reliable information for individual journeys considering all aspects a travel medicine physician would do. Within docQuery we introduced its different case bases and the way they can be queried and combined. Based on this model we then described the requirements that arise from our architecture and its underlying knowledge domain with regard to knowledge modeling, similarity measures, system performance, and accessibility.

We then presented e:IAS, a knowledge management tool, which is based on CBR and developed by empolis introduced its underlying concepts and basic usage, and evaluated it, according to the earlier stated requirements. Following this, we illustrated how we realized docQuery's modular, cross-linked knowledge model using e:IAS and explained how we implemented the novel functionalities using IAS's on-board utilities. Further on, we illustrated how this implementation fits in the general context of our work on SEASALT and CoMES.

docQuery is still work in progress. We are currently building up our knowledge base and plan to further expand it, both manually and by mining knowledge from the corresponding community of experts. Apart from the knowledge acquisition part our next steps will be finding a sensible way of switching similarity measures during run time in order to realize contextual similarities. We also plan to launch a first prototype soon in order to optimize our request strategies based on realistic queries.

References

1. Althoff, K.D., Bach, K., Deutsch, J.O., Hanft, A., Mänz, J., Müller, T., Newo, R., Reichle, M., Schaaf, M., Weis, K.H.: Collaborative multi-expert-systems – realizing knowlegde-product-lines with case factories and distributed learning systems. Technical report, University of Osnabrück, Osnabrück (September 2007)
2. Bach, K., Reichle, M., Althoff, K.D.: A domain independent system architecture for sharing experience. In: Proceedings of LWA 2007, Workshop Wissens- und Erfahrungsmanagement, September 2007, pp. 296–303 (2007)
3. Plaza, E., McGinty, L.: Distributed case-based reasoning. The Knowledge engineering review 20(3), 261–265 (2006)
4. Leake, D., Sooriamurthi, R.: When two case bases are better than one: Exploiting multiple case bases. In: Aha, D.W., Watson, I. (eds.) ICCBR 2001. LNCS (LNAI), vol. 2080, pp. 321–335. Springer, Heidelberg (2001)
5. Althoff, K.D., Reichle, M., Bach, K., Hanft, A., Newo, R.: Agent based maintenance for modularised case bases in collaborative multi-expert systems. In: Proceedings of AI2007, 12th UK Workshop on Case-Based Reasoning, December 2007, pp. 7–18 (2007)
6. Richter, M.M.: Introduction. In: Lenz, M., Bartsch-Spörl, B., Burkhard, H.D., Wess, S. (eds.) Case-Based Reasoning Technology. LNCS (LNAI), vol. 1400. Springer, Heidelberg (1998)
7. ISTM - International Society of Travel Medicine: Introduction to travel medicine. ISTM Slide Lecture Kit (February 2006)
8. Schmidt, T., Herda, L.R., Ly, M.P.: First aid in tourism medicine. Curriculum. Continuing Education Scheme (August 2006)
9. Leake, D.B., Sooriamurthi, R.: Dispatching cases versus merging case-bases: When mcbr matters. In: Proceedings of the Sixteenth International Florida Artificial Intelligence Research Society Conference, FLAIRS-2003, May 2003, pp. 129–133 (2003)
10. Avesani, P., Hayes, C., Cova, M.: Language games: Solving the vocabulary problem in multi-case-base reasoning. In: Muñoz-Avila, H., Ricci, F. (eds.) ICCBR 2005. LNCS (LNAI), vol. 3620, pp. 35–49. Springer, Heidelberg (2005)

11. Perner, P.: An architecture for a cbr image segmentation system. In: Althoff, K.-D., Bergmann, R., Branting, L.K. (eds.) ICCBR 1999. LNCS (LNAI), vol. 1650, pp. 525–534. Springer, Heidelberg (1999)

12. Song, X., Petrovic, S., Sundar, S.: A case-based reasoning approach to dose planning in radiotherapy. In: Wilson, D.C., Khemani, D. (eds.) Proceedings of the 7th International Conference on Case-Based Reasoning (ICCBR) 2007, Workshop on Case-Based Reasoning in the Health Sciences, Belfast, Northern Ireland, August 2007, pp. 348–357 (2007)

13. empolis GmbH: Technical white paper e:information access suite. Technical report (September 2005)

14. Lenz, M., Busch, K.H., Hübner, A., Wess, S.: The simatic knowledge manager. In: Aha, D.W., Becerra-Fernandez, I., Maurer, F., Muñoz-Avila, H. (eds.) Exploring Synergies of Knowledge Management and Case-Based Reasoning (Technical Report WS99 -10), pp. 40–45. AAAI Press, Menlo Park (1999)

15. Schaaf, M.: Managing Experience Items Indexed by Declarative Workflow Descriptions. PhD thesis, Institute of Computer Science, University of Hildesheim (February 2008)

GEP-Induced Expression Trees as Weak Classifiers

Joanna Jędrzejowicz[1] and Piotr Jędrzejowicz[2]

[1] Institute of Computer Science, Gdańsk University,
Wita Stwosza 57, 80-952 Gdańsk, Poland
jj@inf.univ.gda.pl
[2] Department of Information Systems, Gdynia Maritime University,
Morska 83, 81-225 Gdynia, Poland
pj@am.gdynia.pl

Abstract. The paper proposes applying Gene Expression Programming (GEP) to induce expression trees used subsequently as weak classifiers. Two techniques of constructing ensemble classifiers from weak classifiers are investigated in the paper. The working hypothesis of the paper can be stated as follows: given a set of classifiers generated through applying gene expression programming method and using some variants of boosting technique, one can construct the ensemble producing effectively high quality classification results. A detailed description of the proposed GEP implementation generating classifiers in the form of expression trees is followed by the report on AdaBoost and boosting algorithms used to construct an ensemble classifier. To validate the approach computational experiment involving several benchmark datasets has been carried out. Experiment results show that using GEP-induced expression trees as weak classifiers allows for construction of a high quality ensemble classifier outperforming, in terms of classification accuracy, many other recently published solutions.

Keywords: gene expression programming, classification, ensemble methods.

1 Introduction

Appropriately combining information sources to form a more effective output than any of the individual sources is an effective approach used in a variety of areas. In machine learning there are several situations motivating combining multiple learners. For example, Benett [4] argues that "it may not be possible to train using all the data because data privacy and security concerns prevent sharing the data. However, a classifier can be trained over different data subsets and the predictions they issue may be shared. In other cases, the computation burden of the base classifier may motivate classifier combination. When a classifier with a nonlinear training or prediction cost is used, computational gains can be realized by partitioning the data and applying an instance of the classifier to each subset. In other situations, combining classifiers can be seen as a

P. Perner (Ed.): ICDM 2008, LNAI 5077, pp. 129–141, 2008.
© Springer-Verlag Berlin Heidelberg 2008

way of extending the hypothesis space or relaxing the bias of the original base classifier". Among methods based on classifier combination one can mention cascade generalization [12], stacking [27], and boosting [22], [3]. In this paper we propose an ensemble method integrating the gene expression programming with the AdaBoost algorithm. Ensemble methods are methods that first solve a classification or regression problem by creating multiple learners each attempting to solve the task independently, then use a procedure specified by the particular ensemble method for selecting or weighing the individual learners. Apart from the already mentioned stacking and boosting, other example combination methods include composite classifiers [5], voting pool of classifiers [2], combination of multiple classifiers [13] and [17], classifier ensembles [6], [16], and many others. An excellent review of the classifier combination methods can be found in [16].

As Polikar [21] observes there are two types of classifier combination: classifier selection and classifier fusion. In classifier selection, each classifier is trained to become an expert in some local area of the feature space. In classifier fusion all classifiers are trained over entire feature space. In this case, the classifier combination process involves merging individual (weaker) classifiers to obtain a single (stronger) expert of superior performance. The ensemble method proposed in this paper belongs to the classifier fusion category.

Although theoretical results [7] indicate there is no a priori choice of algorithm which will perform best over all problems, experience has shown that some algorithms can dominate large classes of problems. However, even when an algorithm outperforms another algorithm across a problem set, combining the algorithms can lead to better results than either alone. There are many concrete situations where weak classifiers can help improve the performance of a strong classifier [4].

In the research reported in this paper, ensemble of weak classifiers is induced using gene expression programming (GEP). Hence, the working hypothesis of the paper can be stated as follows: given a set of classifiers generated through applying gene expression programming method and using some variants of boosting technique, one can construct an ensemble classifier producing effectively high quality classification results.

The paper is organized in sections. Section 1 contains the introduction and the hypothesis studied in the paper. Section 2 provides a short overview of the gene expression programming application to classification problems and a more detailed description of the proposed GEP implementation generating classifiers in the form of expression trees. Section 3 reports on AdaBoost and boosting algorithms used to construct the ensemble system as well as on its proposed modification. Section 4 shows results of the computational experiment. The final section contains conclusions and a proposal for future research.

2   Using Gene Expression Programming to Induce Classifiers

Gene expression programming introduced by Ferreira [9] is an automatic programming approach. In GEP computer programs are represented as linear character strings of fixed-length called chromosomes which, in the subsequent fitness evaluation, can be expressed as expression trees of different sizes and shapes. The approach has flexibility and power to explore the entire search space, which comes from the separation of genotype and phenotype. As it has been observed by Ferreira [10] GEP can be used to design decision trees, with the advantage that all the decisions concerning the growth of the tree are made by the algorithm itself without any human input, that is, the growth of the tree is totally determined and refined by evolution. Ferreira in her paper [10] proposed two different algorithms to grow the trees. The first one induces decision trees with nominal attributes, and the second one was developed for handling numeric attributes but, in fact, can handle all kinds of attributes.

The ability of GEP to generate decision trees makes it a natural tool for solving classification problems. Ferreira [10] showed several example applications of GEP including classification. The approach was based on the tree induction algorithm proposed by the author. Weinert and Lopes [26] apply GEP to the data mining task of classification by inducing rules. The authors proposed a new method for rule encoding and genetic operators that preserve rule integrity. They also implemented a system, named GEPCLASS which allows for the automatic discovery of flexible rules, better fitted to data. Duan with co-authors [7] claimed to improve efficiency of GEP used as a classification tool. Their contribution includes proposing new strategies for generating the classification threshold dynamically and designing a new approach called Distance Guided Evolution Algorithm. Zeng with co-authors [29] proposed a novel Immune Gene Expression Programming as a tool for rule mining. Another approach to evolving classification rules with gene expression programming was proposed in [30]. A different example of GEP application to classification problems was proposed by Li and co-authors [19]. They proposed a new representation scheme based on prefix notation which brings some advantages as compared with the traditional approach. Wang and co-authors [25] proposed a GEP decision tree system. The system can construct decision tree for classification without prior knowledge about the distribution of data. Karakasis and Stafylopatis [15] proposed a hybrid evolutionary technique for data mining tasks, which combines the Clonal Selection Principle with Gene Expression Programming. The authors claim that their approach outperforms GEP in terms of convergence rate and computational efficiency.

In this paper two-class classification of data with numeric and categorical attributes is considered. Gene expression programming is used to induce expression trees which correspond to rules. The data that satisfy the rule are classified as the first class and those for which the rule does not work - as the second class.

As usual when applying GEP methodology, the algorithm uses a population of chromosomes, selects them according to fitness and introduces genetic variation using several genetic operators. Each chromosome is composed of a single gene divided into two parts as in the original head-tail method [9]. The size of the head (h) is determined by the user with the suggested size not less than the number of attributes in the dataset. The size of the tail (t) is computed as

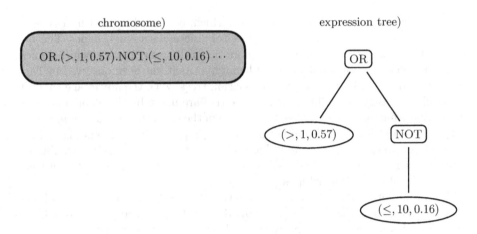

Fig. 1. One chromosome and a corresponding expression tree

$t = h(n - 1) + 1$ where n is the largest arity found in the function set. In the computational experiments the functions are: logical AND, OR and NOT. Thus $n = 2$ and the size of the chromosome is $h + t = 2h + 1$. The terminal set contains triples $(op, attrib, const)$ where op is one of relational operators $<, \leq, >, \geq, =, \neq$, $attrib$ is the attribute number, and finally $const$ is a value belonging to the domain of the attribute $attrib$. As usual in GEP, the tail part of a gene always contains terminals and head can have both, terminals and functions. Observe that in this model each chromosome is syntactically correct and corresponds to a valid expression. Each attribute can appear once, many times or not at all. This allows to define flexible characteristics like for example $(attribute1 > 0.57)$ AND $(attribute1 < 0.80)$. On the other hand, it can also introduce inconsistencies like for example $(attribute1 > 0.57)$ AND $(attribute1 < 0.40)$. This does not cause problems since a decision subtree corresponding to such a subexpression would evaluate it to $false$. Besides, when studying the structure of the best classifiers in our experiments the above inconsistencies did not appear.

The expression of this kind of genes is done in exactly the same manner as in all GEP systems. In Fig. 1 an example of a gene and a corresponding expression tree is given. The start position (position 0) in the chromosome corresponds to the root of the expression tree (OR, in the example). Then, below each function branches are attached and there are as many of them as the arity of the function - 2 in our case. The following symbols in the chromosome are attached to the branches on a given level. The process is complete when each branch is completed with a terminal. The number of symbols from the chromosome to form the expression tree is denoted as the termination point. For the discussed example, the termination point is 4 therefore further symbols are not meaningful and are denoted by $\cdots$ in Fig.1. Subsequent symbols in the gene are separated by dots and in the tree the terminals are drawn as ovals. The rule corresponding to the expression tree from Fig. 1 is:

IF $(attribute1 > 0.57)$ OR NOT $(attribute10 \leq 0.16)$ THEN Class1.

In fact, the above rule was taken as an example from the solution to the Sonar dataset described later in the paper. It was used in one of the shots of the 10 cross validation experiment. It classifies the learning set with accuracy 0.80 and 0.85 for the test set.

The algorithm for learning the best classifier using GEP works as follows. Suppose that a training dataset is given and each row in the dataset has a correct label representing one of two classes $\{0, 1\}$, where 0 represents Class0 - *false*, and 1 represents Class1 - *true*. In the initial step the minimal and maximal value of each attribute is calculated and a random population of chromosomes is generated. For each chromosome the symbols in the head part are randomly selected from the set of functions AND, OR, NOT and the set of terminals of type $(op, attrib, const)$, where the value of *const* is in the range of *attrib*. The symbols in the tail part are all terminals. To introduce variation in the population the following genetic operators are used:

- mutation,
- transposition of insertion sequence elements (IS transposition),
- root transposition (RIS transposition),
- one-point recombination,
- two-point recombination.

Mutation can occur anywhere in the chromosome. We consider one-point mutation which means that with a probability, called mutation rate, one symbol in a chromosome is changed. In case of a functional symbol it is replaced by another randomly selected function, otherwise for $g = (op, attrib, const)$ a random relational operator op', an attribute $attrib'$ and a constant $const'$ in the range of $attrib'$ are selected. Note that mutation can change the respective expression tree since a function of one argument may be mutated into a function of two arguments, or vice versa.

Transposition stands for moving part of a chromosome to another location. Here we consider two kinds of transposable elements. In case of transposition of insertion sequence (IS) three values are randomly chosen: a position in the chromosome (start of IS), the length of the sequence and the target site in the **head** - a bond between two positions. For example consider the chromosome C with head=6, defined below. The termination point is 7.

0	1	2	3	4	5	6	7	8	9
OR	AND	AND	$(>, 1, 0)$	$(=, 2, 05)$	$(>, 1, 2)$	$(<, 1, 10)$	$(>, 3, 0)$	$(<, 3, 5)$	$\cdots$

Suppose that IS is defined as: start position=6, length=2, target=0. Then a cut is made in the bond defined by the target site (in the example between symbol 0 and 1), and the insertion sequence (the symbols from positions 6 and 7) is copied into the site of the insertion. The sequence downstream from the copied IS element loses, at the end of the head, as many symbols as the length of the transposon. The resulting chromosome is shown below:

0	1	2	3	4	5	6	7	8	9
OR	$(<,1,10)$	$(>,3,0)$	AND	AND	$(>,1,0)$	$(<,1,10)$	$(>,3,0)$	$(<,3,5)$	$\cdots$

Observe that since the target site is in the head, the newly created individual is always syntactically correct though it can reshape the tree quite dramatically as in the above case. The termination point is 3 for the new chromosome.

In case of root transposition, a position in the head is randomly selected, the first function following this position is chosen - it is the start of the RIS element. If no function is found then no change is performed. The length of the insertion sequence is chosen. The insertion sequence is copied at the root position and at the same time the last symbols of the head (as many as RIS length) are deleted. For the chromosome C defined before and RIS sequence starting with the function AND at the position 2, of length 2 the resulting chromosome is defined as:

0	1	2	3	4	5	6	7	8	9
AND	$(>,1,0)$	OR	AND	AND	$(>,1,0)$	$(<,1,10)$	$(>,3,0)$	$(<,3,5)$	$\cdots$

Again the change has quite an effect since the termination point is now 9.

For both kinds of recombination two parent chromosomes P_1, P_2 are randomly chosen and two new child chromosomes C_1, C_2 are formed. In case of one-point recombination one position is randomly generated and both parent chromosomes are split by this position into two parts. Child chromosomes C_1 (respectively, C_2) is formed as containing the first part from P_1 (respectively, P_2) and the second part from P_2 (and P_1). In two-point recombination two positions are randomly chosen and the symbols between recombination positions are exchanged between two parent chromosomes forming two new child chromosomes. Observe that again, in both cases, the newly formed chromosomes are syntactically correct no matter whether the recombination positions were taken from the head or tail.

During GEP learning, the individuals are selected and copied into the next generation based on their fitness and the roulette wheel sampling with elitism which guarantees the survival and cloning of the best chromosome in the next generation. The fitness of a chromosome C is calculated as follows. The expression tree T corresponding to C is generated. For each row from the dataset the outcome of T is compared with the label. The ratio of correctly classified rows to the size of the dataset is the fitness of chromosome C. The algorithm of GEP learning is shown below.

Algorithm 1 (GEP weak learning)
Input: training data with correct labels representing two classes, integer NG specifying the number of generations,
Output: classifier C

```
create chromosomes of the initial population,
repeat NG  times the following:
     1. express chromosomes as expression trees,
        calculate fitness of each chromosome,
```

2. keep best chromosome,
3. select chromosomes,
4. mutation,
5. transposition of insertion sequence elements,
6. root transposition,
7. one-point recombination,
8. two-point recombination.

calculate fitness and keep the best chromosome - classifier C

3 Ensemble System Constructed from GEP Induced Expression Trees

As suggested in [22] a weak classifier can be considerably improved by *boosting*. The general idea is to create an ensemble of classifiers by resampling the training dataset and creating a classifier for each sample. In each step the most informative training data is provided - for example those for which the previous classifier misclassified. Then an ensemble of generated classifiers together with an intelligent combination rule proves often to be a more efficient approach. Freund and Schapiro [11] suggested a refinement of a boosting algorithm called AdaBoost. For a predefined number of iterations T, the following procedures are performed. In the ith step, according to the current distribution - which is uniform in the first iteration, a sample is drawn from the dataset. The best classifier C_i is found for the sample and using the whole dataset the error of the current classification is calculated. The distribution is updated so that the weights of those instances that are correctly classified by the current classifier C_i are reduced by the factor depending on the error, and the weights of misclassified instances are unchanged. Once T classifiers are generated 'weighted majority voting' is used to classify the test set. The idea is to promote those classifiers that have shown good performance during training - they are rewarded with a higher weight than the others. The details are shown below.

Algorithm 2 (AdaBoost using GEP as a weak learner)
Input: training dataset
$$LD = \{(\mathbf{x}_i, y_i) : y_i \in \{0,1\}, i = 1, \ldots, N\}$$
with N examples and correct labels representing two classes, test dataset TD, integer T specifying the number of iterations,
Output: qc- quality of the AdaBoost classifier.

1. Initialize the distribution

$$D_1(i) = \tfrac{1}{N}, i = 1, \ldots, N$$

2. for t=1 to T do
2.1   select a training dataset St, a subset of LD drawn for the current distribution,
2.2   call Algorithm 1 for the dataset St,

```
   receive the classifier Ct,
2.3  calculate the error of the classifier Ct
```

$\epsilon_t = \sum_{Ct(\mathbf{x}_i) \neq y_i} D_t(i),$
set $\beta_t = \epsilon_t/(1 - \epsilon_t)$

```
2.4 update the distribution
```

if $Ct(\mathbf{x}_i) = y_i$ then $D_t(i) := D_t(i) \times \beta_t$

```
2.5 normalize the distribution
```

$D_{t+1}(i) = D_t(i)/Z_t, \ Z_t = \sum_i D_t(i)$

```
3. test the ensemble classifier {C1, C2, ...,CT}
      in the test dataset TD
3.1 qc:=0
3.2 for each instance (x,y) from TD
3.2.1 calculate
```

$Vi = \sum_{Ct(\mathbf{x})=i} \log(1/\beta_t), \ i = 0,1$

```
3.2.2 if V0 > V1  then   c:=0   else c:=1
3.2.3 if c=y then qc:=qc+1
3.3  qc:=qc/size of TD
```

As follows from the above in each iteration t the distribution weights of those instances that were correctly classified are reduced by a factor β_t and the weights of the misclassified instances stay unchanged. After the normalization the weights of instances misclassified are raised and they add up to 1/2, and the weights of the correctly classified instances are lowered and they also add up to 1/2. What is more, since it is required that the weak classifier has an error less than 1/2, it is guaranteed to correctly classify at least one previously misclassified instance. In the ensemble decision those classifiers which produced small error and β_t is close to zero, have a large voting role since $1/\beta_t$ and logarithm of $1/\beta_t$ are large. In our second boosting algorithm, which is called MV-boosting in the sequel, the AdaBoost algorithm is modified so that simple majority voting is used by the ensemble.

Algorithm 3 (MV-boosting using GEP as a weak learner)
Input: training dataset
$LD = \{(\mathbf{x}_i, y_i) : \ y_i \in \{0,1\}, i = 1, \ldots, N\}$ with N examples and correct labels representing two classes, test dataset TD, integer T specifying the number of iterations,
Output: qc- quality of the classifier.

```
1. Initialize the distribution
```

$D_1(i) = \frac{1}{N}, i = 1, \ldots, N$

```
2. for t=1 to T do
2.1   select a training dataset St,
       a subset of LD drawn for the current distribution,
2.2   call Algorithm 1 for the dataset St,
       receive the classifier Ct,
2.3   calculate the error of the classifier Ct
```

$\epsilon_t = \sum_{Ct(\mathbf{x}_i) \neq y_i} D_t(i)$,

set $\beta_t = \epsilon_t/(1 - \epsilon_t)$

```
2.4 update the distribution
```

if $Ct(\mathbf{x}_i) = y_i$ then $D_t(i) := D_t(i) \times \beta_t$

```
2.5 normalize the distribution
```

$D_{t+1}(i) = D_t(i)/Z_t, \ Z_t = \sum_i D_t(i)$

```
3. test the ensemble classifier {C1, C2, ...,CT}
    in the test dataset TD
3.1 qc:=0
3.2 for each instance (x,y) from TD
3.2.1 calculate for i=0,1
Vi= the number of classifiers which classified x as the class i
3.2.2 if V0>V1 then c:=0 else c:=1
3.2.3 if   c=y   then qc:=qc+1
3.3 qc:=qc/size of TD
```

The time complexity of Algorithm 1 is $O(N \cdot NG \cdot psize)$ where N is the size of the training set, NG is the number of generations, and $psize$ is the size of the population. Thus the time complexity of Algorithm 2 and 3 is $O(T \cdot N \cdot NG \cdot psize)$ where T is the number of iterations in, respectively, AdaBoost and MV-boosting.

4   Computational Experiment Results

To evaluate the proposed approach computational experiment has been carried out. The experiment involved the following datasets from the UCI Machine Learning Repository [1]: Wisconsin Breast Cancer (WBC), Australian Credit Approval (ACredit), German Credit Approval (GCredit), Heart Disease, and Sonar. Basic characteristics of these sets are shown in Table 1.

In the reported experiment the following classification tools, described in details in the previous Sections, have been used:

- Expression tree induced by gene expression programming (GEP-only)
- Ensemble of weak classifiers induced by gene expression programming combined through the AdaBoost algorithm (AdaBoost)

Table 1. Datasets used in the computational experiment

Name	Data type	Attribute types	# of instances	# of attributes
WBC	Multivariate	Integer	699	11
ACredit	Multivariate	Categorical, Integer, Real	690	15
GCredit	Multivariate	Categorical, Integer	1000	21
Heart	Multivariate	Categorical, Real	303	14
Sonar	Multivariate	Real	208	61

Table 2. Computation parameter settings

Tool	# of individuals in GEP	# of iteration in GEP	# of trees in the ensemble
GEP-only	800	100	-
AdaBoost	100	50	100
MV	100	50	50

Table 3. Percent of the correct classifications and its standard deviations

Dataset	GEP-only	AdaBoost	MV-boosting
WBC	95,4 +/- 3,92	97,7 +/- 2,15	**98,6 +/- 1,79**
ACredit	84,7 =/- 2,39	**89,3 +/- 2,98**	86,4 +/- 3,89
GCredit	77,7 +/- 3,09	80,5 +/- 4,27	**80,07 +/- 3,80**
Heart	83,0 +/- 3,31	85,7 +/- 3,86	**88,0 +/- 6,12**
Sonar	85,9 +/- 9,67	90,6 +/- 7,32	**92,3 +/- 6,68**

– Ensemble of weak classifiers induced by gene expression programming combined through boosting with the majority voting (MV-boosting).

Computations have been run with the parameter settings shown in Table 2. Probabilities of genetic operations have been set as follows: mutation rate - 0,5; root transposition, IS transposition, one-point recombination and two-point recombination - all 0,2.

Performance measures (% of the correct classifications and its standard deviation) have been averaged over ten tenfold independent cross-validation runs for each of the investigated datasets. The respective experiment results are shown in Table 3.

Table 4 contains comparisons of the best out of the two - AdaBoost and MV-boosting called GEP-ensemble, with several literature-reported results. It can be easily seen that the proposed ensembles outperform all recently published classifiers, which we have been able to identify. There are, of course some drawbacks. The approach requires setting values of numerous parameters (numbers of iteration, population size, probabilities of genetic operations). This requires

Table 4. GEP-ensemble versus literature-reported results (% correct classifications and its standard deviation)

Dataset	GEP-ensemble	Reported	Method & Source
WBC	**98,6** +/- 1,79	98,5 +/- n.a.	GEPCLASS (5CV)[26]
		96,5 +/- 2,00	GEP rules [30]
		97,0 +/- n.a.	AdaBoost with LGG [24]
ACredit	**89,3** +/- 2,98	86,9 +/- n.a.	SVM + GA[14]
		86,1 +/- n.a.	Globoost [24]
		85,9 +/- n.a.	C4.5 [17]
GCredit	**80,7** +/- 3,80	79,5 +/- n.a.	LS-SVM [20]
		79,0 +/- n.a.	Bay. F. D. [20]
		77,9 +/- 3,97	SVM+Ga [14]
Heart	**88,0** +/- 6,12	85,1 +/- 0,50	28-NN [28]
		82,2 +/- 7,60	GEP rules [30]
		77,5 +/- n.a.	C4.5 [17]
Sonar	**92,3** +/- 6,68	90,5 +/- n.a.	Fuzzy C-Means [23]
		87,5 +/- 0,80	1-NN Euclid [28]
		84,9 +/- n.a.	Boostexter [24]

an extensive fine-tuning phase. On the other hand, we have observed that apart from the numbers of iteration the remaining parameters do not play a decisive role from the point of view of the efficiency of computations and classifiers performance.

5 Conclusions

The presented research allows to draw the following conclusions:

- Gene expression programming is a versatile and useful tool to automatically induce expression trees.
- Using GEP-induced expression trees as weak classifiers allows for construction of a high quality ensemble classifiers outperforming, in terms of classification accuracy, many other recently published solutions.
- High quality of the ensemble classifier performance can be attributed to the diversity of weak classifiers induced by GEP.
- The computational experiment carried out does not allow to decide which of the two ensemble constructing techniques used is more effective although boosting with the majority voting seems to have slight advantage over the AdaBoost.

Future research should focus on extending the approach to an arbitrary number of classes as well as to considering multigenic chromosomes with each gene representing single rule as a decision subtree.

References

1. Asuncion, A., Newman, D.J.: UCI Machine Learning Repository, University of California, School of Information and Computer Science (2007),
   http://www.ics.uci.edu/~mlearn/MLRepository.html
2. Battiti, R., Colla, A.M.: Democracy in neural nets: Voting schemes for classification. Neural Networks 7(4), 691–707 (1994)
3. Bauer, E., Kohavi, R.: An empirical comparison of voting classification algorithms: Bagging, boosting, and variants. Machine Learning 36(1-2), 105–139 (1999)
4. Bennett, P.N.: Building Reliable Metaclassifiers for Text Learning, Ph.D. Thesis, School of Computer Science, Carnegie Mellon University, Pittsburgh (2006)
5. Dasarathy, B.V., Sheela, B.V.: Composite classifier system design: Concepts and Mathodology. Proceedings of the IEEE 67(5), 708–713 (1979)
6. Drucker, H., Cortes, C., Jackel, L.D., LeCun, Y., Vapnik, V.: Boosting and other ensemble methods. Neural Computation 6(6), 1289–1301 (1994)
7. Duan, L., Tang, C., Zhang, T., Wei, D., Zhang, H.: Distance Guided Classification with Gene Expression Programming. In: Li, X., Zaïane, O.R., Li, Z. (eds.) ADMA 2006. LNCS (LNAI), vol. 4093, pp. 239–246. Springer, Heidelberg (2006)
8. Duda, R., Hart, P., Stork, D.: Pattern Classification. John Wiley & Sons, Inc., New York (2001)
9. Ferreira, C.: Gene Expression Programming: a New Adaptive Algorithm for Solving Problems. Complex Systems 13(2), 87–129 (2001)
10. Ferreira, C.: Gene Expression Programming. Studies in Computational Intelligence 21, 337–380 (2006)
11. Freund, Y., Schapire, R.E.: Decision-theoretic generalization of on-line learning and application to boosting. Journal of Computer and System Science 55(1), 119–139 (1997)
12. Gama, J.: Local cascade generalization. In: Proceedings of the 15th International Conference on Machine Learning, pp. 206–214 (1998)
13. Ho, T.K., Hull, J.J., Srihari, S.N.: Decision Combination in Multiple Classifier Systems. IEEE Transactions on Pattern Recognition and Machine Intelligence 16(1), 66–75 (1994)
14. Huang, C.-L., Chen, M.-C., Wang, C.-J.: Credit scoring with a data mining approach based on support vector machines. Expert Systems with Applications 33, 847–856 (2007)
15. Karakasis, V.K., Stafylopatis, A.: Data Mining based on Gene Expression Programming and Clonal Selection. In: Proc. IEEE Congress on Evolutionary Computation, pp. 514–521 (2006)
16. Kuncheva, L.I.: Classifier ensembles for changing environments. In: Roli, F., Kittler, J., Windeatt, T. (eds.) MCS 2004. LNCS, vol. 3077, pp. 1–15. Springer, Heidelberg (2004)
17. Lam, L., Suen, C.Y.: Optimal combination of pattern classifiers. Pattern Recognition Letters 16(9), 945–954 (1995)
18. Last, M., Maimon, O.: A Compact and Accurate Model for Classification. IEEE Transactions on Knowledge and Data Engineering 16(2), 203–215 (2004)
19. Li, X., Zhou, C., Xiao, W., Nelson, P.C.: Prefix Gene Expression Programming. In: Proc. Genetic and Evolutionary Computation Conference, Washington, pp. 25–31 (2005)
20. Pena Centeno, T., Lawrence, N.D.: Optimising Kernel Parameters and Regularisation Coefficients for Non-linear Discriminant Analysis. Journal of Machine Learning Research 7, 455–491 (2006)

21. Polikar, R.: Ensemble Based Systems in Decision Making. IEEE Circuits and Systems Magazine 3, 22–43 (2006)
22. Schapire, R.E., Freund, Y., Bartlett, P., Lee, W.S.: Boosting the margin: A new explanation for the effectiveness of voting methods. The Annals of Statistics 26(5), 1651–1686 (1998)
23. Srinivasa, K.B., Singh, A., Thomas, A.O., Venugopal, K.R., Patnoik, L.M.: Generic Feature Extraction for Classification Using Fuzzy C-Means Clustering. In: Proc. Intelligent Sensing and Information Processing Conference, pp. 33–38 (2005)
24. Torre, F.: Boosting Correct Least General Generalizations, Technical Report GRAppA-0104, Grenoble (2004)
25. Wang, W., Li, Q., Han, S., Lin, H.: A Preliminary Study on Constructing Decision Tree with Gene Expression Programming. In: Proc. First International Conference on Innovative Computing, Information and Control, vol. 1, pp. 222–225 (2006)
26. Weinert, W.R., Lopes, H.S.: GEPCLASS: A Classification Rule Discovery Tool Using Gene Expression Programming. In: Li, X., Zaïane, O.R., Li, Z. (eds.) ADMA 2006. LNCS (LNAI), vol. 4093, pp. 871–880. Springer, Heidelberg (2006)
27. Wolpert, D.H.: Stacked generalization. Neural Networks 5, 241–259 (1992)
28. Statlog Datasets: comparison of results (accessed on 27 December 2007), www.is.umk.pl/projects/datasets.html#Cleveland
29. Zeng, T., Xiang, Y., Chen, P., Liu, Y.: A Model of Immune Gene Expression Programming for Rule Mining. Journal of Universal Computer Science 13(7), 1239–1252 (2007)
30. Zhou, C., Xiao, W., Tirpak, T.M., Nelson, P.C.: Evolving Accurate and Compact Classification Rules with Gene Expression Programming. IEEE Transactions on Evolutionary Computation 7(6), 519–531 (2003)

Projection with Double Nonlinear Integrals for Classification

JinFeng Wang[1], KwongSak Leung[1], KinHong Lee[1], and ZhenYuan Wang[2]

[1] Department of Computer Science & Engineering, The Chinese University of Hong Kong,
Shatin, NT, Hong Kong SAR
[2] Department of Mathematics, University of Nebraska at Omaha, Omaha, NE 68182, USA
{jfwang,ksleung,khlee}@cse.cuhk.edu.hk,
zhenyuanwang@mail.unomaha.edu

Abstract. In this study, a new classification model based on projection with Double Nonlinear Integrals is proposed. There exist interactions among predictive attributes towards the decisive attribute. The contribution rate of each combination of predictive attributes, including each singleton, towards the decisive attribute can be re presented by a fuzzy measure. We use Double Nonlinear Integrals with respect to the signed fuzzy measure to project data to 2-Dimension space. Then classify the virtual value in the 2-D space projected by Nonlinear Integrals. In our experiments, we compare our classifier based on projection with Double Nonlinear Integrals with the classical method- Naïve Bayes. The results show that our classification model is better than Naïve Bayes.

Keywords: Nonlinear Integrals, Projection, Classification.

1 Introduction

Many classification methods have been proposed based on various approaches [1]. Due to nonlinearity existing in the real world, some linear methods can not satisfy the requirement with high classification accuracy. However, the contribution rate of each combination of predictive attributes including each singleton towards the decisive attribute can be represented by a fuzzy measure. The nonadditivity of the fuzzy measure reflects the interactions among the feature attributes. Recently, many methods which attempt to use nonlinear integrals as aggregation tools [3],[4],[5],[6] has obtained quite encouraging results. In these existing methods, if there are m classes and n predictive attributes, then m sets of fuzzy measures are used and $m(2^n - 2)$ values of fuzzy measures are needed to be determined.

Unlike the methods above, another method called WCIPP (Weighted-Choquet-Integral based Projection Pursuit) use a weighted Choquet Integral as a projection tool [7]. In WCIPP, only one fuzzy measure defined on the power set of the set of all feature attributes is used to describe the importance of each feature attribute as well as their interactions[8],[9],[10] towards the classification of the records. The original classification problem in n-dimensional space is transformed to a one-dimensional space problem through the optimal projection based on fuzzy integrals. But plenty of information may

P. Perner (Ed.): ICDM 2008, LNAI 5077, pp. 142–152, 2008.
© Springer-Verlag Berlin Heidelberg 2008

be missed in the projecting process. In some special cases which will be described in following sections, there exists projection overlapping when the data to be classified have special distribution in the data space, such as one group of the data is surrounded by the data of another group, or the number of classes for the data is large. This will lead to lower classification accuracy. So we propose a new classification model based on Double Fuzzy Integrals in this paper. Double Fuzzy Integrals can lessen loss of information due to the intersection of different classes on real axis in WICPP. Accuracy will be increased accordingly. Computation complexity will be linearly increased, but be acceptable still.

This paper is organized as follows. Section 2 describes the construction of the classification model based on projection with Double Nonlinear Integrals in detail. The experimental results of the new classification model for each dataset are presented in Section 3. Section 4 gives the conclusions and some directions for future work.

2 Classification Based on Nonlinear Integrals

In classification, we are given a data set consisting of l example records, called training set, where each record contains the value of a decisive attribute, Y, and the value of predictive attributes $x_1, x_2, ..., x_n$. Positive integer l is the data size. The classifying attribute indicates the class to which each example belongs, and it is a categorical attribute with values coming from an unordered finite domain. The set of all possible values of the classifying attribute is denoted by $C = c_1, c_2, ..., c_m$, where each c_k, $k = 1, 2, ..., m$, refers to a specified class. The feature attributes are numerical, and their values are described by an n-dimensional vector, ($f(x_1), f(x_2), ..., f(x_n)$). The range of the vector, a subset of n-dimensional Euclidean space, is called the feature space. The jth observation consists of n feature attributes and the classifying attribute can be denoted by $(f_j(x_1), f_j(x_2), ..., f_j(x_n), Y_j)$, $j = 1, 2, ..., l$.

In this section, a method of classification based on nonlinear integrals will be presented. It can be viewed as the idea of projecting the points in the feature space onto a virtual space by double nonlinear integral, and then using a linear classifier to classify these points according to a certain criterion optimally. The parameters are obtained by using an adaptive genetic algorithm. Good performance of this method comes from the use of the fuzzy measure and the relevant nonlinear integral, since the nonadditivity of the fuzzy measure reflects the inherent interactions of the feature attributes towards the discrimination of the points. In fact, each feature attribute has respective important index reflecting their amounts of contributions towards the decision. Furthermore, the global contribution of several feature attributes to the attribute of classification is not just the simple sum of the contribution of each feature to the decision, but may vary nonlinearly. A combination of the feature attributes may have mutually restraining or a complementary synergy effect on their contributions towards the classification. So the fuzzy measure defined on the power set of all feature attributes is a proper representation of the respective importance of the feature attributes and the interactions among them, and a relevant nonlinear integral is a good fusion tool to aggregate the information coming from the feature attributes for the classification. The following are the details of these basic concepts and the mathematical model for the classification problem.

2.1 Fuzzy Measures and Nonlinear Integrals[11]

Let $X = x_1, x_2, \ldots, x_n$, be a nonempty finite set of feature attributes and $P(X)$ be the power set of X.

Definition 2.1. A fuzzy measure, μ, is a mapping from $P(X)$ to $[0, \infty)$ satisfying the following conditions:

1) $\mu(\phi) = 0$;
2) $A \subset B \Rightarrow \mu(A) \leq \mu(B), \forall A, B \in P(X)$.

Set function μ is nonadditive in general. If $\mu(X) = 1$, then μ is said to be regular. The monotonicity and non-negativity of fuzzy measure are too restrictive for real applications. Thus, signed fuzzy measure, which is a generalization of fuzzy measure, has been defined [12] and applied.

Definition 2.2. A set function $\mu : \mathcal{P}(X) \to (-\infty, +\infty)$ is called a signed (non-monotonic) fuzzy measure provided that $\mu(\varnothing) = 0$.

A signed fuzzy measure allows its value to be negative and frees monotonicity constraint. Thus, it is more flexible to describe the individual and joint contribution rates from the predictive attributes in a universal set towards some target. To be convenient, $\mu(\{x_1\}), \mu(\{x_2\}), \ldots, \mu(\{x_n\}), \mu(\{x_1, x_2\}), \ldots, \mu(\{x_1, x_2, \cdots, x_n\})$ are sometimes abbreviated by $\mu_1, \mu_2, \ldots \mu_n, \mu_{12}, \ldots, \mu_{12\ldots n}$, respectively.

Definition 2.3. Let μ be a non-monotonic fuzzy measure on $P(X)$ and f be a real-valued function on X. The Choquet integral of f with respect to μ is obtained by

$$\int f d\mu = \int_{-\infty}^{0} [\mu(F_\alpha) - \mu(X)] d\alpha + \int_{0}^{\infty} \mu(F_\alpha) d\alpha,$$

Where $F_\alpha = \{x | f(x) \geq \alpha\}$, for any $\alpha \in (-\infty, \infty)$, is called the $\alpha - cut$ of f.

To calculate the value of the Nonlinear Integral of a given real-valued function f, usually the values of f, i.e., $f(x_1), f(x_2), \cdots, f(x_n)$, should be sorted in a nondecreasing order so that $f(x_1') \leq f(x_2') \leq \ldots \leq f(x_n')$, where $(x_1', x_2', \ldots x_n')$ is a certain permutation of $(x_1, x_2, \ldots x_n)$. So the value of Nonlinear Integral can be obtained by

$$\int f d\mu = \sum_{i=1}^{n} [f(x_i') - f(x_{i-1}')] \mu(\{x_i', x_{i+1}' \ldots x_n'\}),$$

where $f(x_0') = 0$.

The Choquet integral is based on linear operators to deal with nonlinear space.

Definition 2.4. Let μ, υ be two fuzzy measures on $P(X)$. The double Nonlinear integral of a function $f : X \to [-\infty, \infty]$ with respect to μ and υ is defined by

$$\int f d\mu d\upsilon = < \int f d\mu, \int f d\upsilon >.$$

Where v is determined after the values of μ.The value of Double Nonlinear Integral is the coordinates of the virtual data in the 2-D space projected by Nonlinear Integrals. In fact, the double nonlinear Integrals is the superposed version of classical Nonlinear Integrals.

2.2 Projection Based on Nonlinear Integrals

Based on nonlinear integrals, we can build an aggregation tool that projects the feature space onto a virtual space which maybe 1-dimenstional, 2-Dimensional or more dimensional. Under the projection, each point in the feature space becomes a value of the virtual variable.

A point $(f(x_1), f(x_2), ..., f(x_n))$ is projected to be $\hat{Y}$, the value of the virtual variable, on a real axis through a nonlinear integral defined by $\hat{Y} = \int f d\mu$. Once the value of μ are determined, we can calculate virtual value $\hat{Y}$ from f. Figure 1 illustrates the projection from a 2-D feature space onto a real axis, L, by the nonlinear integral. The contours being broken are due to the nonaditivity of the fuzzy measure. The points on the same projection line have the same set of fuzzy measure values, so they can be projected onto the same location. In our model, we used the signed fuzzy measure in Nonlinear Integrals, so the direction of projection can show differently due to different signs of fuzzy measures.

Projection to 1-D makes the original information simple. But some useful information for classification may be left out, which leads to overlapping situation as fig.2. That star in the right position represents a point misclassified by projection to 1-D. We can't classify it with the other points around it very well.

As described above, overlapping of classification on 1-D space exists in real world problems indeed. When this situation comes up, we need add more information to classify. So the 1-D space is stretched to 2-D space. Except that the first fuzzy measure μ is learned, another fuzzy measure v must be introduced into the classification model. The learning process of the second fuzzy measure is similar to the previous one. The real axis on the 1-Dimension space is used as one axis of the 2-Dimension space.

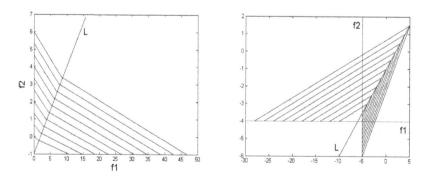

Fig. 1. Projection onto axis by nonlinear integrals

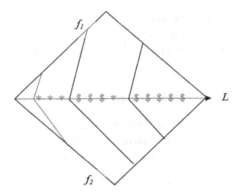

Fig. 2. Overlapping in Projection to real axis

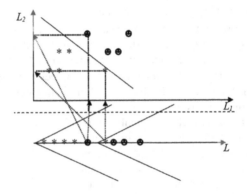

Fig. 3. Projection to 2-D due to overlapping in Projection in 1-D

Then we learn the second fuzzy measure v using GA and the value of the fuzzy integral respect to v is distributed on the other direction in 2-Dimension space. The linear classifier is used as fitness function on to classify the points projected to 2-Dimension space with fuzzy integrals. The graphical representation of projection with Fuzzy Integrals to 2-D is shown to fig. 3. The vague one in fig. 2 can be projected again into the 2-Demension space in fig. 3 and separated from the other class.

We can classify the cases according to the virtual values on the axis projected by nonlinear integrals. But overlapping situations among classes may exist in data on real axis projected by fuzzy integrals. To solve this problem, we need to retain more information of the original space by adding the degree of dimensionality. Then more information needs to be added into the model. So a mapping can be denoted by $M : R^n \to R^2$. To reflect the intersection among the predictive attributes towards the classification better, another signed fuzzy measure v, defined on the power set of X, is used for measuring the strength of contribution from each individual predictive attributes in different point of view. The values of fuzzy integrals with respect to μ and

v constitute the coordinates of original data in 2-Dimension space. Then we can classify these points in 2-D using a linear classifier. The two fuzzy measures can be learned by Genetic Algorithm to get optimal values in order to obtaining high accuracy.

2.3 GA-Based Adaptive Classifier

Now, based on the Choquet integral, we want to find an appropriate formula that projects the n-dimensional feature space onto a real axis, L, such that each point $f = (f(x_1), f(x_2), ..., f(x_n))$ which can be written in brief format $f = (f_1, f_2, ..., f_n)$ becomes a value of the virtual variable that is optimal with respect to the classification. In such a way, each classification boundary is just a point on real axis L.

The classification process can be divided into two parts to implement.

1) The nonlinear integral classifier depends on the fuzzy measures μ and v, so determining the optimal values of μ and v is in the first place of our work;

2) When the fuzzy measures μ and v are determined, the virtual values y' can be obtained by using fuzzy integral. So, we must decide how to classify these virtual values on 2-Dimension space.

The following subsections focus on the above problems respectively.

GA Based Learning Fuzzy Measure. Here we discuss the optimization of the fuzzy measure μ under the criterion of minimizing the corresponding global misclassification rate which is obtained in the second part above.

In our GA model, we use a variant of original integrand f, $f' = \vec{a} + \vec{b}f$, to substitute, where $\vec{a}$ is a vector to shift the coordinates of data and $\vec{b}$ is a vector to scale the values of predictive attributes. Here, $\vec{a}$ and $\vec{b}$ attempt to balance the scales of the predictive attributes in case that they have different measurement units. Each chromosome represents fuzzy measure μ, shifting vector $\vec{a}$ and scaling vector $\vec{b}$. A signed fuzzy measure is 0 at empty set. If there is n attributes in training data, a chromosome has $2^n - 1 + 2n$ genes. Genetic operations are traditional methods. At each generation, for each chromosome, all variables are fixed and the virtual values of all training data are calculated using the fuzzy integral with respect to the signed fuzzy measure, so the classification function used in the second part is used as the fitness function and the misclassification rate is used as fitness value.

In our model, projection to 2-D based on fuzzy integrals is adopted for higher accuracy. So we must repeat above described optimal process. In first step, we get the optimal value of the first fuzzy measure and the feature space is projected to 1-Dimension space. The real axis on the 1-Dimension space is used as one axis of the 2-Dimension space in the step 2. Then we learn the second fuzzy measure v using GA and the value of the fuzzy integral respect to v is distributed on the other direction in 2-Dimension space. The linear classifier is used as fitness function on 2-Dimension space.

Linear Classifier for the Virtual Values. After determining the fuzzy measure μ and ν, shifting vector $\vec{a}$, scaling vector $\vec{b}$ and the respective classification function from the training data in GA, original data in the n-dimensional feature space are projected onto 2-Dimension space using fuzzy integrals. One linear classifier is needed to classifying the virtual data $\hat{Y} = (y'_1, y'_2, \cdots, y'_l)$. Discriminant analysis is introduced in details [13].

We use Fisher's linear discriminant[14] function to perform classification in projected space. Positive and negative centroids for projected data are determined by the following formulas.

$$m_+ = \frac{\sum_{i:y_i=1} y'_i}{\sum_{i:y_i=1} 1}. \qquad m_- = \frac{\sum_{i:y_i=-1} y'_i}{\sum_{i:y_i=-1} 1}.$$

Ronald Fisher defined Scatter Matrices as $S_\pm \equiv \sum_{y'_i:y_i=\pm 1}(y'_i - m_\pm)(y'_i - m_\pm)'$.

$S_W = S_+ + S_-$ is called the Within Class Scatter Matrix. Similarly, the Between Class Scatter Matrix can be defined as $S_B \equiv (m_+ - m_-)(m_+ - m_-)'$.

So this results in an equivalent expression for Fishers discriminant criterion as a ratio between two quadratic forms is

$$J(w) = \frac{w' S_B w}{w' S_W w}.$$

in which w represents the direction of projection space. We can solve the programming problem maximizing $J(w)$. The optimal w can be represented as $w = S_w^{-1} * (m_+ - m_-)$. So the Fisher's discriminant function is formulated as:

$$y = w * (y' - n_+ * m_+ - n_- * m_-).$$

in which $n_\pm$ is the sum of observations in each class respectively. Finally, a threshold needs to be fixed in order to define a complete classifier.

3   Experiments

In this section, we present the results of Classification model by projection with Nonlinear Integral to classify two kinds of datasets. One is synthetic datasets which is produced randomly and contains two sets with 100 cases and 200 cases respectively. All data belong to two classes which show ying-yang distribution as figure 4. The other dataset which is selected from UCI[15] is Monk Dataset which includes 3 sub-problems. They come from the same event space with 6 attributes and 2 classes[16]. All sub-problems, i.e. Monk1, Monk2 and Monk3, differ in type of the target concept to be learned. Especially, 5% of the examples in Monk3 are noise. The detailed information is listed in Table 1.

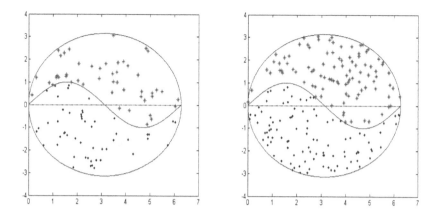

Fig. 4. The synthetic data distribution

Table 1. Description of Datasets

Datasets	Examples	Attributes	Classes
Syn_Data1	100	2	2
Syn_Data2	200	2	2
Monk1	556	6	2
Monk2	601	6	2
Monk3	554	6	2

For implementing our new classifier based on nonlinear integrals, GA tool in Matlab v7.2 Programming is called combining with Fisher's discriminant function. In our experiments, all parameters in GA are set with the default values. We design the generation limit which is 100 as the stopping criteria.

We adopt 10-fold cross validation method to make sure that the testing data can cover the whole dataset. That means that we randomly break the data into 10 sets of size of n/10, train on 9 sets and test on 1, and repeat 10 times in turn and take the mean result. After ten iterations, all data are used for testing and the average can be computed to evaluate the performance of the classifier based on nonlinear integrals.

Figure 5 illustrates the classification situation for the synthetic data in 2-D space by projection with Double Nonlinear Integrals. The comparison results of projection with classical Nonlinear Integrals to 1-D and projection with Double Nonlinear Integrals to 2-D are listed in table 2. Apparently, the Projection with Double Nonlinear Integrals can classified the data better than Projection to 1-D.

Table 3 show the results of Projection with Double Nonlinear Integrals to 2-D compared with the classical method, Naïve Bayes[17]. The best results are highlighted in bold. We can see that our new algorithm has higher accuracy than the Naïve Bayes for the most cases.

Table 2. Comparison Results between 2-D Projection and Classical Projection

Algorithm / Datasets		NIC with 1-D	NIC with 2-D
Dataset1	Train accuracy	0.963	**0.968**
	Test accuracy	0.836	**0.914**
Dataset2	Train accuracy	0.958	**0.958**
	Test accuracy	0.909	**0.924**
Monk1	Train accuracy	0.899	**0.948**
	Test accuracy	0.820	**0.862**
Monk2	Train accuracy	0.711	**0.712**
	Test accuracy	0.667	**0.692**
Monk3	Train accuracy	0.966	**0.966**
	Test accuracy	0.950	**0.959**

Table 3. Comparison Results between 2-D Projection and Other Methods

Algorithm / Datasets		NIC with 2-D	NB
Dataset1	Train accuracy	**0.968**	0.871
	Test accuracy	**0.914**	0.858
Dataset2	Train accuracy	**0.958**	0.892
	Test accuracy	**0.924**	0.891
Monk1	Train accuracy	**0.948**	0.606
	Test accuracy	**0.862**	0.661
Monk2	Train accuracy	**0.712**	0.659
	Test accuracy	**0.692**	0.657
Monk3	Train accuracy	**0.966**	0.928
	Test accuracy	**0.959**	0.921

Note: NIC with 1-D stands for Nonlinear Integrals with Projection to 1-Dimension space; NIC with 2-D stands for Nonlinear Integrals with Projection to 2-Dimension space; and NB stands for Naïve Bayes.

We introduce the new algorithm for solving the problem existing in projection to 1-Dimension space with Nonlinear Integrals. From the table 2 we can see the aim for improving has gotten. The performance of NIC with 2-D has greatly enhanced for NIC with 1-D. Except for that, it is enough to prove that our new algorithm can win over some classical method although we just compare it with NB.

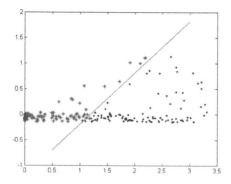

Fig. 5. The graphical representation of classification for the synthetic dataset with 200 cases

4 Conclusions and Future Work

In this paper, a new classification model based on projection with Double Nonlinear Integrals has been proposed. This method has good performance using the signed fuzzy measure and the nonlinear integral, since the nonadditivity of the fuzzy measure reflects the importance of the feature attributes, as well as their inherent interactions. Projection to 2-Dimension based on Double Nonlinear Integrals enhances the performance due to solving the intersection situations in projection onto 1-Dimension space. However, the computation complexity is just increased linearly, which is acceptable.

In future work, we will compare our new model with more classical methods to evaluate its performance and extend projection to 2-D based on fuzzy integrals to more dimensional space if necessary so that better performance can be obtained.

References

1. Hutter, M.: Universal Artificial Intelligence: Sequential Decisions based on Algorithmic Probability. Springer, Berlin (2004)
2. Leung, K.S., Ng, Y.T., Lee, K.H., Chan, L.Y., Tsui, K.W., Mok, T., Tse, C.H., Sung, J.: Data Mining on DNA Sequences of Hepatitis B Virus by Nonlinear Integrals. In: Proceedings Taiwan-Japan Symposium on Fuzzy Systems & Innovational Computing, 3rd meeting, Japan, pp. 1–10 (2006)
3. Grabisch, M.: The representation of importance and interaction of features by fuzzy measures. Pattern Recognition Letters 17, 567–575 (1996)
4. Grabisch, M., Nicolas, J.M.: Classification by fuzzy integral: Performance and tests. Fuzzy Sets and Systems 65, 255–271 (1994)
5. Keller, J., Yan, M.B.: Possibility expectation and its decision making algorithm. In: 1st IEEE Int. Conf. On Fuzzy Systems, San Diago, pp. 661–668 (1992)
6. Mikenina, L., Zimmermann, H.-J.: Improved feature selection and classification by the 2-additive fuzzy measure. Fuzzy Sets and Systems 107, 197–218 (1999)
7. Kebin, X., Zhenyuan, W., Heng, P.-A., Kwong-Sak, L.: Classification by nonlinear integral projections. IEEE Transactions on Fuzzy System 11(2), 187–201 (2003)

8. Wang, W., Wang, Z.Y., Klir, G.J.: Genetic algorithm for determining fuzzy measures from data. Journal of Intelligent and Fuzzy Systems 6, 171–183 (1998)
9. Wang, Z.Y., Klir, G.J.: Fuzzy Measure Theory. Plenum, New York (1992)
10. Wang, Z.Y., Leung, K.S., Wang, J.: A genetic algorithm for determining nonadditive set functions in information fusion. Fuzzy Sets and Systems 102, 463–469 (1999)
11. Murofushi, T., Sugeno, M., Machida, M.: Non monotonic fuzzy measures and the Choquet integral. Fuzzy Sets and Systems 64, 73–86 (1994)
12. Grabisch, M., Murofushi, T., Sugeno, M. (eds.): Fuzzy Measures and Integrals: Theory and Applications. Physica-Verlag (2000)
13. McLachlan, G.J.: Discriminant Analysis and Statistical Pattern Recognition. Wiley, New York (1992)
14. Mika, S., Smola, A.J., Schölkopf, B.: An Improved Training Algorithm for Fisher Kernel Discriminants. In: Jaakkaola, T., Richardson, T. (eds.) Proc. Artifical Intelligence and Statistics (AISTATS 2001), pp. 98–104 (2001)
15. Merz, C., Murphy, P.: UCI repository of machine learning databases[Online] (1996), ftp://ftp.ics.uci.edu/pub/machine-learning-databases
16. The MONK's Problems, A Performance Comparison of Different Learning Algorithms (1991)
17. Borgelt, C.: Bayes Classifier Induction, http://fuzzy.cs.uni-magdeburg.de/~borgelt/bayes.html

Local Modelling in Classification

Gero Szepannek[1], Julia Schiffner[1], Julie Wilson[2], and Claus Weihs[1]

[1] Department of Statistics, University of Dortmund
44227 Dortmund, Germany
[2] Department of Mathematics and Chemistry, University of York
Heslington, York YO1 5DD, UK
szepannek@statistik.uni-dortmund.de

Abstract. In classification tasks it may sometimes not be meaningful to build single rules on the whole data. This may especially be the case if the classes are composed of several subclasses. Several common as well as recent issues are presented to solve this problem. As it can e.g. be seen in Weihs et al. (2006) there may result strong benefit from such *local modelling*. All presented methods are evaluated and compared on four real-world classification problems in order to obtain some overall ranking of their performance following an idea of Hornik and Meyer (2007).

1 Introduction

In the context of *local modelling* it has to be distinguished between *global classification models* generating one single classification rule that holds for the entire data and *local classification models* that do not hold for the entire population but rather for a subsample (see e.g. Morik et al., 2004). To give an idea of the problem, three simple examples of different artificial data sets in the two dimensional space are shown in Figure 1. Both data sets on the left hand side can be correctly classified using single discrimination rules as they result from e.g. linear or polynomial svms or linear or quadratic discriminant analysis. Contrariwise, the right plot shows data where a combination of two rules is necessary to obtain a good classification model.

Proposed classification methods in literature that perform local modelling may be split into several groups according to their way of performing the local modelling task. For some of the methods the classes are assumed to be composed of several subclasses each. A first distinction might be whether these subclasses have to be given in the training data or not.

Section 2 describes different approaches to local modelling: methods that assume the subclasses to be specified (section 2.1) as well as methods that perform an implicit subclass detection (section 2.2). Furthermore, in section 2.3 some common classification methods are mentioned that implicitly perform some sort of local modelling.

In section 3 all methods are evaluated. Four real-world data sets are introduced in this paper. Methods to evaluate the performance of the different methods are presented in section 3.2: a statistical test to judge significant dominance of some

P. Perner (Ed.): ICDM 2008, LNAI 5077, pp. 153–164, 2008.
© Springer-Verlag Berlin Heidelberg 2008

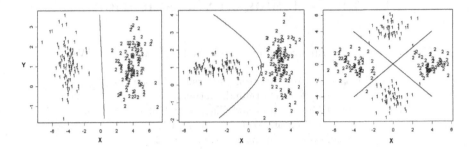

Fig. 1. Examples of classification rules for different data sets. Left: rule generated by a linear svm, middle: rule resulting from quadratic discriminant analysis, right: example of data where one single rule is not sufficient for discrimination.

of the methods as well as a method to derive a consensus ranking of them as a combination from their results on the four different data sets. The results of the study are shown in section 3.3.

2   Local Modelling in Classification

2.1   Known Subclasses in the Training Data

Combining local hypotheses
To motivate the task of combining several local hypotheses in case of *known subclasses* of the training set let us introduce some first example:

Example 1: music data
The classification problem concerning the *music* data set consists in register classification by means of timbre characteristics, i.e. without any information about the underlying fundamental frequency (for details see e.g. Weihs et al., 2006). The variables are formed by masses and widths of the fundamental frequency and the first twelve harmonics (see Figure 2), summing up to 26 variables in total. The data set consists of 432 tones (= observations) played by 9 different instruments/voices. The goal was to predict the correct register (i.e. high or low). Using the well-known linear discriminant analysis (Fisher, 1936) which is known to perform well and robust under a lot of data situations (see e.g. Hastie et al. 2001, p.89 or Michie et al., 1994) resulted in a (cross validated) unsatisfyingly high error rate of 0.352.

The idea of local modelling may result here in building local classification models for each instrument (denoted by l) separately. We can consider the population to be the union $\Omega = \cup_{l=1}^{L} \Omega_l$ of *subpopulations*. The problem consists in prediction of a new observation if the instrument (and thus the choice of the local model) is not known. The resulting task can be formulated as some *globalization of local classification rules*.

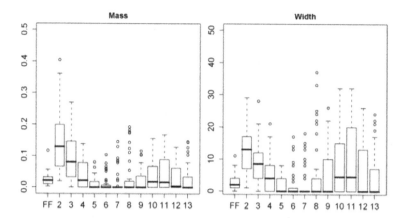

Fig. 2. Variables of the register classification problem: voiceprint of a male singer (boxplots of tones over an entire song)

Imagine all local (subpopulation-)classifiers returning local class posterior probabilities

$$P(k|l, x) \tag{1}$$

where $k = 1, \ldots, K$ denotes the class, x the actual observation and $l = 1, \ldots, L$ is the index of the local model, i.e. the instrument.

Among several issues that were investigated by Weihs et al. (2006) the Bayes rule

$$\hat{k} = \arg\max_k \sum_l P(k|l, x)P(l|x) \tag{2}$$

showed best performance for the musical register classification problem. To implement equation 3 an additional classifier has to be built to predict the presence of each local model l for a given new observation x. All classification models, the local ones from equation 4 as well as the one in equation 3 can be built by any classifier returning posterior probability membership values, e.g. using linear discriminant analysis. Doing so improved the error rates on the music data set up to 0.263. Note that – since only posterior probabilities are used to build the classification rule – all models can be build on different subsets of variables. A variable selection can be performed e.g. using the stepclass algorithm, minimizing cross-validated error rates or also using Wilk's Λ statistic. Both methods are implemented in the statistical software package klaR of the statistical programming language R (see Weihs et al., 2005).

Unequal subclasses of the classes
The data situation as it is described in example 1 is not of a very general nature. Thus, we introduce a second data set:

Example 2: crystallization data

The *crystallization data set* consists of 2746 observations. 37 features were obtained from image analysis and the goal is to automatically classify correctly whether an object is of crystalline nature or not, specified by three different classes i.e. different types of crystallization (for details, see Wilson 2006). Each of the classes is composed of several subclasses. Figure 3 shows an example image of each of seven subclasses (with the corresponding class given in parentheses).

Fig. 3. Exemplary images from (sub)classes of the crystallization data set. Subclasses (and corresponding classes) from left to right: empty image (negative, i.e. poor results), rubbish (negative), precipitate (possible, i.e. results could be optimized to give crystals), quite interesting (possible), very interesting (possible), crystal clusters (positive, i.e. crystalline results), crystal (positive).

It can be seen that each of the subclasses (let them be denoted by l_i^k, $i = 1, \ldots, n_k$ where k is the class index) can be mapped to one and only one of the three classes. From a classification rule for the subclasses that yields subclass membership posterior probabilities $P(l_i^k|x)$ for any new observation x one can derive a Bayes classification rule as

$$\hat{k} = \arg\max_k \sum_{i=1}^{n_k} P(l_i^k|x). \tag{3}$$

Szepannek and Weihs (2006) show that this approach can be considered as a generalization of the upper case of subpopulations as they are given by the instruments in the register classification task. Each class (register) can be considered to be composed of subclasses of some register played by the different instruments.

Note that a classification model for prediction of the correct subclass l_i^k has to be build on one explicit choice of a subset of variables. Szepannek and Weihs (2006) therefore suggest to perform (sub)class wise modelling on possibly different variable subspaces using the pairwise coupling algorithm of Hastie and Tibshirani (1998) to construct a classification rule from the $K(K+1)/2$ pairwise decisions. Going back to example 1, due to this kind of subclass pair specific variable selection it has been possible to further improve the error rate to 0.243. This method will be referred to as *LocPVS*.

2.2   Unknown Subclasses in the Training Data

Mixture discriminant analysis
A different situation is given when the subclass memberships are not specified
in the training data. In such situations some kind of preliminary clustering will
be necessary. A classic algorithm of this sort of methods is *mixture discriminant
analysis (MDA)* (Hastie and Tibshirani, 1996). Here, each class k is modelled as
a mixture of Gaussians

$$P(X|k) = \sum_{j=1}^{n_k} \pi_{l_j^k} \phi(X; \mu_{l_j^k}, \Sigma). \tag{4}$$

The mixture weights $\pi_{l_j^k}$ as well as the distribution parameters are trained by
the EM algorithm (Dempster et al., 1977).

Common components
In the common components model (Titsias and Likas, 2001) the conditional
density of class k is

$$P(X|k) = \sum_l \pi_{lk} \phi(X; \mu_l, \Sigma_l) \quad \text{for } k = 1, \ldots, K, \tag{5}$$

where π_{lk} represents the conditional prior $P(l|k)$ of the lth local model given
class k.

The densities $\phi(X; \mu_l, \Sigma_l)$ do not depend on k. Therefore all class conditional
densities are explained by the same mixture component densities. This motivates
the name *common components*.

In order to calculate the class posterior probabilities, the parameters μ_l and
Σ_l as well as the priors π_{lk} and P_k are estimated based on maximum likelihood
and the EM algorithm. A derivation of the EM steps is given in Titsias and
Likas, 2001.

Hierarchical approach
The hierarchical mixture classifier (Titsias and Likas, 2002) can be considered
as an extension of the common components classifier. We still assume that the
data are generated by L local models. But additionally, we suppose that the
data generated by each local model l consist of class-labeled subgroups that are
modeled by the densities $\phi(X; \mu_{kl}, \Sigma_{kl})$ for $k = 1, \ldots, K$.

Then the class conditional densities take the form

$$P(X|k) = \sum_l \pi_{lk} \phi(X; \mu_{kl}, \Sigma_{kl}) \quad \text{for } k = 1, \ldots, K. \tag{6}$$

Here, the mixture component densities $\phi(X; \mu_{kl}, \Sigma_{kl})$ depend on the class labels k and hence, in contrast to the common components approach, each class
conditional density is described by a separate mixture.

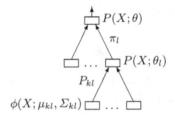

Fig. 4. Illustration of the hierarchical mixture model. The density $P(X; \theta_l)$ is indexed by the parameter $\theta_l = (\mu_{1l}, \ldots, \mu_{Kl}, \Sigma_{1l}, \ldots, \Sigma_{Kl})$. The unconditional density $P(X; \theta)$ depends on $\theta = (\theta_1, \ldots, \theta_L)$.

Figure 4 shows the differences between the common components and the hierarchical approach. In the common components model $P(X; \theta)$ is a mixture of the component densities $\phi(X; \mu_l, \Sigma_l)$ that do not depend on the class labels whereas in the hierarchical approach $P(X; \theta_l)$ is described by another mixture of $\phi(X; \mu_{kl}, \Sigma_{kl})$.

The parameters and priors are estimated by two different algorithms (referred to as HM1 and HM2) that are both based on EM. Details can be found in Titsias and Likas (2002).

Localized discriminant analysis
In *localized discriminant analysis (LLDA)* (Czogiel et al., 2007) observation specific classification models are built, depending on the explicit observation to be classified: for any new observation x normal distribution parameters $\hat{\mu}_k$ and $\hat{\Sigma}$ are calculated by weighting the training observations x_i according to their distance $||x - x_i||$. The weight w_i for training observation x_i are determined using some kernel function

$$w_i = K(||x - x_i||). \tag{7}$$

In empirical studies a RBF kernel of type

$$K(||x - x_i||) = \exp(-\gamma ||x - x_i||) \tag{8}$$

showed good results where the free parameter γ has to be chosen according to the explicit problem (e.g. using cross validated optimization). Principally, the principle can be applied using any classifier not only LDA.

2.3   Methods Performing Implicit Local Modelling

Apart from the methods mentioned above performing an explicit local modelling there are also several common methods that implicitly do so. They all share the property that the nature of their resulting rules of class assignment do allow for more than one single rule per class. *k nearest neighbours* classifiers do represent an extreme case where each training observation represents such a local rule.

support vector machines using *RBF kernels* also allow for different local models of the classes as well as *decision trees* (Breiman et al., 1984) and their bagged cousins *random forests* (Breiman, 2001). Also, *learning vector quantization* (Kohonen, 1995) should be mentioned here.

3 Comparative Study

3.1 Data Sets

Two data sets have already been introduced in examples 1 and 2: the *music* data set and the *crystallization data*. For the empirical study of this paper two more data sets that are assumed to posses local structures are used:

Example 3: medical diagnosis data
The data set comprises 794 patients recruited at the general practitioner's office with the aim to screen for a disease with a complex profile. The features that should be used to allow for the disease status of a patient are 6 biomarkers selected from an initial biomarker pool as the most promising ones with respect to this diagnostic task. The cases stand now for diseased and the controls for non-diseased patients. However, the control group has in the screening frame a natural subclass structure, being composed of 4 different diseases with a similar symptomatic with the disease in case.

Example 4: vowel recognition
The fourth data set comes from the domain of automatic speech recognition: 25ms segments from the center of 389 vowel pronunciations are extracted from the TIMIT speech data base (see Garofolo et al., 1993). Each data entry is of one of six different phonemes: $\{/ah/, /ax/, /ae/, /ih/, /ix/, /uh/\}$. According to Lee and Hon (1989) some of the original phoneme transcriptions can be summarized due to their similar sound. These are $/ih/$ (like 'bit') and $/ix/$ (like 'roses') as well as $/ah/$ and $/ax/$ so that there finally remain the four vowel classes $\{/a/, /ae/, /i/, /uh/\}$. These vowel classes are chosen to represent the four extreme positions of articulation. 13 standard MFCCs (see e.g. Davies and Mermelstein, 1980) are extracted from the sound signal and used as features for classification.

3.2 Methodology

Developing consensus rankings
The idea is to use the experimental results of the different methods on the four data sets to derive a *consensus method ranking* following a concept of Hornik and Meyer (2007).

Consider latent scores π_i denoting the probability of method i to yield the best results among the set of proposed classifiers for any classification problem. Then $\pi_{ij} = \frac{\pi_i}{\pi_i + \pi_j}$ is the probability that method i will outperform method j.

Let r_{ijn} be 1 if method i has been observed to outperform method j on data set $n = 1, \ldots, N$ and 0 otherwise. Then (assuming independence of all observed paired comparisons which is of course not true since all comparisons are made on the same data sets) one can derive the likelihood for a given set of $\{\pi_1, \ldots, \pi_M\}$:

$$L(\pi_1, \ldots, \pi_M) = \prod_{i<j} \binom{N}{\sum_{n=1}^{N} r_{ijn}} \frac{\prod_{i=1}^{M} \pi_i^{(\sum_{j \neq i} \sum_{n=1}^{N} r_{ijn})}}{\prod_{i \neq j} (\pi_i + \pi_j)^N}. \tag{9}$$

Maximization results in solving

$$\hat{\pi}_i = \frac{\sum_{i \neq j} \sum_{n=1}^{N} r_{ijn}}{N(\sum_{j \neq i} (\hat{\pi}_i + \hat{\pi}_j)^{-1})} \tag{10}$$

which can be done iteratively (see David, 1988). This model is called *Bradley/ Terry model* (Bradley and Terry, 1952).

Testing performance

To be able to judge whether one method outperforms another on a specific (real world) data set a test suggested by Dietterich (1995) is used, performing 5x2 fold cross validation.

The reason for using 2 fold cross validation are non-overlapping training sets. Moreover, a large test set is necessary for convergence of the distribution of the errors towards normality.

Let $\Delta_{ij,q}^{(m)} = err_{i,q}^{(m)} - err_{j,q}^{(m)}$ with $err_{i,q}^{(m)}$ being the test error of method i on the mth half of the training set in the qth repetition of 2 fold cross validation.

Furthermore,

$$s_{ij,q}^2 := \sum_m (\Delta_{ij,q}^{(m)} - \bar{\Delta}_{ij,q})^2. \tag{11}$$

Assuming independence $s_{ij,q}^2 / \sigma_{ij}^2 \sim \chi_1^2$ with σ_{ij}^2 being the 'true' variance of the approximative error difference distribution. Dietterich (1995) observed instable variance estimates $s_{ij,q}^2$ and thus proposed a stabilization over five repetitive cross validations:

$$s_{ij}^2 := \frac{1}{5} \sum_{q=1}^{5} s_{ij,q}^2. \tag{12}$$

Finally, he proposes the test statistic

$$T_{ij} := \frac{\Delta_{ij,1}^{(1)}}{s_{ij}^2} \sim_{approx.} t_5 \tag{13}$$

to test whether one of both methods outperforms the other one.

3.3   Results

All methods were used to classify all $N = 4$ real data sets both on the entire set of features and on a feature subset resulting from application of Wilk's Λ feature subset selection (see Weihs et al., 2005). The better of both results has been chosen for each method. Free parameters are specified based on optimizing error rates on a separate test-training split on the data in order to avoid overfitting. All computations were implemented using the statistical programming language R (Ihaka and Gentleman, 1996). Linear discriminant analyses (Fisher, 1936) are also implemented to obtain a reference baseline for classification results if no local modelling is carried out.

A consensus ranking of the methods is derived from estimates $\hat{\pi}_i$ of the Bradley/Terry model of paired comparisons where $r_{ijn} = 1$ if method i significantly ($\alpha = 0.05$) outperforms method j on data set n according to the test by Dietterich (1995) and $r_{ijn} = 0.5$ if neither method i nor method j performed significantly better.

The results are given in Table 1. The first four columns show the averaged 5x2 cross validated test error rates. All four data sets illustrate quite different data situations among the results of the different methods:

Table 1. Mean 5x2cv error rates of the different methods. In **bold**: methods not performing significantly worse than the best method.

Method	music	crystal	diagnosis	vowel	**consensus** $\hat{\pi}_i$
LocLDA	**0.257**	**0.187**	**0.107**	**0.344**	0.237
LocPVS	**0.268**	**0.188**	0.132	**0.348**	0.118
RBF SVM	0.281	0.203	**0.106**	**0.362**	0.112
Random forests	**0.238**	0.205	**0.095**	**0.376**	0.107
MDA	**0.293**	0.217	**0.111**	**0.373**	0.102
HM2	0.287	0.218	**0.108**	0.442	0.070
kNN	0.336	0.224	**0.114**	**0.393**	0.067
HM1	0.288	0.221	**0.108**	0.467	0.051
LDA	0.345	0.210	0.130	**0.345**	0.051
LVQ	0.325	0.236	0.117	0.416	0.038
CC	**0.296**	0.337	0.116	0.545	0.025
CART	0.310	0.280	**0.109**	0.494	0.020

For the music data strong improvements are observed for some of the local modelling approaches, namely random forests, localized discriminant analysis and the Bayes approach for known subclasses (LocPVS). Local modelling shows to be strongly beneficial compared to the 'global' LDA rule.

Classification of the crystallization data shows the best results for the same methods as in the musical register classification problem: LLDA, LocPVS, random forests and RBF SVMs are the top four classifiers.

On the medical diagnosis data set a bunch of classifiers shows similarly good results. Random forests perform best but not significantly better than the other methods. The worst error rates are obtained using the global LDA classifier or LocPVS. This result indicates that the prespecified natural subclass information

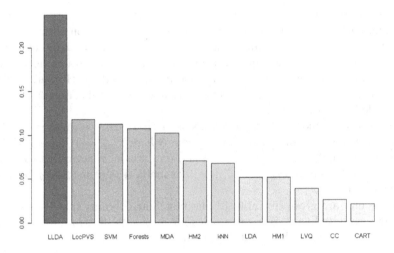

Fig. 5. Resulting consensus ranking of the methods: $\hat{\pi}_i$ of the Bradley/Terry model

is not helpful in this case for the classification task. It might be worth to perform unsupervised local modelling.

For the vowel data set the best results are obtained using the standard LDA rule or its localized alternative LLDA. This indicates that the subclass-grouping as it is proposed by Lee and Hon (1989) is meaningful and local modelling can not improve the performance in this case.

The last column of Table 1 shows the final consensus ranking (see also Figure 5): as it could have been expected random forests and RBF-SVMs which turned out to yield good results in a couple of other benchmarking studies (see e.g. Meyer et al., 2003) and which perform an implicit local modelling are also among the top classifiers here.

The most stable results are observed for localized discriminant analysis. According to the estimates, better results are between four and five times more probable using localized LDA compared to standard LDA classification rules on our data sets. Nevertheless, random forests have the lowest error rates for two of the data sets.

Quite good results are also observed for the known-subclasses Bayes rule (LocPVS), RBF-SVMs and standard mixture discriminant analysis, all being twice as probable to outperform standard LDA.

Finally one has to notice that the observed results only represent a survey of our experiences on local modelling. A new data set may of course be of different structure und thus the performance of the methods need not necessarily be related to our observations. The study has to be understood rather as an approach to give a hint towards the general performance of different methods following the approach of local modelling.

4   Summary

In this paper the idea of local modelling of classification problems is presented on four real-world data sets. Several methods are presented that take into account local modelling, among them both standard methods and recently developed ones as well as some common methods that do implicitly perform local modelling.

All methods are compared using statistical tests of performance on the real world data examples. In order to summarize the results a consensus ranking of the classification methods is derived from the results giving a hint towards their general performance.

It turned out that local modelling in classification may lead to significant improvements in terms of the error rates in some but not all situations.

Nevertheless, the presented data do by no means represent an exhaustive sample of all possible data sets and thus interpretation of the results has to be made carefully. Additional results of a simulation study for different data situations and some of the presented methods are given in Schiffner(2008).

Acknowledgment. This work has been supported by the Collaborative Research Center 'Reduction of Complexity in Multivariate Data Structures' (SFB 475) of the German Research Foundation (DFG). Further thanks should be announced to Raluca Ilinca Schmitt for her cooperation on analyzing the medical diagnosis data set.

References

Bradley, R., Terry, M.: The rank analysis of incomplete block designs, i. the method of paired comparisons. Biometrics, 324–345 (1952)

Breiman, L.: Random forests. Machine Learning 45(1), 5–32 (2001)

Breiman, L., Friedman, J., Olshen, R., Stone, C.: Classification and Regression Trees. Chapman & Hall, New York (1984)

Czogiel, I., Luebke, K., Zentgraf, M., Weihs, C.: Localized linear discriminant analysis. In: Decker, R., Lenz, H., Gaul, W. (eds.) Advances in Data Analysis, pp. 133–140. Springer, Heidelberg (2007)

Davis, K., Mermelstein, P.: Comparison of parametric representation for monosyllabic word recognition in continously spoken sentences. IEEE Trans.Acoust.Speech Signal Process 28(4), 357–366 (1980)

Dempster, A., Laird, N., Rubin, D.: Maximum likelihood from incomplete data via the EM algorithm. Journal of the Royal Statistical Society B 39(1), 1–22 (1977)

Dietterich, T.: Approximate statistical tests for comparing supervised classification learning algorithms. Neural Computation 10(7), 1895–1923 (1995)

Fisher, R.: The use of multiple measures in taxonomic problems. Annals of Eugenics 7, 179–188 (1936)

Garofolo, J., Lamel, L., Fiesher, W., Fiscus, J., Pallet, D., Dahlgren, N.: DARPA TIMIT acoustic-phonetic continuous speech corpus. Tech. Rep. NISTIR 4930, NIST, Gaithersburgh, MD (1993)

Gold, L., Morgan, N.: Speech and Audio Signal Processing. Wiley, New York (1999)

Hastie, T., Tibshirani, R.: Classification by pairwise coupling. Annals of Statistics 26(1), 451–471 (1998)

Hastie, T., Tibshirani, R.: The Elements of Statistical Learning - Data Mining, Inference and Prediction. Springer, NY (2001)

Hastie, T., Tibshirani, R., Friedman, J.: Discriminant analysis by Gaussian mixtures. Journal of the Royal Statistical Society B 58, 158–176 (1996)

Herbert, D.: The Method of Paired Comparisons, 2nd edn. Charles Griffin, London (1988)

Hornik, K., Meyer, D.: Consensus rankings from benchmarking experiments. In: Decker, R., Lenz, H., Gaul, W. (eds.) Advances in Data Analysis, pp. 163–170. Springer, Heidelberg (2007)

Ihaka, R., Gentleman, R.: R: a language for data analysis and graphics. Journal of Computational and graphical statistics 5(3), 299–314 (1996)

Kohonen, T.: Self-Organizing Maps. Springer, Berlin (1995)

Lee, K., Hon, H.: Speaker-independent phone recognition using Hidden Markov Models. IEEE Transactions on Speech and Signal Processing 37(11), 1641–1648 (1989)

Meyer, D., Leisch, F., Hornik, K.: The support vector machine under test. Neurocomputing 55, 169–186 (2003)

Michie, D., Spiegelhalter, D., Taylor, C.: Machine Learning, Neural and Statistical Classification. Ellis Horwood Limited, Hertfordshire (1994)

Morik, K., Siebes, A., Boulicault, J.: Preface. In: Morik, K., Siebes, A., Boulicault, J. (eds.) Local Pattern Detection. Springer, Heidelberg, V-IX (2004)

Schiffner, J., Weihs, C.: Comparison of local classification methods. In: Preisach, C., Burkhardt, H., Schmidt-Thieme, L., Decker, R. (eds.) Data Analysis, Machine Learning, and Applications. Springer, Heidelberg (to appear, 2008)

Szepannek, G., Weihs, C.: Local modelling in classification on different feature subsets. In: Perner, P. (ed.) ICDM 2006. LNCS (LNAI), vol. 4065, pp. 226–238. Springer, Heidelberg (2006)

Titsias, M.K., Likas, A.: Shared kernel models for class conditional density estimation. IEEE Transactions on Neural Networks 12(5), 987–997 (2001)

Titsias, M.K., Likas, A.: Mixtures of experts classification using a hierarchical mixture model. Neural Computation 14, 2221–2244 (2002)

Weihs, C., Ligges, U., Luebke, K., Raabe, N.: klaR - analyzing German business cycles. In: Baier, D., Becker, R., Schmidt-Thieme, L. (eds.) Data Analysis and Decision Support, pp. 335–343. Springer, Berlin (2005)

Weihs, C., Szepannek, G., Ligges, U., Luebke, K., Raabe, N.: Local models in register classification by timbre. In: Batagelij, V., Bock, H., Ferligoj, A., Ziberna, A. (eds.) Data Science and Classification, pp. 315–322. Springer, Heidelberg (2006)

Wilson, J.: Automated classification of images from crystallisation experiments. In: Perner, P. (ed.) ICDM 2006. LNCS (LNAI), vol. 4065, pp. 459–473. Springer, Heidelberg (2006)

Improving Imbalanced Multidimensional Dataset Learner Performance with Artificial Data Generation: Density-Based Class-Boost Algorithm

Ladan Malazizi, Daniel Neagu, and Qasim Chaudhry

Abstract. Improving the learner performance over imbalanced and multidimensional datasets raises a challenging task for machine learning community. Although a salient characteristic in data modeling is the amount of data provided for the learner, the proportional distribution of that data in each class has also direct relationship with the classifier performance. In imbalanced datasets when data is distributed into different classes, various in size, understanding of data structure and characteristics plays an important role in improving the learner accuracy. In this paper we introduce a new approach that combines the information gained from traditional classification algorithms, confusion matrix parameters and density-based clustering to generate artificial data in order to increase the learner performance. First a classification algorithm is run on training data. Then the confusion matrix is studied and the True Positive (TP) rate of each class is measured. The class with the lowest TP rate is selected. Using density-based clustering we identify the centroid of the class and measure the samples distribution in multidimensional space in the next step. With the values gained from Probability Density Function estimations for clusters, extra samples are generated and added to the original dataset to rebalance the class proportion and the weight of different classes in the whole training set. Our method has been evaluated in terms of TP, F-Measure and also overall accuracy against a number of Demetra (toxicology) and UCI datasets. Our method provides an insight view of the data structure and characteristics in order to identify how much and where the data need to be added for increasing the classification accuracy of the learner.

Keywords: Imbalanced multidimensional dataset, Class-Boost, Density-based Clustering.

1 Introduction

In data mining, classification learning is a supervised learning scheme that uses knowledge gained through the training process of classified instances for classification of unseen examples. One of the main issues for classifier during this process is the samples distribution of classes or class balance. Imbalanced or skewed [1] dataset, affect the performance of classification algorithms. The over represented classes provide enough information for training the classifier because

P. Perner (Ed.): ICDM 2008, LNAI 5077, pp. 165–176, 2008.

of their sufficient number of samples against the under represented class. Real world scientific applications often face this problem for a number of reasons [2].

For instance, in toxicology domain this problem is severe. When the chemical compounds need to be tested on different species, high toxicity chemicals cannot be sampled as many as low toxicity compounds. In these datasets the important task of classification has to focus on high toxic chemical compounds since misclassification of high toxic chemicals may lead to disastrous consequences.

In this paper we propose a new approach, which combines the supervised classification task with unsupervised clustering in order to maximize the knowledge gained from the data characteristics. Firstly selected datasets from Demetra project [3] and UCI [4] repository are trained using a classification algorithm. At the second stage the poorly classified samples are identified by studying the produced confusion matrix of classification task. Then TP rate for these samples is measured and compared with other samples belonging to classes with higher classification accuracy or TP. The class with lowest prediction accuracy produced on its samples is separated and used for the density-based clustering task study. This task is performed on the selected class in order to identify the samples distribution density inside its clusters. The cluster, which contains more samples or with higher prior probability would be identified as the representative set.

Based on the class population and also cluster density, artificial data are generated. The generated data are added to the original dataset and a new training dataset is constructed. With this method we increase the classification accuracy of the less represented class and in most cases with effect on learner accuracy on other classes and also the overall prediction accuracy.

This paper is organized as follows. Section two describes related work in this field. Section three introduces the Probability Density Clustering method, it provides a brief description of statistical measures used in this paper and also the artificial data generation process. The description of the Density-based Class-Boost Algorithm (DCBA) comes at the end of section 3. The proposed method is evaluated in section four. Conclusions and further work finalize the paper.

2   Related Work

Various approaches and methods have been proposed to tackle imbalanced data problem. One of these methods is one-sided selection [5] in which the borderline/negative examples or the ones overlapping in two class dimensional space are removed.

Another method is DataBoost-IM approach [6]. According to this method the hard examples from minority and majority class are identified. Then the synthetic samples are generated using the hard samples and added to the original dataset. The class distribution and the total weights of the different classes in the new training set are re-balanced at the last stage.

Guided re-sampling technique [7] is another solution which first determines the subcomponents within each class. The element in each subcomponent is re-sampled until each subcomponent has the same number of examples as biggest subcomponent. Then the between-class imbalance is eliminated by randomly selecting and duplicating members of the minority class.

SMOTHBoost [8] is another method which increases the learner performance in classification of minority class with creating synthetic instances by operating in the feature space rather than data space. Using this method a new minority class sample is created in the neighborhood of the minority class target.

There are also some methods which down-size the majority class in order to equalize the distribution of two classes [9] [10]. All these methods concentrate on the two-class problem with minority and majority class: either over-sampling or under-sampling presentation by overlooking the distribution of the class sub-components [7]. The statistical relationships between these elements is not addressed in detail. This could be very important in terms of how the new samples are generated in order to improve this relationship and help the learner in the classification process.

3   Density-Based Class-Boost Algorithm(DCBA)

The Density-based Class-Boost Algorithm applies to multi-class domain problem and is based on insight view of class characteristics in order to determine the distribution density of class samples. The idea is based on boosting the core of the hard recognizable class in order to highlight class influence zones [11] or boundaries. The algorithm is presented in Figure 2.

3.1   Probability Density Clustering

Clustering is based on a statistical model called finite mixture. A mixture is a set of k probability distributions of k clusters. The distribution gives the probability that an instance has a certain set of attribute values if it was identified to be a member of that cluster [12].

With Probability Density Clustering there are few parameters measured for each attribute in the data set and also each cluster within a class. For each attribute, mean, standard deviation and sampling probability are produced. For each cluster S with mean (μ_S) and a standard deviation (σ_S), if the classification is already determined for each sample then:
mean(average):

$$\mu = \frac{1}{n}\sum_{1}^{n} X_i \tag{1}$$

standard deviation:

$$\sigma^2 = \frac{1}{n-1}\sum_{1}^{n}(X_i - \mu)^2 \tag{2}$$

Sampling Probability for the class (S), P(S)= the estimation of the number of instances belonging to the class.

With these parameters already identified, the probability that instance x belonging to cluster S is:

$$P(S|x) = \frac{P(x|S)P(S)}{P(x)} \tag{3}$$

where $P(x|S)$ is the density function for:

$$S, f(x; \mu_S, \sigma_S) = \frac{1}{\sqrt{2\pi}\sigma_S} e^{\frac{-(x-\mu_S)^2}{2\sigma_S^2}} \tag{4}$$

Finally the joint probability of an instance is calculated as a sum of the probabilities of all its attributes which is produced as prior probability of instances distribution for each cluster [12] [13]. Figure 1 shows the Density-based clustering for class 3 in Demetra Trout dataset [3].

3.2   ROC Analysis/Evaluation Measures

In this paper we use a number of measures which are produced by confusion matrix during classification process. A brief description is provided below.

- **Confusion Matrix:** In a binary dataset when the classification is performed the prediction for each sample has four possible outcomes: True Positive, False Positive, True Negative and False Negative. They are produced in the form of Confusion Matrix.
- **Overall Accuracy**$= \frac{TP+TN}{TP+TN+FP+FN}$
- True Positives are the members of the class that have been predicted correctly for which the predicted and actual value for class membership are equal.
- **True Positive Rate**$= \frac{TP}{TP+FN} = Recall$
- **Recall:** Shows the proportional relationship between TP and FN rate.
- **Precision:** Shows the proportional relationship between TP and FP.
- Precision $= \frac{TP}{TP+FP}$
- **F-Measure:** This statistical figure simply produces the relationship between Precision and - Recall as follows: F$= \frac{2PR}{(R+P)}$

3.3   Artificial Data Generation

As the weak class is identified after first training with the classifier (the class with lowest TP rate), unsupervised Density-based Clustering is performed.

For every single attribute, mean, standard deviation and sampling prior probability are calculated. The determining facts for the size of additional artificial data generation are:

- the whole class proportion (the number of samples in target class). In some cases the target class members are as little as four in OralQuail data set. In order to affect the learner performance, enough artificial data need to be generated. Table 1 shows the number of classes and also samples in each class in all data sets.

- the cluster size inside the class (the number of samples in the cluster). After performing Density-based Clustering task, the class is divided into two clusters with different proportion. When the original class members size is small consequently the constructed clusters would be smaller. For every cluster the following is applied: if $S(x) = S_1(x) + S_2(x)$ when a cluster $S(x)$ is consist of cluster $S_1(x)$ and cluster $S_2(x)$.
- the effect that the additional data is caused (increase in the classification accuracy).

The data (numerical values) is generated based on the normal distribution/ mean values of each attribute based on following: If the frequency distribution has k attributes/features intervals with midpoints: $m_1, m_2, ..., m_k$ and corresponding frequencies $f_1, f_2, ..., f_k$, then:

Grouped mean $\bar{x} = \frac{\sum_{i=1}^{k} m_i f_i}{n} = \frac{\sum_{Cells}(Midpoint \times Frequency)}{Total frequency}$ [14].

In general in our work the added artificial data size is between 10 to 100 percent. If the constructed training data set (data set with added artificial data) is identified as T_{nm} and artificial training data as T_m then:

$$T_{nm} = T_n \cup T_m \qquad (5)$$

Then the artificial data are added to the class. This resized class would replace the original class in the training data set and the new training set is constructed and retrained. The stopping point for generation of more artificial data is when the classification accuracy start decreasing and also no more than the original set size.

Cluster 0: Prior probability: 0.7353
Cluster1: Prior probability: 0.2647

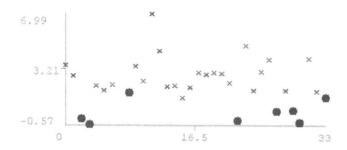

Fig. 1. Density-based clustering on class 3 in Trout dataset: x shows the number of instances in the class against y which is the value for an attribute ACD1 (-0.57 to 6.99)

Table 1. Datasets class distribution; note: in Glass dataset class4 had no samples and it has been deleted. The label for other classes has been shifted accordingly.

Datasets	No. of Classes	Class1	Class2	Class3	Class4	Class5	Class6
Wine	3	51	79	48			
Iris	3	50	50	50			
Vehicle	4	199	217	212	218		
Ecolio	5	143	77	52	35	20	
Glass	6	70	76	17	13	9	29
Trout	4	117	84	34	27		
Daphnia	4	107	64	50	23		
OralQuail	4	4	21	21	58		
Bee	5	14	19	12	38	12	
DietaryQuail	5	8	35	32	22	10	

3.4 Algorithm Description

Firstly using a Meta learner classification algorithm [12] [15] in Weka [16] the accuracy for each class is measured (Figure 2 steps 1 and 2). The confusion matrix is presented after the process, the class with lowest TP rate is selected as target class (Figure 2 step 3). Sometimes there are two classes with the same TP rate and both are targeted.

The targeted class or classes are then analyzed and with the help of unsupervised Density-based Clustering (Figure 2 steps 4 and 5) the prior probability of each cluster is measured. Then within that class, the cluster with highest prior probability is selected (Figure 2 step 6). At this stage the value of normal distribution mean of each attribute within the cluster and the frequency of the midpoint value for the samples (Figure 2 step 7) are used for generating artificial data. Proportion or sample size in each cluster determines how much data need to be added. The numerical data is generated in a way that satisfies the cluster mean. For instance if the mean value for an attribute is about 0.7, with highest frequency distribution of 0.5-0.6 then the generated numerical values would fall in this range.

The generated data are added to the cluster in target class and finally to the original data set and the new training set is constructed (Figure 2 step 8). The training data is balanced and weights are updated (step 9). The error of classification is calculated. This computed error is used to update the weights distribution of the samples (Figure 2 steps 10,11). The data is retrained using the same classification algorithm and result is presented (Output). The stopping criteria for adding more artificial data would be determined by overall classification accuracy of the whole training set in the consequent modeling.

TP rates of all classes are measured at every step of modeling since our experimental work proves that sometimes the increase in TP rate of one class affects the decrease of the same statistical measure in another class. Although initially the less representative class with lowest TP rate is targeted, the performance of the classification algorithm on all the other classes is also measured and watched in order to assure the effectiveness of the method.

Density-based Class Boost Algorithm
Input: T_n, set of n examples $x_1 = (x_1^1, x_1^2, ..., x_1^m), x_2 = (x_2^1, x_2^2, ..., x_2^m), x_n = (x_n^1, x_n^2, ..., x_n^m)$, with labels $c_i \in C^*$
-m_i , midpoint of distribution of class/cluster intervals(attribute values)
-f_i , corresponding frequencies
-L , number of iterations
For $l = 1$ to L

1. Initialize distribution weights on samples: $D_i(x_i) = \frac{1}{n}$ for all $x_i \in T_n$
2. Train data with meta learner
3. Identify target class S_i if $TP(S_i) = TP_{min}$
4. Calculate $P(s|x) = \frac{p(x|s)p(s)}{p(x)}$
5. Produce: S, $f(x; \mu_S, \sigma_S) = \frac{1}{\sqrt{2\pi}\sigma_S} e^{\frac{-(x-\mu_S)^2}{2\sigma_S^2}}$ in which
6. Calculate: $P(s|x) where P(s|x) = p(s|x)_{min} + p(s|x)_{max}$
7. For $p(s|x)_{max}$, generate $T_m = \{x_i, i = (1...m)\}$ when $\bar{x} = \frac{\sum_{i=1}^{k} m_i f_i}{n}$
8. Add artificial data T_m to original data set: $T_n \cup T_m = T_{nm}$
9. Balance training data and update weights
10. Train given the distribution $D_l, S_l = learner(T, D_l)$
11. Set $\beta_l = \varepsilon_l/(1 - \varepsilon_l)$ where ε_l error of S_l is: $S_l, \varepsilon_l = \sum_{x_i \in T, S_l(x_i) \neq y_i} D_l(x_i)$

Output: $S^*(x) = \underset{y \in Y}{\mathrm{argmax}} \sum_{l:S_l(x)=y} log \frac{1}{\beta_l}$

Fig. 2. DCBA algorithm

The experimental results show that the implementation of the method (adding data to one class) highlights the boundaries and border lines of the other classes which causes the increase in the overall classification accuracy and also each individual class (in most cases).

4   Method Evaluation

As it has been mentioned in abstract for the purpose of the evaluation, we have chosen Trout, Bee, Daphnia, DietaryQuail and OralQuail data sets from real-world applications provided by Demetra project [3]. We have also selected Glass, Iris, Wine, Ecolio and Vehicle from UCI Repository [4]. All these datasets are multi-class and imbalanced (Table 1) except Iris which is multi-class but balanced dataset.

The results of experiment show that the method has been effective in all data sets in terms of increase in overall classification accuracy. In the case of Iris data set although the data set is not imbalanced but the original classification accuracy (on original data set with no artificial data) for class2 was much lower than other classes, so we tested the method to see if it is effective in order to increase the TP rate for this class which was successful.

Table 2. Classification Accuracy for Demetra Datasets; target classes are in bold

Dataset	Class1	Class2	Class3	Class4	Class5
Bee					
TP	0.286	0.053	**0**	0.684	0.167
F-Measure	0.286	0.071	**0**	0.52	0.2
ROC area	0.765	0.511	**0.494**	0.605	0.702
Bee+artificial					
TP	0.571	0.105	**0.375**	0.658	0.25
F-Measure	0.516	0.143	**0.462**	0.543	0.286
ROC area	0.863	0.559	**0.571**	0.635	0.788
DietaryQuail					
TP	**0.1**	0.457	0.375	0.273	**0.1**
F-Measure	**0.2**	0.41	0.353	0.267	**0.154**
ROC area	**0.487**	0.543	0.581	0.569	**0.656**
DQ+artificial					
TP	**0.417**	0.457	0.406	0.409	**0.286**
F-Measure	**0.476**	0.421	0.366	0.439	**0.381**
ROC area	**0.668**	0.604	0.536	0.695	**0.719**
Trout					
TP	0.615	0.595	**0.147**	0.593	
F-Measure	0.643	0.549	**0.172**	0.533	
ROC area	0.71	0.72	**0.702**	0.847	
Trout+artificial data					
TP	0.667	0.583	**0.295**	0.556	
F-Measure	0.69	0.547	**0.329**	0.5	
ROC area	0.741	0.737	**0.751**	0.825	
Daphnia					
TP	0.664	0.313	0.38	**0.261**	
F-Measure	0.617	0.336	0.376	**0.316**	
ROC area	0.723	0.656	0.71	**0.852**	
Daphnia+artificial data					
TP	0.682	0.281	0.38	**0.407**	
F-Measure	0.635	0.324	0.365	**0.431**	
ROC area	0.7	0.659	0.699	**0.878**	
OralQuail					
TP	**0**	0.048	**0**	0.931	
F-Measure	**0**	0.074	**0**	0.701	
ROC area	**0.463**	0.456	**0.41**	0.482	
OQ+artificial					
TP	**0.143**	0.16	**0.276**	0.931	
F-Measure	**0.2**	0.235	**0.41**	0.697	
ROC area	**0.695**	0.62	**0.546**	0.669	

As the results show for Demetra data sets (Table 2) and for UCI data sets (Table 3) the method not only increased the classification accuracy for the target class and the overall accuracy of classification but also TP rate, F-Measure and ROC area of other classes as well (in most cases). In the case of Glass, Bee, DietaryQuail and Iris data sets after adding artificial data TP, F-Measure and ROC area increased for all the classes. In the Vehicle data set all the statistical measured for all the classes have improved except there is a slight decrease in TP rate of class4 after addition of artificial data. For Daphnia data set although decrease in values of TP rate and F-Measure for class2 and class3 occurred all the other statistical measures have improved. In OralQuail except the decrease in F-Measure for class4 after addition of artificial data the other measures show good improvement.

Table 3. Classification Accuracy for UCI Datasets; target classes are in bold

Dataset	Class1	Class2	Class3	Class4	Class5	Class6
Glass						
TP	0.786	0.737	**0.118**	0.769	0.667	0.828
F-Measure	0.738	0.723	**0.19**	0.769	0.6	0.842
ROC area	0.891	0.867	**0.832**	0.939	0.99	0.919
Glass+artificial data						
TP	0.814	0.737	**0.56**	0.769	0.889	0.828
F-Measure	0.792	0.737	**0.636**	0.741	0.727	0.873
ROC area	0.921	0.87	**0.929**	0.942	0.989	0.938
Ecolio						
TP	0.986	0.831	0.788	**0.514**	0.75	
F-Measure	0.956	0.8	0.837	**0.554**	0.833	
ROC area	0.979	0.95	0.942	**0.912**	0.985	
Ecolio+artificial data						
TP	0.972	0.87	0.788	**0.745**	0.7	
F-Measure	0.949	0.832	0.82	**0.792**	0.778	
ROC area	0.978	0.952	0.923	**0.953**	0.979	
Vehicle						
TP	0.94	0.507	**0.481**	0.963		
F-Measure	0.874	0.525	**0.523**	0.923		
ROC area	0.986	0.839	**0.864**	0.982		
Vehicle+artificial data						
TP	0.945	0.512	**0.575**	0.959		
F-Measure	0.87	0.534	**0.605**	0.937		
ROC area	0.988	0.845	**0.878**	0.992		
Wine						
TP	0.966	**0.915**	0.979			
F-Measure	0.958	**0.935**	0.959			
ROC area	0.991	**0.982**	0.997			
Wine+artificial data						
TP	0.949	**0.938**	0.979			
F-Measure	0.949	**0.943**	0.969			
ROC area	0.997	**0.991**	0.998			
Iris						
TP	0.917	**0.793**	0.923			
F-Measure	0.88	**0.821**	0.911			
ROC area	0.923	**0.908**	0.986			
Iris+artificial data						
TP	1	**0.96**	1			
F-Measure	1	**0.98**	0.96			
ROC area	1	**0.997**	0.995			

As it is shown in the result table (Table 2) for this data set, the class1 and class3 in the first run classification had zero TP rate. None of the samples belonging to these two classes have been classified correctly. The method shows good implication for such data sets. In the case of Wine data set all the parameters have improved except the TP rate and F-Measure in class1. For Ecolio data set the effect is different. For class1 and 5 the result is not satisfactory also for class3 the F-Measure and ROC area shows decrease but all other parameters for the rest of the data set is good.

The process of adding artificial data to datasets has been done in one iteration. The method can be applied and data can be added until the overall classification start decreasing. The confusion matrix has to be studied after every iteration in order to target classes with lowest TP rate.

Table 4. Classification Accuracy for all data sets after testing models

Datasets	Cross-Validation: Overall Accuracy (%)		
	Original	Data added to class 3 (model1)	
Trout	54.6		57
Tested with model1	75.5		
		Data added to class 4 (model1)	
Daphnia	47.54		49
Tested with model1	73		
		Data added to class1 &5 (model1)	
DietaryQuail	33.6		41
Tested with model1	63.5		
		Data added to class1 &3 (model1)	
OralQuail	52.8		56.3
Tested with model1	68.2		
		Data added to class3 (model1)	
Bee	34.7		45.45
Tested with model1	61		
		Data added to class 2 (model1)	
Iris	87.5		98.5
Tested with model1	96.2		
		Data added to class3 (model1)	
Vehicle	71.9		74.04
Tested with model1	91.72		
		Data added to class4 (model1)	
Ecolio	85.3		87.17
Tested with model 1	89.9		
		Data added to class3 (model1)	
Glass	71.49		76
Tested with model 1	86.4		
		Data added to class2 (model1)	
Wine	94		95.2
Tested with model 1	96.06		

Although the application of the method show decrease in few parameters(TP, F-Measure or ROC area) in some cases, the overall result is promising and the method is very effective in severe imbalanced data sets such as Bee and OralQuail. Table 4 shows the result of the testing gained models (from added artificial data)on the original data sets (using Cross-Validation). There is a good increase in classification accuracy after this process. There are three values for each dataset in this table. For examples in the case of Trout dataset the first value (54.6) is the overall classification accuracy for the original dataset with no artificial data. After adding artificial data to class3 the accuracy increased to (57). Then the dataset with artificial data was tested against the original dataset which caused the improvement in classification accuracy to (75.5).

5   Conclusion

In this paper we proposed a hybrid algorithm. We combined the supervised classification process with unsupervised clustering in order to get insight view of the classes internal components and characteristics. We focused our study and implementation of our method on imbalanced and multi-class data sets in which the severe class samples distribution exists.

We have shown that as long as we understand how class members are constructed in dimensional space in each cluster we can reform the distribution and provide more knowledge domain for classifier. Our results are promising and show the affect of the method even in special cases such as Demetra data sets where data is highly imbalanced with very low overall classification accuracy. Our process of data generation and the way the numerical values are produced proved to be effective.

The future work will focus on evaluation measures of an algorithm before and after addition of artificial data in order to see how the additional data could affect the whole data set characteristics not only the target class. The number of artificial samples added to the set will also be considered.

Acknowledgement

This work is partially funded by Central Science Laboratory York, SeedCorn project. We also thank the DEMETRA project for the use of toxicity data sets.

References

[1] Maloof, M.A.: Learning When Data Sets are Imbalanced and When Costs are Unequal and Unknown. In: ICML-2003: Workshop on Learning from imbalanced data sets II (2003)

[2] Ertekin, S., Huang, J., Bottou, L., Giles, C.L.: Learning on the Border: Active Learning In Imbalanced Data Classification. In: CIKM 2007, Lisbon, Portugal (2007)

[3] Demetra Project, http://www.demetra-tox.net

[4] UCI Data Repository, http://kdd.ics.uci.edu

[5] Kubat, M., Matwin, S.: Addressing the Curse of Imbalanced Training Sets: One-Sided Selection. In: 14th International Conference on Machine Learning, pp. 179–186. Morgan Kaufmann, San Francisco (1997)

[6] Guo, H., Viktor, H.L.: Learning From Imbalanced Data Sets with Boosting and Data Generation:The DataBoost-IM Approach. Sigkdd Explorations 6, 30–39 (2007)

[7] Nickerson, A.S., Japkowicz, N., Milios, E.: Using Unsupervised Learning to Guide Re-sampling in Imbalanced Data Sets. In: 8th International Workshop on Artificial Intelligence and Statistics

[8] Chawla, N.V., Lazarevic, A., Hall, L.O., Bowyer, K.: SMOTEBoost: Improving Prediction of the Minority Class in Boosting. In: 7th European Conference on Principles and Practice of Knowledge Discovery in Databases, pp. 107–119. Cavtat-Dubrovnik, Croatia (2003)

[9] Japkowicz, N.: Learning From Imbalanced Data Sets: A Comparison of Various Strategies. Technical Report, The AAAI Workshop (2000)

[10] Japkowicz, N.: The Class Imbalance Problem: Significance and Strategies. In: 2000 International Conference on Artificial Intelligence, pp. 111–117. IC-AI (2000)

[11] Herbin, M., Bonnet, N., Vautrot, P.: Estimation of the Number of Clusters and Influence Zones. Pattern Recognition Letters 22, 1557–1568

[12] Witten, I.H., Frank, E.: Practical Machine Learning Tools and Techniques with Java Implementations. Morgan Kaufmann, San Francisco (2005)
[13] Frank, E., Hall, M.: Visualizing Class Probability Estimators (2003)
[14] Bhattacharyya, G.K., Johnson, R.A.: Statistical Concepts and Methods. John Wiley and Sons, Chichester (1977)
[15] Melville, P., Mooney, R.J.: Creating Diversity In Ensembles Using Artificial Data. Elsevier Science, Amsterdam (2004)
[16] The University of Waikato New Zealand. Weka Data Mining System, http://www.cs.waikato.ac.nz/ml/weka

CPL Clustering with Feature Costs

Leon Bobrowski[1,2]

[1] Faculty of Computer Science, Bialystok Technical University
[2] Institute of Biocybernetics and Biomedical Engineering, PAS, Warsaw, Poland

Abstract. The convex and piecewise linear (*CPL*) criterion functions can be specified for the purposes of data clustering or unsupervised learning. The data set is in this case composed of feature vectors without additional knowledge in the form of vectors categories. The minimisation of the *CPL* criterion functions allows for discovering linear dependence among feature vectors from a given data set. Introducing feature costs to the *CPL* criterion functions allows to combine linear dependence examination with feature selection process.

Keywords: convex and piecewise linear (*CPL*) criterion functions, feature selection, linear dependence of feature vectors.

1 Introduction

The main goal of data exploration methods is discovering regularities (patterns) in sets of feature vectors of the same structure. Pattern recognition methodology offers variety of tools used for realisation of data exploration goals [1], [2], [3]. Many of these tools have originated from cluster analysis approach. In accordance with this approach, there is an assumed availability of some set of feature vectors, but without additional knowledge about vectors categories. Clusters (groups) of feature vectors resulting from application of particular clustering algorithm are supposed to reflect an objective similarity between these vectors. Data exploration goals also includes discovering dependencies among feature vectors. Depending feature vectors can be treated as specific clusters.

Patterns in streams of feature vectors can be discovered during self-learning process in formal neurons [4]. Self-learning process is based on a sequence of feature vectors with an unknown category. The self-learning algorithm in a stationary environment can lead to the formation of novelty detector or detector of typical patterns in a given formal neuron [5].

The minimisation of convex and piecewise linear (*CPL*) criterion functions can be used instead of self-learning algorithms. Such criterion functions can be specified on a given data set for variety of tasks of exploratory data analysis [6]. In particular, the minimisation of the *CPL* criterion functions gives the possibility for discovering and evaluating linear dependencies among feature vectors from a given data set. The *CPL* criterion functions can be also used for designing novelty detectors or detectors of typical patterns in formal neurons. The convexity of the criterion functions allows to

P. Perner (Ed.): ICDM 2008, LNAI 5077, pp. 177–188, 2008.

avoid problems with local solutions. The enlargement of the *CPL* criterion functions with feature costs components gives the possibility to combining the detectors designing with feature selection process.

Discovering linear dependencies among feature vectors from a given data set or detectors designing can be realised via using the basis exchange algorithms [7]. These algorithms are similar to linear programming and allow to perform the minimization task in an efficient manner, even in the case of large amount of multidimensional feature vectors.

2 Self-learning Sequences and Data Sets

Let us take into considerations the self-learning sequence $\{x[n]\}$ of feature vectors $x[n]$, where $n = 1,2,3,....$:

$$\mathbf{x}[1], \mathbf{x}[2], \mathbf{x}[3], \ldots \tag{1}$$

It is assumed here that the sequence $\{x[n]\}$ (1) is generated independently in each step n, in accordance with constant probabilities p_j:

$$(\forall n \in \{1,2,3,....\}) \quad P\{\mathbf{x}[n] = \mathbf{x}_j\} = p_j \tag{2}$$

where $\mathbf{x}_j = [x_{j1},..,x_{jN}]^T$ $(\mathbf{x}_j \in R^N)$ is the j-th feature vector $(j = 1,...., m)$. Components x_{ji} of the vector $\mathbf{x}_j$ could be numerical results of n standardized examinations of given objects O_j ($x_{ji} \in \{0,1\}$ or $x_{ji} \in R$). Each vector $\mathbf{x}_j$ can be treated as the point of the N-dimensional feature space $F[N] \subset R^N$.

The probabilities p_j are usually unknown, but they can be estimated from given sequence $\{x[n]\}$ (1) of a finite length n_0 in the below manner ($n = 1,2,......, n_0$):

$$(\forall j \in \{1,2,...., m\}) \quad p_j \approx n_j / n_0 \tag{3}$$

where n_j is the number of the events $\{\mathbf{x}[n] = \mathbf{x}_j\}$ in the sequence $\{x[n]\}$ (1) of the length n_0.

The self-learning sequence (1) can be aggregated as the data matrix X or as the set C of the below form:

$$X = [\mathbf{x}_1,........, \mathbf{x}_m]^T \tag{4}$$

or

$$C = \{\mathbf{x}_j, \alpha_j\} \tag{5}$$

where $\mathbf{x}_j$ is the j-th feature vector $(j = 1,...., m)$, and α_j is the nonnegative parameter (*price*). The price α_j of the vector $\mathbf{x}_j$ can be equal to $\alpha_j = 1 / m$ if there is no frequency information about the sequence (1) or can be equal to the probability p_j estimator $\alpha_j = n_j / n_0$ (3). Let us remark, that the self-learning set C (1) contains no information about category related to particular feature vectors $\mathbf{x}_j$.

The formal neurons $FN(w,\theta)$ can be defined by the below decision rule:

$$1 \quad if \quad \mathbf{w}^{\mathrm{T}}\mathbf{x} \geq \theta \tag{6}$$

$$r = \mathrm{r}(\mathbf{w},\theta; \mathbf{x}) =$$

$$0 \quad if \quad \mathbf{w}^{\mathrm{T}}\mathbf{x} < \theta$$

where r is the binary output ($r \in \{0,1\}$), $\mathbf{w} = [w_1,........,w_N] \in R^N$ is the weight vector, θ is the threshold ($\theta \geq 0$), and $\mathbf{x}$ ($\mathbf{x} \in R^N$) is the input vector.

The formal neuron $FN(\mathbf{w},\theta)$ divides the feature space $F[N]$ into two regions (half-spaces) by the following hyperplane $H(\mathbf{w},\theta)$ depending on the weight vector $\mathbf{w}$ and the threshold θ:

$$H(\mathbf{w},\theta) = \{\mathbf{x}: \mathbf{w}^{\mathrm{T}}\mathbf{x} = \theta\} \tag{7}$$

The feature vector $\mathbf{x}_j$ activates ($r = 1$) the formal neuron $FN(\mathbf{w},\theta)$ (5) if and only if $\mathbf{x}_j$ is situated on the positive side of the hyperplane $H(\mathbf{w},\theta)$ ($\mathbf{w}^{\mathrm{T}}\mathbf{x}_j \geq \theta$). In other words, the hyperplane $H(\mathbf{w},\theta)$ separates such vectors $\mathbf{x}_j$ which activate ($r = 1$) the neuron $FN(\mathbf{w},\theta)$ from the non-activating vectors. Such separation depends on the value of the weight vector $\mathbf{w}$.

The weights $\mathbf{w}$ of the neuron $FN(\mathbf{w},\theta)$ (5) can be formed as a result of self-learning process. During such process the sequence of weights $\{\mathbf{w}[n]\}$ is generated in accordance with the selected self-learning algorithm on the basis of the sequence $\{\mathbf{x}[n]\}$ (1). The self-learning algorithms can be represented and analysed as the Robbins-Monro procedure of the stochastic approximation [5].

The self-learning process based on the random sequence $\{\mathbf{x}[n]\}$ (1) of input vectors $\mathbf{x}[n]$ could lead to formation of novelty detector or detector of typical patterns in formal neuron $FN(\mathbf{w},\theta)$ [4]. The novelty detector is formed if the neuron $FN(\mathbf{w},\theta)$ is activated ($r = 1$) mainly by rare feature vectors $\mathbf{x}_j$ (vectors with low probabilities p_j (3)). The detector of typical patterns is formed in the neuron $FN(\mathbf{w},\theta)$ if this neuron is activated mainly by the feature vectors $\mathbf{x}_j$ with the highest probabilities p_j (3).

Variety of functionalities of the neuron $FN(\mathbf{w},\theta)$ (5) can be formed also through minimisation of the convex and piecewise linear (*CPL*) criterion functions $\Phi(\mathbf{w})$ [6].

3 Convex and Piecewise Linear (CPL) Penalty and Criterion Functions

Let us consider the convex and piecewise linear (*CPL*) penalty functions $\varphi_j(\mathbf{w})$ defined on the feature vectors $\mathbf{x}_j$ from the self-learning set C (4) [4]:

$$(\forall \mathbf{x}_j \in C) \qquad\qquad \delta - \mathbf{w}^{\mathrm{T}}\mathbf{x}_j \quad if \quad \mathbf{w}^{\mathrm{T}}\mathbf{x}_j \leq \delta \tag{8}$$

$$\varphi_j(\mathbf{w}) =$$

$$\mathbf{w}^{\mathrm{T}}\mathbf{x}_j - \delta \quad if \quad \mathbf{w}^{\mathrm{T}}\mathbf{x}_j > \delta$$

where δ is some parameter (*margin*) ($\delta \in R$).

The penalty functions $\varphi_j(\mathbf{w})$ are equal to the absolute values $|\delta - \mathbf{w}^{\mathrm{T}}\mathbf{x}_j|$ (Fig. 1).

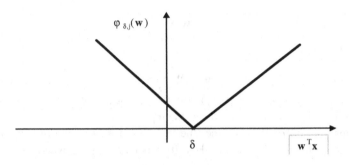

Fig. 1. The penalty function $\varphi_j(\mathbf{w})$ (7)

Let us define the criterion function $\Phi(\mathbf{w})$ as the weighted sum of the penalty functions $\varphi_j(\mathbf{w})$ (8):

$$(\forall\,(j,j') \in I^+)\quad \mathbf{x}_j \prec \mathbf{x}_{j'} \tag{9}$$

The positive parameters α_j in the function $\Phi(\mathbf{w})$ and can be treated as the *prices* of particular vectors $\mathbf{x}_j$ and can be equal, for example, to the estimators $\alpha_j = n_j\,/\,n_0$ (3) of the probabilities p_j .

The criterion function $\Phi(\mathbf{w})$ (9) is convex and piecewise linear (*CPL*) as the sums of such type of functions $\alpha_j\,\varphi_j(\mathbf{w})$.

Each feature vector $\mathbf{x}_j$ from the self-learning set C (5) defines the hyperplane h_j in the parameter (*weight*) space:

$$(\forall \mathbf{x}_j \in C)\qquad h_j = \{\mathbf{w}\colon (\mathbf{x}_j)^T\mathbf{w} = \delta\} \tag{10}$$

The hyperplanes h_j (10) are linked to the penalty functions $\varphi_j(\mathbf{w})$ (83). The function $\varphi_j(\mathbf{w})$ (8) is equal to zero if and only if, the vector $\mathbf{w}$ is situated on the hyperplane h_j ($\mathbf{x}_j^T\mathbf{w} = \delta$).

Any set of N linearly independent feature vectors $\mathbf{x}_j$ ($j \in I_k$) can be used for designing the non-singular matrix (*basis*) $\mathbf{B}_k = [\mathbf{x}_{j(1)},\ldots.,\mathbf{x}_{j(N)}]$ with the columns composed from these vectors. The vectors $\mathbf{x}_j$ ($j \in I_k$) from this set define such N hyperplanes h_j (10) which pass through the below point (*vertex*) $\mathbf{w}_k$:

$$\mathbf{B}_k^T\mathbf{w}_k = \delta = [\delta,\ldots,\delta]^T = \delta\,[1,\ldots\ldots,1]^T = \delta\,\mathbf{1} \tag{11}$$

or

$$\mathbf{w}_k = (\mathbf{B}_k^T)^{-1}\,\delta = \delta\,(\mathbf{B}_k^T)^{-1}\mathbf{1} \tag{12}$$

In the case of "short" vectors $\mathbf{x}_j[N]$, when the number m of the vectors $\mathbf{x}_j[N]$ is much greater than the vectors dimensionality N (m >> N), there may exist many bases $\mathbf{B}_k$ (11) and many vertices $\mathbf{w}_k$ (12). It can be proved that the minimal value Φ^* of the criterion functions $\Phi(\mathbf{w})$ (9) is situated in one of the vertices $\mathbf{w}_k$ (12) [6]:

$$(\exists \mathbf{w}_k^*)\ (\forall \mathbf{w})\ \ \Phi(\mathbf{w}) \geq \Phi(\mathbf{w}_k^*) = \Phi^* \tag{13}$$

Theorem 1. The minimal value $\Phi(\mathbf{w_k}^*)$ (13) of the criterion function $\Phi(\mathbf{w})$ (9) with $\delta \neq 0$ is equal to zero $(\Phi(\mathbf{w_k}^*) = 0)$ if and only if all the feature vectors $\mathbf{x_j}$ from the C (5) are situated on some hyperplane $H(\mathbf{w},\theta)$ (9) with $\theta \neq 0$.

Proof. Let assume that the feature vectors $\mathbf{x_j}$ from the set C (5) are situated on some hyperplane $H(\mathbf{w},\theta)$ (7) with $\theta \neq 0$. In this case, the following equations are fulfilled:

$$(\forall \mathbf{x_j} \in C) \quad \mathbf{w}^T \mathbf{x_j} = \theta \quad or \quad (\delta / \theta) \mathbf{w}^T \mathbf{x_j} = \delta \tag{14}$$

or

$$(\forall \mathbf{x_j} \in C) \quad \varphi_j((\delta / \theta) \mathbf{w}) = 0 \tag{15}$$

On the other hand, if the conditions $\varphi_j(\mathbf{w}) = 0$ (8) are fulfilled for all the feature vectors $\mathbf{x_j}$ from the set C (5), then these vectors have to be situated on the hyperplane $H(\mathbf{w}, \delta)$ (7).

If all the feature vectors $\mathbf{x_j}$ from the set C (5) are situated on some hyperplane $H(\mathbf{w},\theta)$ (7) with $\theta = 0$, then the minimal value $\Phi(\mathbf{w_k}^*)$ (13) of the criterion function $\Phi(\mathbf{w})$ (9) with $\delta = 0$ is equal to zero $(\Phi(\mathbf{w_k}^*) = 0)$.

Let us now examine properties of the criterion function $\Phi(\mathbf{w})$ (9) when feature vectors $\mathbf{x_j}$ are linearly transformed.

$$(\forall \mathbf{x_j} \in C) \quad \mathbf{x_j}' = A \, \mathbf{x_j} \tag{16}$$

where A is a non-singular matrix of dimension $(N \times N)$ $(A^{-1}$ exists).

Theorem 2. The minimal value $\Phi(\mathbf{w_k}^*)$ (13) of the criterion function $\Phi(\mathbf{w})$ (9) does not depend (is invariant) on linear, non-singular data transformations (16).

Proof. The minimal value $\Phi(\mathbf{w_k}^*)$ (13) of the criterion function $\Phi(\mathbf{w})$ (9) is located in one of vertices $\mathbf{w_k}^*$ (13). Let us consider the following transformation of the vector $\mathbf{w_k}^*$:

$$\mathbf{w_k}' = (A^{-1})^T \mathbf{w_k}^* \tag{17}$$

The below equations result directly from (16)

$$(\forall \mathbf{x_j} \in C) \quad (\mathbf{w_k}')^T \mathbf{x_j}' = (\mathbf{w_k}^*)^T (A^{-1}) A \, \mathbf{x_j} = (\mathbf{w_k}^*)^T \mathbf{x_j} \tag{18}$$

In accordance with the above relations the value of all products $(\mathbf{w_k}^*)^T \mathbf{x_j}$ can be preserved by replacement of the optimal vertex $\mathbf{w_k}^*$ (13) by the point $\mathbf{w_k}'$ (17). As a result, the minimal value $\Phi(\mathbf{w_k}^*)$ (13) of the criterion functions $\Phi(\mathbf{w})$ (9) is equal to the minimal value $\Phi'(\mathbf{w_k}')$ of the criterion functions $\Phi'(\mathbf{w})$ (9) determined on the transformed vectors $\mathbf{x_j}'$ (16):

$$\Phi'(\mathbf{w_k}') = \Phi(\mathbf{w_k}^*) \tag{19}$$

where the minimal point $\mathbf{w_k}'$ of the criterion functions $\Phi'(\mathbf{w})$ is given by (17).

It is also possible to assure the invariancy of the minimal value $\Phi(\mathbf{w_k}^*)$ (13) of the criterion function $\Phi(\mathbf{w})$ (9) in respect to data translations:

$$(\forall \mathbf{x}_j \in C) \quad \mathbf{x}_j' = \mathbf{x}_j + \mathbf{b} \tag{20}$$

where $\mathbf{b}$ is N-dimensional vector.

Remark 1. The minimal value $\Phi(\mathbf{w}_k^*)$ (13) does not depend (19) on data translations (20), if the criterion function $\Phi(\mathbf{w})$ (9) is defined on the centred feature vectors $\mathbf{x}_j''$:

$$(\forall \mathbf{x}_j \in C) \quad \mathbf{x}_j'' = \mathbf{x}_j - \mathbf{m}_x \tag{21}$$

where $\mathbf{m}_x$ is the mean vector with the prices α_j (5):

$$\mathbf{m}_x = \sum_{j=1,\ldots,m} \alpha_j \mathbf{x}_j \tag{22}$$

where is the mean vector $\mathbf{m}_x$.

4 Feature Selection Problem

Feature selection procedures are aimed at the maximal reduction of the features x_i which are unimportant for a given problem. Unimportant features x_i can be characterized by the weights w_i equal to zero ($w_i = 0$).

Let the symbol $F = \{x_1,\ldots,x_N\}$ stand for the set of all features x_i used in defining the criterion function $\Phi(\mathbf{w})$ (9). The feature subset F' is obtained from the set F by neglecting (reduction) of some features x_i.

Definition 1. The function $\Phi(\mathbf{w})$ (9) is *monotonical* in respect to features x_i neglecting if and only if the below implication holds:

$$(F' \subset F) \Rightarrow \Phi_{F'}^* \geq \Phi_F^* \tag{23}$$

where Φ_{F*}^* is the minimal value (13) of the criterion function $\Phi(\mathrm{w})$ (9) defined on the features x_i from the set F'.

Theorem 3. The criterion function $\Phi(\mathbf{w})$ (9) is monotonical (23) in respect to neglecting features x_i.

Proof. The feature subset F' is obtained from the set $F = \{x_1,\ldots,x_N\}$ by neglecting some features x_i. It can be assumed that the neglected features x_i are characterized by the weights w_i equal to zero ($w_i = 0$). The features x_i with the weights w_i equal to zero ($w_i = 0$) can be neglected because such features have no influence on the decision rule $r(\mathbf{w},\theta; \mathbf{x})$ (6). Neglecting the feature x_i can be seen as a result of an additional constrain in the below form:

$$(\mathbf{e}_i)^T \mathbf{w} = 0 \Rightarrow w_i = 0 \tag{24}$$

where $\mathbf{e}_i = [0,\ldots,0,1,0,\ldots,0]^T$ is the i-th unit vector, and $\mathbf{w} = [w_1,\ldots,w_N]$.

The minimal value (13) of the criterion function $\Phi(\mathbf{w})$ (9) with an additional constraint '$w_i = 0$' can be not less than the minimal value (13) of the function $\Phi(\mathbf{w})$ (9) without such constraints.

The feature selection problem: For given *margin* γ ($\gamma \geq 0$) find such minimal feature subset F' ($F' \subset F$), that

$$\Phi_{F'}^{*} - \Phi_{F}^{*} \leq \gamma \tag{25}$$

where $\Phi_{F'}^{*}$ is the minimal value (13) of the criterion function $\Phi(\mathbf{w})$ (9) defined on the features x_i from the set F'.

The monotonocity property (23) gives the possibility for an increasing efficiency in the solution of the feature selection problem by using the *branch and bound strategy* [2].

5   CPL Criterion Function with Features Costs

Neglecting unessential features x_i in the cost sensitive manner can be supported by the modified *CPL* criterion function $\Psi_\lambda(\mathbf{w})$ in the below form:

$$\Psi_\lambda(\mathbf{w}) = \Phi(\mathbf{w}) + \lambda \sum_{i \in I} \gamma_i \, \phi_i(\mathbf{w}) \tag{26}$$

where $\Phi(\mathbf{w})$ is given by (9), λ is the *cost level* ($\lambda \geq 0$), γ_i – is the *cost* of the feature x_i ($\gamma_i > 0$), $I = \{1,....,n\}$, and the *cost functions* $\phi_i(\mathbf{w})$ are defined by the unit vectors $\mathbf{e}_i = [0,..,1,..,0]^T$:

$$(\forall i \in \{1,....,n\}) \qquad \phi_i(\mathbf{w}) = |w_i| = \begin{cases} -(\mathbf{e}_i)^T\mathbf{w} & if \ (\mathbf{e}_i)^T\mathbf{w} < 0 \\ (\mathbf{e}_i)^T\mathbf{w} & if \ (\mathbf{e}_i)^T\mathbf{w} \geq 0 \end{cases} \tag{27}$$

The criterion function $\Psi_\lambda(\mathbf{w})$ (26) is the convex and piecewise linear (*CPL*) as the sum of the *CPL* functions $\Phi(\mathbf{w})$ (9) and $\lambda \, \gamma_i \, \phi_i(\mathbf{w})$ (26). Like previously (13), we are taking into account the optimal point $\mathbf{w}_\lambda^{*}$ constituting the minimal value of the criterion function $\Psi_\lambda(\mathbf{w})$:

$$(\exists \mathbf{w}_\lambda^{*}) \ (\forall \mathbf{w}) \ \ \Psi_\lambda(\mathbf{w}) \geq \Psi_\lambda(\mathbf{w}_\lambda^{*}) \tag{28}$$

Each *CPL* cost function $\phi_i(\mathbf{w})$ tends to reach the condition $w_i = 0$ (24) through the minimization of the function $\Psi_\lambda(\mathbf{w})$ (26) and to reducing the feature x_i. The influence of the cost functions $\phi_i(\mathbf{w})$ increases with the value of the parameter λ. The increase of the cost level λ can lead to reducing additional features x_i.

Each unit vector $\mathbf{e}_i$ defines the below hyperplane $h_{0,i}$ in the parameter space:

$$(\forall i \in \{1,2,..., N\}) \qquad h_{0,i} = \{\mathbf{w}: (\mathbf{e}_i)^T\mathbf{w} = 0\} \tag{29}$$

The minimum point $\mathbf{w}_\lambda^{*}$ (46) of the function $\Psi_\lambda(\mathbf{w})$ (26) is situated in one of the vertices $\mathbf{w}_k'$ defined by the equation of the below type (11):

$$\mathbf{B}_k^T \mathbf{w}_k' = \delta' = [\delta,..., \delta, 0,...,0]^T \tag{30}$$

In this case the columns of the basis $\mathbf{B}_k$ can be composed partly of some feature vectors $\mathbf{x}_j$ and partly of some unit vectors $\mathbf{e}_i$. The vertex $\mathbf{w}_k'$ (30) is the point of intersection of hyperplanes $h_{\delta,j}$ (10) defined by some feature vectors $\mathbf{x}_j$ and hyperplanes $h_{0,i}$ (29) defined by unit vectors $\mathbf{e}_i$. The minimum point $\mathbf{w}_\lambda^*$ (28) of the function $\Psi_\lambda(\mathbf{w})$ (28) is situated in one of such vertices $\mathbf{w}_k'$, which is the point of intersecting of N hyperplanes $h_{\delta,j}$ (10) and $h_{0,i}$ (29). Such features x_i, which are linked to the unit vectors $\mathbf{e}_i$ in the optimal basis $\mathbf{B}_k^*$ (30) fulfil the equation (24) and can be reduced. The number of the reduced features x_i can be increased by increasing the cost level λ value in the criterion function $\Psi_\lambda(\mathbf{w})$ (26).

6   Feature Selection Based on the *CPL* Criterion Function $\Psi_\lambda(\mathbf{w})$

Let us consider the case of *"long"* vectors $\mathbf{x}_j[N]$ of dimensionality N ($j = 1,.....,m$). In this case, the dimensionality N of the vectors $\mathbf{x}_j[N]$ is much greater then the number m of these vectors $(N \gg m)$. Each basis $\mathbf{B}_k$ (30) in the N-dimensional feature space $F[N]$ ($\mathbf{x}_j[N] \in F[N]$) contains m' vectors $\mathbf{x}_j[N]$, where $0 \le m' \le m$, and $N - m'$ unit vectors $\mathbf{e}_i$. It can be proved that the minimal value $\Psi_\lambda(\mathbf{w}_k^*[N])$ (28) of the criterion functions $\Psi_\lambda(\mathbf{w}[N])$ (26) is situated in one of the vertices $\mathbf{w}_k'$ [N] (30):

$$(\exists \mathbf{w}_k^*[N]) \ (\forall \mathbf{w}[N]) \ \Psi_\lambda(\mathbf{w}[N]) \ge \Psi_\lambda(\mathbf{w}_k^*[N]) \tag{31}$$

Theorem 4. If all the feature vectors $\mathbf{x}_j[N]$ from the C (5) are situated on some hyperplane $H(\mathbf{w}[N], \theta)$ (9) with $\theta \ne 0$ in the feature space $F[N]$, then there exists such small value of the parameter λ ($0 \le \lambda \le \lambda_{max}$), that the below relations hold in the optimal point $\mathbf{w}_k^*[N] = [w_{k1}^*,......,w_{kN}^*]^T$ (31):

$$(\forall \lambda \in (0, \lambda_{max})) \ \Phi(\mathbf{w}_k^*[N]) = 0 \tag{32}$$

where $\Phi(\mathbf{w}_k^*[N])$ is the minimal value (13) of the criterion function $\Phi(\mathbf{w})$ (9) with $\delta \ne 0$, and

$$(\forall \lambda \in (0, \lambda_{max})) \quad \Psi_\lambda(\mathbf{w}_k^*[N]) = \lambda \sum_{i \in \{1,....,N\}} \chi_i |w_{ki}^*| \tag{33}$$

The above theorem can be proved by using the *Theorem* 1 by an analysis similar to this given in the paper [8].

If the features prices χ_i have a constant value χ, then

$$(\forall \lambda \in (0, \lambda_{max})) \quad \Psi_\lambda(\mathbf{w}_k^*[N]) = \lambda \chi \sum_{i \in \{1,....,N\}} |w_{ki}^*| = \|\mathbf{w}_k^*[N]\|_{L1} \tag{34}$$

Remark 2. If the assumptions of the *Theorem* 4 are fulfilled, then the minimization (31) of the criterion function $\Psi_\lambda(\mathbf{w}[N])$ (26) with equal features prices χ_i ($\chi_i = \chi$) leads to such optimal vertex $\mathbf{w}_k^*[N]$ (30) which has the minimal L_1 norm (34) among all vertices $\mathbf{w}_k'[N]$ (30) fulfilling the condition $\Phi(\mathbf{w}_k'[N]) = 0$ (32).

The vertex $\mathbf{w}_k^*$ [N] (30) defines the hyperplane $H(\mathbf{w}_k^*$ [N],θ) (7) in the feature space $F[N]$:

$$H(\mathbf{w}_k^*[N],\theta) = \{\mathbf{x}[N]: \mathbf{w}_k^* [N]^T\mathbf{x}[N] = \theta\} \tag{35}$$

The Euclidean distance $\rho_{L2}(\mathbf{w}_k{}'[N],\theta)$ between the hyperplane $H(\mathbf{w}_k{}'[N],\theta)$ (7) and the origin of the coordinate system is given by the below expression:

$$\rho_{L2}(\mathbf{w}_k{}^*[N],\theta) = \theta / (\mathbf{w}_k{}^*[N]^T \mathbf{w}_k{}^*[N])^{1/2} = \theta / \|\mathbf{w}_k{}^*[N]\|_{L2} \tag{36}$$

The L_1 distance $\rho_{L1}(\mathbf{w}_k{}'[N],\theta)$ between the hyperplane $H(\mathbf{w}_k{}'[N],\theta)$ (35) and the origin is defined similarly

$$\rho_{L1}(\mathbf{w}_k{}^*[N],\theta) = \theta / (\sum_{i\in\{1,...,N\}} |w_{ki}{}^*|) = \theta / \|\mathbf{w}_k{}^*[N]\|_{L1} \tag{37}$$

As it results from the *Theorem* 1, the condition $\Phi(\mathbf{w}_k{}^*[N]) = 0$ (32) means that all the feature vectors $\mathbf{x}_j$ from the set C (5) are situated on the hyperplane $H(\mathbf{w}_k{}^*[N],\theta)$ (35) with $\theta \neq 0$.

Remark 3. If the assumptions of the *Theorem* 4 are fulfilled, then the minimization (31) of the criterion function $\Psi_\lambda(\mathbf{w}[N])$ (26) with equal features prices χ_i ($\chi_i = \chi$) leads to the hyperplane $H(\mathbf{w}_k{}^*[N],\theta)$ (35) with the maximal L_1 distance $\rho_{L1}(\mathbf{w}_k{}^*[N],\theta)$ (31) among the hyperplanes $H(\mathbf{w}_k{}^*[N],\theta)\mathbf{w}_k{}^*[N]$ (35) containing all the feature vectors $\mathbf{x}_j$ from the set C (5).

The *Remark* 3 makes precise the geometrical meaning of the optimal hyperplane $H(\mathbf{w}_k{}^*[N],\theta)$ (35) defined by the minimum of the criterion function $\Psi_\lambda(\mathbf{w}[N])$ (26) with equal features prices χ_i. This geometrical meaning has some similarity to the optimality criteria used in the *Support Vector .Machines* approach (*SVM*) [8]. The optimality criteria used in *SVM* are based on the Euclidean L_2 norms.

The minimization (31) of the criterion function $\Psi_\lambda(\mathbf{w}[N])$ (26) leads to the optimal vertex $\mathbf{w}_k{}^*[N]$ with the basis $\mathbf{B}_k{}^*$ (30). The basis $\mathbf{B}_k{}^*$ may contain some unit vectors $\mathbf{e}_i$. Such features x_i which are related to the unit vectors $\mathbf{e}_i$.in the basis $\mathbf{B}_k{}^*$ (30) can be neglected without changing the decision rule (6), because they have the weights w_i equal to zero (24). In other words, the feature selection problem can be solved through minimization (31) of the *CPL* criterion function $\Psi_\lambda(\mathbf{w})$ (28).

7 Clustering with the *CPL* Criterion Function $\Psi_\lambda(\mathbf{w})$

The minimization (31) of the *CPL* criterion function $\Psi_\lambda(\mathbf{w})$ (28) can be also used for clustering data represented by the matrix X (4) or by the set C (5). Let us consider for this purpose the case of *"short"* vectors $\mathbf{x}_j[N]$, when the number m of these vectors is much greater then the vectors dimensionality N ($m \gg N$). Usually in this case it is impossible to locate all the feature vectors on the optimal hyperplane $H(\mathbf{w}_k{}^*[N],\theta)$ (35) in the feature space $F[N]$. A family of the optimal hyperplanes $H(\mathbf{w}_k{}^*[N],\theta)$ (35) can be designed and used for this purpose. We will use the monotonicity property of the function $\Psi_\lambda(\mathbf{w})$ (28) in designing procedure.

Definition 2. The criterion function $\Psi_\lambda(\mathbf{w}[N])$ (26) is *monotonical* in respect to features vectors $\mathbf{x}_j[N]$ neglecting if and only if the below implication holds:

$$(C'' \subset C) \Rightarrow \Psi_{C''}{}^{*} \leq \Psi_{C}{}^{*} \tag{38}$$

where $\Psi_{C''}{}^{*} = \Psi_{\lambda}(\mathbf{w}_k{}^{*}[N])$ (28) is the minimal value of the criterion function $\Psi_{\lambda}(\mathbf{w})$ (26) defined on the features vectors $\mathbf{x}_j[N]$ from the set C''.

Theorem 5. The criterion function $\Psi_{\lambda}(\mathbf{w}[N])$ (26) is monotonically in respect to features vectors $\mathbf{x}_j[N]$ neglecting.

Proof. The set C'' (23) is obtained from the set C by means of neglecting some feature vectors $\mathbf{x}_j[N]$. The neglecting feature vector $\mathbf{x}_k[N]$ means that adequate penalty function $\varphi_k(\mathbf{w})$ (8) is reduced from the sum (9). The minimal value (31) of the reduced criterion function can not be greater than the minimal value (31) of the original criterion function $\Psi_{\lambda}(\mathbf{w})$ (28).

Let us consider the below multistage procedure of the optimal hyperplanes $H(\mathbf{w}_k{}^{*}[N],\theta)$ (35) designing on the basis of given set $C_1 = \{\mathbf{x}_j[N]\}$.

    i.   The minimum value $\Psi_{\lambda}(\mathbf{w}_1{}^{*}[N])$ (31) and the optimal vector $\mathbf{w}_1{}^{*}[N]$ of the criterion function $\Psi_{\lambda}(\mathbf{w}[N])$ (26) defined on all vectors $\mathbf{x}_j[N]$ from the set C_1 are found.

    ii.   If $\Phi(\mathbf{w}_1{}^{*}[N]) > 0$, then such feature vectors $\mathbf{x}_j[N]$ which are maximally distanced from the optimal hyperplane $H(\mathbf{w}_1{}^{*}[N],\theta)$ (35) are successively removed from the set C_1 until the below condition is met for given a priori margin δ ($\delta > 0$):

$$\Phi(\mathbf{w}_{C'}{}^{*}[N]) \leq \delta \tag{39}$$

where $\mathbf{w}_{C'}{}^{*}[N]$ is the optimal vector (31) of the criterion function $\Psi_{\lambda}(\mathbf{w}[N])$ (26) defined on the reduced set C'' (38).

    iii.   The feature vectors $\mathbf{x}_j[N]$ reduced during the stage *ii.* are collected as the set C_2.

    iv.   The minimum value $\Psi_{\lambda}(\mathbf{w}_2{}^{*}[N])$ (31) and the optimal vector $\mathbf{w}_2{}^{*}[N]$ of the criterion function $\Psi_{\lambda}(\mathbf{w}[N])$ (26) defined on all vectors $\mathbf{x}_j[N]$ from the set C_2 are found.

    v.   If $\Phi(\mathbf{w}_2{}^{*}[N]) > 0$, then the stage *ii.* is repeated with the set C_1 replaced by C_2 (or C_k replaced by C_{k+1}).

In accordance with the above procedure, the initial set $C_1 = \{\mathbf{x}_j[N]\}$ is reduced to zero ($\emptyset$) through K successive subsets $C_1 \supset C_2 \supset \ldots\ldots \supset C_K = \emptyset$. After finite number of $K - 1$ stages, a family of optimal hyperplanes $H(\mathbf{w}_k{}^{*}[N],\theta)$ (35) will be designed to allow to divide an arbitrary set $C_1 = \{\mathbf{x}_j[N]\}$ into K clusters S_k. The cluster S_k based on the k-th optimal hyperplanes $H(\mathbf{w}_k{}^{*}[N],\theta_k)$ (35) can be defined as some *slice* of the thickness ρ ($\rho \geq 0$):

$$S_k = \{\ \mathbf{x}_j[N]\colon |\ \mathbf{w}_k{}^{*}[N]^{T}\mathbf{x}_j[N] - \theta_k| \leq \rho\ \mathbf{w}_k{}^{*}[N]^{T}\ \mathbf{w}_k{}^{*}[N]^{1/2} \tag{40}$$

The shape and the number of the clusters S_k depend on the structure of data set $C_1 = \{\mathbf{x}_j[N]\}$. In particular, the clusters S_k can overlap ($S_k \cap S_{k'} \neq \emptyset$) (Fig. 2)

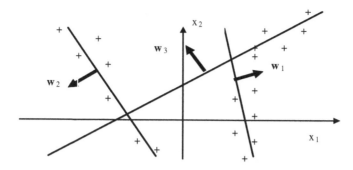

Fig. 2. An example of three clusters S_k defined on the plane by the hyperplanes $H(\mathbf{w}_1, \delta)$, $H(\mathbf{w}_2, \delta)$ and $H(\mathbf{w}_3, \delta)$ (35)

Let us remark that each cluster S_k (40) can be characterized by its own linear dependency among features x_{ji}, where:

$$w_{k1}^* x_{j1} + w_{k2}^* x_{j2}. + \ldots\ldots + w_{kN}^* x_{jN} = \theta_k \qquad (41)$$

where $\mathbf{x}_j = [x_{j1},..,x_{jN}]^T$ and $\mathbf{w}_k^*[N] = [w_{k1}^*, \ldots\ldots, w_{kN}^*]^T$.

Part of the optimal weights w_{ki}^* can be equal to zero ($w_{ki}^* = 0$). The number r_k o nonzero weights w_{ki}^* ($w_{ki}^* \neq 0$) can be used for determining the *rank* of the subset C_k of feature vector $\mathbf{x}_j[N]$.

The above procedure can be specified for designing novelty detectors or detectors of typical patterns on the basis of the self-learning set C (5). The subset C_{R1} ($C_{R1} \subset C$) of rare vectors $\mathbf{x}_j[N]$ and the subset C_{T1} ($C_{T1} \subset C$) of typical vectors can be selected from the set C (5) on the basis of the vectors frequencies $\alpha_j = n_j / n_0$ (3). The subset C_{R1} of rare vectors is used for the purpose of designing novelty detectors. Similarly, the subset C_{T1} is used for designing detectors of typical patterns in accordance with the described earlier procedure.

8 Concluding Remarks

The minimization (31) of the *CPL* criterion function $\Psi_\lambda(\mathbf{w})$ (28) can be used for designing   detectors of typical patterns or novelty detectors on the basis of the self - learning set C (5). The subset C_T of detected typical patterns $\mathbf{x}_j[N]$) and the subset C_R of rare patterns depends not only on the prices α_j (5) of particular feature vectors $\mathbf{x}_j$, but also on the structure of the set C (5), on the feature costs γ_i, and on the cost level λ (26). An increase of the feature costs γ_i (26) allows to reduce particular features x_i from the detected dependencies (41). An increase of the cost level λ (26) also allows to decrease the number of the regarded features x_i.

The *CPL* criterion functions $\Psi_\lambda(\mathbf{w})$ (28) can be minimised by using the basis exchange algorithms, which are similar to the linear programming [7]. Such algorithms give the possibility of finding the minimum of a given piecewise linear function $\Psi_\lambda(\mathbf{w})$ (28) efficiently for different values of feature costs γ_i, and for different values of the cost level λ even in the cases of large, multidimensional data set C (4).

The self learning set C (4) can be explored efficiently and the most interesting clusters can be detected in this way.

Aknowledgements

The work was partially financed by the KBN grant 3T11F01130, and by the grant S/WI/2/08 from the Białystok University of Technology.

References

1. Duda, O.R., Hart, P.E., Stork, D.G.: Pattern Classification. J. Wiley, Newyork
2. Fukunaga, K.: Introduction to Statistical Pattern Recognition. Academic Press, London (1972)
3. Johnson, R.A., Wichern, D.W.: Applied Multivariate Statistical Analysis. Prentice-Hall, Inc., Englewood Cliffs (1991)
4. Bobrowski, L.: Rules of forming receptive fields of formal neurons during unsupervised learning processes. Biological Cybernetics 43, 23–28 (1982)
5. Kushner, H.J., Clark, D.S.: Stochastic approximation Methods for Constrained and Unconstrained Systems. Springer, Berlin (1978)
6. Bobrowski, L.: Eksploracja danych oparta na wypukłych i odcinkowo-liniowych funkcjach kryterialnych (Data mining based on convex and piecewise linear (CPL) criterion functions) (in Polish), Technical University Białystok (2005)
7. Bobrowski, L.: Design of piecewise linear classifiers from formal neurons by some basis exchange technique. Pattern Recognition 24(9), 863–870 (1991)
8. Bobrowski, L., Huntsinger, R.C.: Exploring the linearity of models on the basis of ranked data. In: SCSC 2007, Summer Simulation Multiconference, SCS, San Diego, California, USA, pp. 411–418 (2007)
9. Vapnik, V.N.: Statistical Learning Theory. J. Wiley, New York (1998)

Relative Linkage Disequilibrium: A New Measure for Association Rules

Ron Kenett[1] and Silvia Salini[2]

[1] KPA Ltd., Raanana, Israel and University of Torino, Torino, Italy
ron@kpa.co.il
[2] Department of Economics Business and Statistics, University of Milan, Italy
silvia.salini@unimi.it

Abstract. Association rules are one of the most popular unsupervised data mining methods. Once obtained, the list of association rules extractable from a given dataset is compared in order to evaluate their importance level. The measures commonly used to assess the strength of an association rule are the indexes of support, confidence, and the lift.

Relative Linkage Disequilibrium (RLD) was originally proposed as an approach to analyse both quantitatively and graphically general two way contingency tables. RLD can be considered an adaptation of the lift measure with the advantage that it presents more effectively the deviation of the support of the whole rule from the support expected under independence given the supports of the LHS (A) and the RHS (B). RLD can be interpreted graphically using a simplex representation leading to powerful graphical display of association relationships. Moreover the statistical properties of RLD are known so that confirmatory statistical tests of significance or basic confidence intervals can be applied.

This paper will present the properties of RLD in the context of association rules and provide several application examples to demonstrate it's practical advantages.

Keywords: contingency table, simplex representation, text mining.

1 Introduction

In evaluating the structure of a 2x2 contingency table we consider four relative frequencies, $x_1, x_2, x_3, x_4, \sum_{i=1}^{4} x_i = 1, 0 \leq x_i, i = 1...4$. In the context of this paper, the two variables we consider are occurrence, in a set of transactions, of items A and B on the Left Hand Side and Right Hand Side of an association rule. The frequencies are described in the table below.

In evaluating the association between the two variables, several measures of association are available such as the cross product or odds ratio, $\alpha = \frac{x_1 x_4}{x_2 x_3}$, the chi square statistic [2] and Cramer's index, among others (see [1]). An inherent advantage to informative graphical displays is that the experience and intuition of the experimenter who collects the data can contribute to the statistician's

P. Perner (Ed.): ICDM 2008, LNAI 5077, pp. 189–199, 2008.

	B	^B
A	x_1	x_2
^A	x_3	x_4

data analysis. In [3] a graphical representation of 2x2 tables is proposed using a simplex and, a measure of association, the Relative Linkage Disequilibrium with an intuitive visual interpretation, is proposed. This paper expands on these ideas with a specific focus on association rules that are used, among other things, for analyzing semantic unstructured data.

In Section 2 some details about Relative Linkage Disequilibrium (RLD) and simplex representation are given. Section 3 briefly describes association rules and the classical measure used to select the rules, moreover the implementation of RLD in the association rules context is described. Section 4 presents two practical application of the RLD; the first example considers a classical market basket analysis data set and compares RLD with the classical measures; the second example shows the simplex representation and its interesting interpretation using a data set related to the event category (EC) of an electronic product under constant monitoring.

2   Relative Linkage Disequilibrium and Simplex Representation

Relative Linkage Disequilibrium (RLD) is an association measure motivated by indices used in population genetics to assess stability of the genetic composition of populations under various forces of natural selection and migration patterns (see[4] , [5], [6] and [7]). One specific such measure is the linkage disequilibrium, $D = x_1 x_4 - x_2 x_3$. Under independence, the odds ratio, $\alpha = 1$ and $D = 0$.

There is a natural one to one correspondence between the set of all possible 2x2 contingency tables and point on a simplex. We exploit this graphical representation to map out association rules derived, for example, from text analysis. The tables that correspond to independence in the occurrence of A and B, correspond to a specific surface within the simplex (see figure 1). On that surface, $\alpha = 1$ and $D = 0$.

Let $f = x_1 + x_3$ and $g = x_1 + x_2$. It can be easily verified that:

$x_1 = fg + D$
$x_2 = (1 - f)g - D$
$x_3 = f(1 - g) - D$
$x_4 = (1 - f)(1 - g) + D$

The geometric interpretation of D makes it an appealing measure of interaction. However points closer to the edges of the simplex will have intrinsically smaller values of D. The Relative Linkage Disequilibrium standardizes D by the distance D_M the maximum distance to the surface of the simplex from the point on the surface D=0 corresponding to the contingency table along the direction (1, -1, -1, 1). RLD is therefore computed as D/D_M.

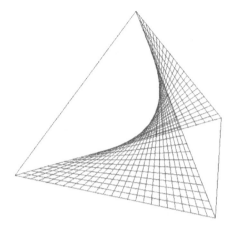

Fig. 1. The surface D=0

The computation of RLD can be performed through the following algorithm:

If $D > 0$
then
    if $x_3 < x_2$
    then $RLD = \frac{D}{D+x_3}$
    else $RLD = \frac{D}{D+x_2}$
else
    if $x_1 < x_4$
    then $RLD = \frac{D}{D-x_1}$
    else $RLD = \frac{D}{D-x_4}$

Some asymptotic properties of RLD are available [3] and can be used for statistical inference.

3   Association Rules and Relative Linkage Disequilibrium

Association rules are one of the most popular unsupervised data mining methods [8]. They were developed in the field of computer science and typically used in applications such as market basket analysis to measure the association between products purchased by each consumer, or in web clickstream analysis, to measure the association between the pages seen (sequentially) by a visitor of a site. Association rules belong to the category of local models, i.e. methods that deal with selected parts of the dataset in the form of subsets of variables or subsets of observations, rather than being applied to the whole database. This element constitutes both the strength and the weak point of the approach. The strength is in that being local, they do not require a large effort from a computational point of view. On the other hand, the locality itself means that a generalization of the results cannot be allowed, not all the possible relations are evaluated at the same time.

Mining frequent *itemsets* and association rules is a popular and well researched method for discovering interesting relations between variables in large databases. Piatetsky-Shapiro in [9] describes analyzing and presenting strong rules discovered in databases using different measures of interest. The structure of the data to be analyzed is typically referred to as transactional in a sense explained below.

Let $I = \{i_1, i_2, \ldots, i_n\}$ be a set of n binary attributes called "items". Let $T = \{t_1, t_2, \ldots, t_n\}$ be a set of transactions called the database. Each transaction in T has a unique transaction ID and contains a subset of the items in I. Note that each individual can possibly appear more than once in the dataset. In market basket analysis, a transaction means a single visit to the supermarket, for which the list of products bought is recorded, while in web clickstream analysis, a transaction means a web session, for which the list of all visited web-pages is recorded. From this very topic specific structure, the more common data matrix can be easily derived, a different transaction (client) for each row, and a product (page viewed) for each column. The internal cells are filled with 0 or 1 according to the presence or absence of the product (page).

A rule is defined as an implication of the form $X => Y$ where $X, Y \in I$ and $X \cap Y = \emptyset$. The sets of items (for short *itemsets*) X and Y are called antecedent (left-hand-side or LHS) and consequent (right-hand-side or RHS) of the rule. In an *itemset*, each variable is binary, taking two possible values only, "1" if a specific condition is true, and "0" otherwise.

Each association rule describes a particular local pattern, based on a restricted set of binary variables, and represents relationships between variables which are binary by nature. In general, however, this does not have to be the case and continuous rules are also possible. In this case, the elements of the rules can be intervals on the real line, that are conventionally assigned a value of TRUE= 1 and FALSE=0. For instance, a rule of this kind can be $X > 0 => Y > 100$.

Once obtained, the list of association rules extractable from a given dataset is compared in order to evaluate their importance level. The measures commonly used to assess the strength of an association rule are the indexes of support, confidence, and lift.

- The **support** for a rule $A => B$ is obtained by dividing the number of transactions which satisfy the rule, N{A=>B}, by the total number of transactions, N

$$support \ \{A=>B\} = N\{A=>B\} \ / \ N$$

  The support is therefore the frequency of events for which both the LHS and RHS of the rule hold true. The higher the **support** the stronger the information that both type of events occur together.

- The **confidence** of the rule $A => B$ is obtained by dividing the number of transactions which satisfy the rule N{A=>B} by the number of transactions which contain the body of the rule, A.

$$confidence \{A=>B\} = N\{A=>B\} / N\{A\}$$

The confidence is the conditional probability of the RHS holding true given that the LHS holds true. A high **confidence** that the LHS event leads to the RHS event **implies causation** or statistical dependence.

– The **lift** of the rule $A => B$ is the deviation of the support of the whole rule from the support expected under independence given the supports of the LHS (A) and the RHS (B).

$$lift \{A=>B\} = confidence\{A=>B\} / support\{B\}$$

$= support\{A=>B\}/support\{A\} support\{B\}$

Lift is an indication of the effect that knowledge that LHS holds true has on the probability of the RHS holding true.

- **when lift is exactly 1:** No effect (LHS and RHS independent). No relationship between events.
- **for lift greater than 1:** Positive effect (given that the LHS holds true, it is more likely that the RHS holds true). Positive dependence between events.
- **if lift is smaller than 1:** Negative effect (when the LHS holds true, it is less likely that the RHS holds true). Negative dependence between events.

Relative Linkage Disequilibrium (RLD) is an alternative measure of association between A and B. For example, take a specific relation A=>B which is observed in 57 cases out of 254 transactions. The LHS, A, is observed without B on the RHS (^RHS) in 40 cases, The RHS B is observed in 109 cases without A on the LHS and, in 48 cases, neither A nor B were observed. In our example we obtain the following table:

	B	^B			B	^B
A	57	40		A	X_1	X_2
^A	109	48		^A	X_3	X_4

and in general
Let $x_i = \frac{X_i}{N}, i = 1...4$
For this data:

– $Support\{A =>B\} = 57/254 = .224$
– $Confidence\{A =>B\} = 57/97 = .588$
– $Support \{B\} = 166/254 = .654$
– $Lift\{A=>B\} = .588/.654 = .90$
– $D = x_1 x_4 - x_2 x_3 = (57*48 - 40*109)/(254*254) = -1624/64516 = -0.0252$
  and since D<0 and $x_1 > x_4$ we have that
– **RLD**$= D/(D - x4) = -0.0252/(-0.0252 - 0.189) = 0.118$

4   Application Examples

The **arules** extension package for R [10] provides the infrastructure needed to create and manipulate input data sets for the mining algorithms and for analyzing the resulting *itemsets* and rules. Since it is common to work with large sets of rules and *itemsets*, the package uses sparse matrix representations to minimize memory usage. The infrastructure provided by the package was also created to explicitly facilitate extensibility, both for interfacing new algorithms and for adding new types of interest measures and associations.

The library **arules** provides the function interestMeasure() which can be used to calculate a broad variety of interest measures for *itemsets* and rules. All measures are calculated using the quality information available from the sets of *itemsets* or rules (i.e., support, confidence, lift) and, if necessary, missing information is obtained from the transactions used to mine the associations. For example, available measures for *itemsets* are:

- All-confidence [11]
- Cross-support ratio [12]

For rules the following measures are implemented:

- Chi square measure [13]
- Conviction [14]
- Hyper-lift and hyper-confidence[15]
- Leverage [9]
- Improvement [16]
- Several measures from [17] (e.g., cosine, Gini index, ϕ- coefficient, odds ratio)

In this paper we implement the Relative Linkage Disequilibrium measure (RLD) in the function InterestMeasure() and we use the function quadplot() and triplot() of the library **klaR** [18] to produce the simplex 3D and 2D representation.

The first example that we consider is an application to a classical market basket analysis data set. The Groceries (provided by [14] data set contains 1 month (30 days) of real-world point-of-sale transaction data from a typical local grocery outlet. The data set contains 9835 transactions and the items are aggregated to 169 categories.

In order to compare the classical measure of association rule with RLD we plot in Figure 2 the measures of the 430 rules obtained with the apriori algorithm [19] setting minimum support 0.01 and minimum confidence 0.1.

The plot shows that RLD, like confidence and lift, is able to identify rules that have similar support. Moreover for low level of confidence, the value of RLD is more variable. The relationship with lift is interesting. It seems that RLD can differentiate between groups of rules with the same level of lift.

In Table 1, the first 20 rules sorted by lift are displayed. For each rule the RLD, the odds Ratio and the Chi Square are reported. Figure 3 shows the value of RLD versus odds ratio and versus Chi Square for the top 10 rules.

As we expect for the relationship between RLD and odds ratio (see [1]) the two measure are coherent but different. The Chi Square appears not correlated with

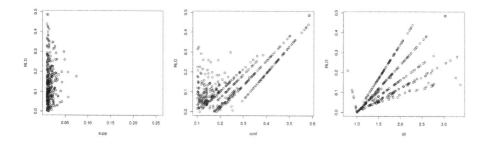

Fig. 2. Plot of Relative Disequilibrium versus Support, Confidence and Lift for the 430 rules of Groceries data set

Table 1. First 20 rules for groceries data, sorted by Lift

lhs	rhs	supp	conf	lift	RLD	odds	chi
{whole milk, yogurt}	{curd}	0.010	0.180	3.372	0.141	4.566	184.870
{citrus fruit, other vegetables}	{root vegetables}	0.010	0.359	3.295	0.281	4.958	188.438
{other vegetables,yogurt}	{whipped/sour cream}	0.010	0.234	3.267	0.175	4.450	177.154
other vegetables}	{root vegetables}	0.012	0.343	3.145	0.262	4.679	206.042
{root vegetables}	{beef}	0.017	0.160	3.040	0.250	4.631	277.341
{beef}	{root vegetables}	0.017	0.331	3.040	0.250	4.631	277.341
{citrus fruit, root vegetables}	{other vegetables}	0.010	0.586	3.030	0.487	6.183	175.058
{tropical fruit,, root vegetables}	{other vegetables}	0.012	0.585	3.021	0.485	6.195	207.203
{other vegetables, whole milk}	{root vegetables}	0.023	0.310	2.842	0.225	4.390	330.231
{other vegetables, whole milk}	{butter}	0.012	0.154	2.771	0.143	3.639	146.317
{whole milk, curd}	{yogurt}	0.010	0.385	2.761	0.286	4.088	132.726
{whipped/sour cream}	{curd}	0.011	0.146	2.742	0.135	3.539	129.718
{curd}	{whipped/sour cream}	0.011	0.197	2.742	0.135	3.539	129.718
{other vegetables, whole milk}	{whipped/sour cream}	0.015	0.196	2.729	0.140	3.702	183.728
{other vegetables, yogurt}	{root vegetables}	0.013	0.297	2.729	0.212	3.791	163.187
{whole milk, yogurt}	{whipped/sour cream}	0.011	0.194	2.709	0.132	3.500	131.650
{other vegetables, yogurt}	{tropical fruit}	0.012	0.283	2.701	0.199	3.688	151.333
{root vegetables, other vegetables}	{citrus fruit}	0.010	0.219	2.645	0.148	3.407	119.391
{other vegetables, rolls/buns}	{root vegetables}	0.012	0.286	2.628	0.199	3.568	141.814
{tropical fruit, whole milk}	{root vegetables}	0.012	0.284	2.602	0.196	3.514	136.436

RLD. The major advantage of this new measure is the fact that it is more intuitive than odds ratio and Chi Square and has a useful graphical representation.

In the following example we present the simplex representation and its interesting interpretation. We consider a data set made available by KPA Ltd. The problem consists of mapping the severity level of problems, and the event category (EC) of an electronic product under constant monitoring. Six variables are considered, as shown in Table 2.

The data are recoded as a binary incidence matrix by coercing the data set to transactions. The new data sets present 3733 transactions (rows) and 124 items (columns). Figure 4 shows the item frequency plot (support) of the item with support major than 0.1.

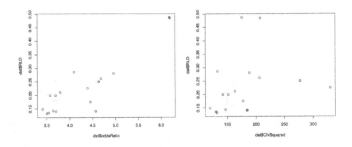

Fig. 3. Plot of Relative Disequilibrium versus Odds Ratio and ChiSquare for the top 10 rules of Groceries data set sorted by RLD

Table 2. Event Category Data Set

PBX No	Severity	Customer Type	EC2	EC1	AL1	AL2	AL3
90009	2	High Tech	SEC08	Security	NO_AL	NP	NP
90009	2	High Tech	NTC09	Network Com	NO_AL	NP	NP
90009	2	High Tech	SEC08	Security	NO_AL	NP	NP
90009	2	High Tech	SEC08	Security	NO_AL	NP	NP
90021	2	Municipalities	SEC08	Security	NO_AL	NP	NP
90033	2	Transportation	SFW05	Software	PCM TS	NP	NP
90033	3	Transportation	INT04	Interface	PCM TS	NP	NP
90033	3	Transportation	SEC05	Security	PCM TS	NP	NP
90038	2	Municipalities	SFW05	Software	NO_AL	NP	NP

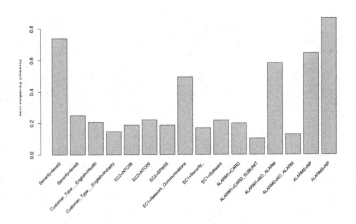

Fig. 4. Item Frequency Plot (Support>0.1) of EC data set

We apply to the data the apriori algorithm setting minimum support 0.1 and minimum confidence 0.8. We obtain 200 rules. The aim of this example is to show the intuitive interpretation of RLD through his useful graphical representation. Figure 5 shows the simplex representation of the contingency tables

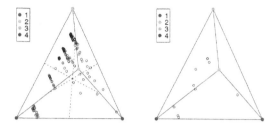

Fig. 5. 3D Simplex Representation for 200 rules of EC data set and for the top 10 rules sorted by RLD

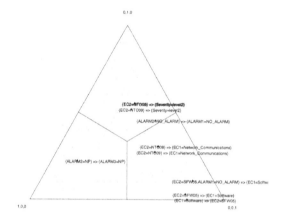

Fig. 6. 2D Simplex Representation for the top 10 rules sorted by RLD of EC data set

corresponding to the rules. The corners represent four tables with relative frequency $(1,0,0,0)$, $(01,0,0)$, $(0,0,1,0)$, $(0,0,0,1)$. We represent the 200 rules obtained from the EC data set and we represent, in the same space, the first 10 rules sorted by RLD.

Figure 5 shows that using a simplex representation, it is possible to immediately have an idea of the rules' structure. In our case, there are 4 groups of rules aligned. Aligned rules imply that they have the same support. In the right part of the Figure 5, the top 10 rules sorted by RLD are plotted. There are some rules in different part of the plot but some of them appear on the same virtual line, only one point is not aligned.

In order to improve the interpretation, we can decide to reduce the dimension and exclude the ˆRHS cell. The 2D representation is shows in Figure 6. In comparison to Table 3, in the left bottom part of the simplex, there are rules with high support, in the right bottom are the rules with low level and in the top are the ones with medium support. The edge in the center represents the middle point of the line, so the point $(0.5, 0.5, 0)$, $(0,0.5,0.5)$ and $(0.5, 0, 0.5)$ obviously assuming ˆRHS equal to 0.

Table 3. Top 10 rules sorted by RLD of EC data set

lhs	rhs	sup	conf	lift	RLD
{EC1=Software}	{EC2=SFW05}	0.1864	0.8593	4.6086	1.0000
{AL2=NO_AL}	{AL1=NO_AL}	0.1286	1.0000	1.7171	1.0000
{EC2=SFW05}	{EC1=Software}	0.1864	1.0000	4.6086	1.0000
{EC2=SFW05}	{Severity=level2}	0.1864	1.0000	1.3550	1.0000
{EC2=NTC08}	{EC1=Network_Com}	0.1878	1.0000	2.0277	1.0000
{EC2=NTC08}	{Severity=level2}	0.1878	1.0000	1.3550	1.0000
{EC2=NTC09}	{EC1=Network_Com}	0.2207	1.0000	2.0277	1.0000
{EC2=NTC09}	{Severity=level2}	0.2207	1.0000	1.3550	1.0000
{AL2=NP}	{AL3=NP}	0.6440	1.0000	1.1543	1.0000
{EC2=SFW05,AL1=NO_AL}	{EC1=Software}	0.1090	1.0000	4.6086	1.0000

5 Summary and Future Work

Relative Linkage Disequilibrium is a useful measure in the context of association rules, especially for its intuitive visual interpretation. An inherent advantage to informative graphical displays is that the experience and intuition of the experimenter who collects the data can contribute to the statistician's data analysis.

The examples proposed in this paper show that RLD, like confidence and lift, is able to identify rules that have similar support. Moreover for low level of confidence, the value of RLD is more variable. The relationship with lift is interesting, it seems that RLD can differentiate between groups of rules with the same level of lift. Moreover RLD is coherent with odds ratio and differ from Chi Square. The second example highlight the major advantage of the new measure: it is more intuitive than odds ratio and Chi Square and has a useful graphical representation that make possible to immediately have an idea of the rules' structure and to identify groups of rules.

In our future intention RLD can be applied to text mining and web mining data and to the analysis of comments in survey questionnaires. The context of application of RLD ranges from cognitive science, ability tests and customer satisfaction surveys, in order to find associated item and to test if there are some redundant items.

Another important future work is the exploration of statistical properties of RLD in the context of association rules. For a partial study of such properties see [3].

Acknowledgment. This work has been partially funded by the FP6 project IST-27097, MUSING (MUlti-industry, Semantic based next generation business INtelliGence).

References

1. Kenett, R., Zacks, S.: Modern Industrial Statistics. Duxbury Press, San Francisco (1998)
2. Shimada, K., Hirasawa, K., Hu, J.: Association Rule Mining with Chi-Squared Test Using Alternate Genetic Network Programming. In: Perner, P. (ed.) ICDM 2006. LNCS (LNAI), vol. 4065, pp. 202–216. Springer, Heidelberg (2006)

3. Kenett, R.: On an Exploratory Analysis of Contingency Tables. The Statistician 32, 395–403 (1983)
4. Fisher, R.A.: The Genetical Theory of Natural Selection. Clarendon, Oxford (1930)
5. Lewontin, R., Kojima, K.: The evolutionary dynamics of complex polymorphisms. Evolution 14, 458–472 (1960)
6. Karlin, S., Feldman, M.: Linkage and selection: Two locus symmetric viability model. Theoretical Population Biology 1, 39–71 (1970)
7. Karlin, S., Kenett, R.: Variable Spatial Selection with Two Stages of Migration and Comparisons Between Different Timings. Theoretical Population Biology 11, 386–409 (1977)
8. Agrawal, R., Imielienski, T., Swami, A.: Mining Association Rules between Sets of Items in Large Databases. In: Proc. Conf. on Management of Data, pp. 207–216. ACM Press, New York (1993)
9. Piatetsky-Shapiro, G.: Discovery, analysis, and presentation of strong rules. In: Knowledge Discovery in Databases, pp. 229–248 (1991)
10. Hahsler, M., Gr¨un, B., Hornik, K.: arules – A computational environment for mining association rules and frequent item sets. Journal of Statistical Software 14(15), 1–25 (2005), http://www.jstatsoft.org/v14/i15/
11. Omiecinski, E.: Alternative interest measures for mining associations in databases. IEEE Transactions on Knowledge and Data Engineering 15(1), 57–69 (2003)
12. Xiong, H., Tan, P.-N., Kumar, V.: Mining strong affinity association patterns in data sets with skewed support distribution. In: Goethals, B., Zaki, M.J. (eds.) Proceedings of the IEEE International Conference on Data Mining, Melbourne, Florida, November 19–22, pp. 387–394 (2003)
13. Liu, B., Hsu, W., Ma, Y.: Pruning and summarizing the discovered associations. In: KDD 1999: Proceedings of the fifth ACM SIGKDD international conference on Knowledge discovery and data mining, pp. 125–134. ACM Press, New York (1999)
14. Brin, S., Motwani, R., Ullman, J., Tsur, S.: Dynamic itemset counting and implication rules for market basket data. In: SIGMOD 1997, Proceedings ACM SIGMOD International Conference on Management of Data, Tucson, Arizona, USA, pp. 255–264 (1997)
15. Hahsler, M., Hornik, K., Reutterer, T.: Implications of probabilistic data modeling for mining association rules. In: Spiliopoulou, M., Kruse, R., Borgelt, C., Nuernberger, A., Gaul, W. (eds.) From Data and Information Analysis to Knowledge Engineering, Studies in Classification, Data Analysis, and Knowledge Organization, pp. 598–605. Springer, Heidelberg (2006)
16. Bayardo, R., Agrawal, R., Gunopulos, D.: Constraint-based rule mining in large, dense databases. Data Mining and Knowledge Discovery 4(2/3), 217–240 (2000)
17. Tan, P.-N., Kumar, V., Srivastava, J.: Selecting the right objective measure for association analysis. Information Systems 29(4), 293–313 (2004)
18. Roever, C., Raabe, N., Luebke, K., Ligges, U., Szepannek, G., Zentgraf, M. (2008), http://www.statistik.tu-dortmund.de
19. Borgelt, C.: Apriori – Finding Association Rules/Hyperedges with the Apriori Algorithm. Working Group Neural Networks and Fuzzy Systems, Otto-von-Guericke-University of Magdeburg, Universit ¨atsplatz 2, D-39106 Magdeburg, Germany (2004), http://fuzzy.cs.uni-magdeburg.de/~borgelt/apriori.html

Weighted Association Rule Mining from Binary and Fuzzy Data

M. Sulaiman Khan[1], Maybin Muyeba[1], and Frans Coenen[2]

[1] School of Computing, Liverpool Hope University, Liverpool, L16 9JD, UK
[2] Department of Computer Science, University of Liverpool, Liverpool, L69 3BX, UK
khanm@hope.ac.uk, muyebam@hope.ac.uk, frans@csc.liv.ac.uk

Abstract. A novel approach is presented for mining weighted association rules (ARs) from binary and fuzzy data. We address the issue of invalidation of downward closure property (DCP) in weighted association rule mining where each item is assigned a weight according to its significance w.r.t some user defined criteria. Most works on weighted association rule mining so far struggle with invalid downward closure property and some assumptions are made to validate the property. We generalize the weighted association rule mining problem for databases with binary and quantitative attributes with weighted settings. Our methodology follows an Apriori approach [9] and avoids pre and post processing as opposed to most weighted association rule mining algorithms, thus eliminating the extra steps during rules generation. The paper concludes with experimental results and discussion on evaluating the proposed approach.

Keywords: Association rules, fuzzy, weighted attributes, apriori, downward closure.

1 Introduction

Association rules (ARs) [11] are a popular data mining technique used to discover behaviour from market basket data. The technique tries to generate association rules (with strong support and high confidence) in large databases. Classical Association Rule Mining (ARM) deals with the relationships among the items present in transactional databases [9, 10]. Typically, the algorithm first generates all large (frequent) itemsets (attribute sets) from which association rule (AR) sets are derived. A large itemset is defined as one that occurs more frequently in the given data set than a user supplied support threshold. To limit the number of ARs generated, a careful selection of the support and confidence thresholds is done. By so doing, care must be taken to ensure that itemsets with low support but from which high confidence rules may be generated are not omitted.

Given a set of items $I = \{i_1, i_2, \dots i_m\}$ and a database of transactions $D = \{t_1, t_2, \dots t_n\}$ where $t_i = \{I_{i1}, I_{i2}, \dots I_{ip}\}$, $p \leq m$ and $I_{ij} \in I$, if $X \subseteq I$ with k = |X| is called a k-itemset or simply an itemset. Let a database D be a multi-set

P. Perner (Ed.): ICDM 2008, LNAI 5077, pp. 200–212, 2008.
© Springer-Verlag Berlin Heidelberg 2008

of subsets of I as shown. Each $T \in D$ supports an itemset $X \subseteq I$ if $X \subseteq T$ holds. An association rule is an expression $X \to Y$, where X, Y are itemsets and $X \cap Y = \varnothing$ holds. The number of transactions T supporting an item X w.r.t D is called support of X, $Supp(X) = |\{T \in D \mid X \subseteq T\}| / |D|$. The strength or confidence (c) for an association rule $X \to Y$ is the ratio of the number of transactions that contain $X \cup Y$ to the number of transactions that contain X, $Conf(X \to Y) = Supp(X \cup Y) / Supp(X)$. For non-binary items, fuzzy association rule mining was proposed using fuzzy sets such that quantitative and categorical attributes can be handled [12]. A fuzzy quantitative rule represents each item as (item, value) pair. Fuzzy association rules are expressed in the following form:

If X is A satisfies Y is B

For example,

if (age is young) ➜ (salary is low)

Given a database T, attributes I with itemsets $X \subset I, Y \subset I$ and $X = \{x_1, x_2, ... x_n\}$ and $Y = \{y_1, y_2, ... y_n\}$ and $X \cap Y = \varnothing$, we can define fuzzy sets $A = \{fx_1, fx_2, ..., fx_n\}$ and $B = \{fx_1, fx_2, ..., fx_n\}$ associated to X and Y respectively. For example (X, A) could be *(age, young), (age, old), (salary, high)* etc. The semantics of the rule is that when the antecedent "X is A" is satisfied, we can imply that "Y is B" is also satisfied, which means there are sufficient records that contribute their votes to the attribute fuzzy set pairs and the sum of these votes is greater than the user specified threshold.

However, classical ARM framework assumes that all items have the same significance or importance i.e. their weight within a transaction or record is the same (weight=1) which is not always the case. For example, from table 1, the rule [printer ➜ computer, 50%] may be more important than [scanner ➜ computer, 75%] even though the former holds a lower support because those items in the first rule usually come with more profit per unit sale. In contrast, standard ARM simply ignores this difference.

The main challenge in weighted ARM is validating the "downward closure property (DCP)" which is crucial for the efficient iterative process of generating and pruning frequent itemsets from subsets.

Table 1. Weighted items database				
ID	**Item**	**Profit**	**Weight**	...
1	Scanner	10	0.1	...
2	Printer	30	0.3	...
3	Monitor	60	0.6	...
4	Computer	90	0.9	...

Table 2. Transactions	
TID	**Items**
1	1,2,4
2	2,3
3	1,2,3,4
4	1,3,4

In this paper we address the issue of DCP in Weighted ARM. We generalize and solve the problem of downward closure property for databases with binary and quantitative items and evaluate the proposed approach with experimental results.

The paper is organised as follows: section 2 presents background and related work; section 3 gives problem definition for weighted ARM with binary and fuzzy data and details weighted downward closure property; section 4 gives frameworks comparison; section 5 reviews experimental evaluation and section 8 concludes paper.

2 Background and Related Work

In literature on association rule mining, weights of items are mostly treated as equally important i.e. weight one (1) is assigned to each item until recently where some approaches generalize this and give items weights to reflect their significance to the user [4]. The weights may be attributed to occasional promotions of such items or their profitability etc. There are two approaches for analyzing data sets with weighted settings: pre- and post-processing. Post processing handles firstly the non-weighted problem (weights=1) and then perform the pruning process later. Pre-processing prunes the non-frequent itemsets after each iteration using weights. The issue in post-processing weighted ARM is that first; items are scanned without considering their weights and later, the rule base is checked for frequent weighted ARs. By doing this, we end up with a very limited itemset pool to check weighted ARs and potentially missing many itemsets.

In pre-processed classical ARM, itemsets are pruned by checking frequent ones against weighted support after every scan. This results in less rules being produced as compared to post processing because many potential frequent super sets are missed. In [2] a post-processing model is proposed. Two algorithms were proposed to mine itemsets with normalized and un-normalized weights. The k-support bound metric was used to ensure validity of the DCP but still there is no guarantee that every subset of a frequent set will be frequent unless the k-support bound value of (k-1) subsets was higher than (k).

An efficient mining methodology for Weighted Association Rules (WAR) is proposed in [3]. A numerical attribute was assigned for each item where the weight of the item was defined as part of a particular weight domain. For example, soda[4,6] → snack[3,5] means that if a customer purchases soda in the quantity between 4 and 6 bottles, he is likely to purchase 3 to 5 bags of snacks. WAR uses a post-processing approach by deriving the maximum weighted rules from frequent itemsets. Post WAR doesn't interfere with the process of generating frequent itemsets but focuses on how weighted AR's can be generated by examining weighting factors of items included in generated frequent itemsets.

Similar techniques for weighted fuzzy quantitative association rule mining are presented in [5, 7, 8]. In [6], a two-fold pre processing approach is used where firstly, quantitative attributes are discretised into different fuzzy linguistic intervals and weights assigned to each linguistic label. A mining algorithm is applied then on the resulting dataset by applying two support measures for normalized and un-normalized cases. The closure property is addressed by using the z-potential frequent subset for each candidate set. An arithmetic mean is used to find the possibility of frequent k+1itemset, which is not guaranteed to validate the valid downward closure property.

Another significance framework that handles the downward closure property (DCP) problem is proposed in [1]. Weighting spaces were introduced as inner-transaction space, item space and transaction space, in which items can be weighted depending on different scenarios and mining focus. However, support is calculated by only considering the transactions that contribute to the itemset. Further, no discussions were made on interestingness issue of the rules produced.

In this paper we present an approach to mine weighted binary and quantitative data (by fuzzy means) to address the issue of invalidation of DCP. We then show that using the proposed technique, rules can be generated efficiently with a valid DCP without any biases found in pre- or post-processing approaches.

3 Problem Definition

The problem definition consists of terms and basic concepts to define item's weight, itemset transaction weight, weighted support and weighted confidence for both binary (boolean attributes) and fuzzy (quantitative attributes) data. Technique for binary data is termed as Binary Weighted Association Rule Mining (BWARM) and technique for fuzzy data is termed as Fuzzy Weighted Association Rule mining (FWARM). Interested readers can see [14] for the formal definitions and more details.

3.1 Binary Weighted Association Rule Mining (BWARM)

Let the input data D have transactions $T = \{t_1, t_2, t_3, \cdots, t_n\}$ with a set of items $I = \{i_1, i_2, i_3, \cdots, i_{|I|}\}$ and a set of positive real number weights $W = \{w_1, w_2, \cdots, w_{|I|}\}$ attached to each item i. Each ith transaction t_i is some subset of I and a weight w is attached to each item $t_i[i_j]$ ("j^{th}" item in the "i^{th}" transaction).

Thus each item i_j will have associated with it a weight corresponding to the set W, i.e. a pair (i, w) is called a weighted item where $i \in I$. Weight for the "j^{th}" item in the "i^{th}" transaction is given by $t_i[i_j[w]]$.

Table 3. Transactional database

T	Items	T	Items
t_1	A B C D	t_6	A B C D E
t_2	B D	t_7	B C E
t_3	A D	t_8	D E
t_4	C	t_9	A C D
t_5	A B D E	t_{10}	B C D E

Table 4. Items with weights

Items i	Weights (IW)
A	0.60
B	0.90
C	0.30
D	0.10
E	0.20

We illustrate the terms and concepts using tables 3 and 4. Table 3 contains 10 transactions for 5 items. Table 4 has corresponding weights associated to each item i in T. We use sum of votes for each itemset by aggregating weights per item as a standard approach.

Item Weight IW is a non-negative real value given to each item i_j ranging $[0..1]$ with some degree of importance, a weight $i_j[w]$.

Itemset Transaction Weight ITW is the aggregated weight of all the items in the itemset present in a single transaction. Itemset transaction weight for an itemset X can calculated as:

$$\text{vote for } t_i \text{ satisfying } X = \prod_{k=1}^{|X|} (v_{[i[w]] \in X}) t_i[i_k[w]] \tag{1}$$

Itemset transaction weight of itemset (A, B) is calculated as: $ITW(A, B) = 0.6 \times 0.9 = 0.54$.

Weighted Support WS is the aggregated sum of itemset transaction weight ITW of all the transactions in which itemset is present, divided by the total number of transactions. It is calculated as:

$$WS(X) = \frac{\sum_{i=1}^{n} \prod_{k=1}^{|X|} (v_{[i[w]] \in X}) t_i[i_k[w]]}{n} \tag{2}$$

WS of itemset (A, B) is calculated as: $\dfrac{1.62}{10} = 0.16$

Weighted Confidence WC is the ratio of sum of votes satisfying both $X \cup Y$ to the sum of votes satisfying X. It is formulated (with $Z = X \cup Y$) as:

$$WC(X \to Y) = \frac{WS(Z)}{WS(X)} = \sum_{i=1}^{n} \frac{\prod_{k=1}^{|Z|} (v_{[z[w]] \in Z}) t_i[z_k[w]]}{\prod_{k=1}^{|X|} (v_{[i[w]] \in X}) t_i[x_k[w]]} \tag{3}$$

Weighted Confidence of itemset (A, B) is calculated as: $WC(A, B) = \dfrac{0.16}{0.30} = 0.54$

3.2 Fuzzy Weighted Association Rule Mining (FWARM)

A fuzzy dataset D' consists of fuzzy transactions $T' = \{t_1', t_2', t_3', ..., t_n'\}$ with fuzzy sets associated with each item in $I = \{i_1, i_2, i_3, \cdots, i_{|I|}\}$, which is identified by a set of

Table 5. Fuzzy transactional database

TID	X		Y	
	Small	**Medium**	**Small**	**Medium**
t_1	0.5	0.5	0.2	0.8
t_2	0.9	0.1	0.4	0.6
t_3	1.0	0.0	0.1	0.9
t_4	0.3	0.7	0.5	0.5

Table 6. Fuzzy items with weights

Fuzzy Items $i[l]$	Weights (IW)
(X, Small)	0.9
(X, Medium)	0.7
(Y, Small)	0.5
(Y, Medium)	0.3

linguistic *labels* $L = \{l_1, l_2, l_3, ..., l_{|L|}\}$ (for example $L = \{small, medium, l \arg e\}$).

We assign a weight w to each l in L associated with i. Each attribute $t'_i[i_j]$ is associated (to some degree) with several fuzzy sets. The degree of association is given by a *membership degree* in the range [0..1], which indicates the correspondence between the value of a given $t'_i[i_j]$ and the set of *fuzzy linguistic labels*. The " k^{th} " weighted fuzzy set for the " j^{th} " item in the " i^{th} " fuzzy transaction is given by $t'_i[i_j[l_k[w]]]$.

We illustrate the fuzzy weighted ARM definition terms and concepts using tables 5 and 6. Table 5 contains transactions for 2 quantitative items discretised into two overlapped intervals with fuzzy values. Table 6 has corresponding weights associated to each fuzzy item *i[l]* in *T*.

Fuzzy Item Weight FIW is a value attached with each fuzzy set. It is a non-negative real number value in $[0..1]$ wrt some degree of importance (table 6). Weight of a fuzzy set for an item i_j is denoted as $i_j[l_k[w]]$.

Fuzzy Itemset Transaction Weight FITW is the aggregated weights of all the fuzzy sets associated with items in the itemset present in a single transaction. Fuzzy Itemset transaction weight for an itemset *(X, A)* can be calculated as:

$$\text{vote for} t'_i \text{ satisfying } X = \prod_{k=1}^{|A|} {}_{(\forall [i[l[w]]] \in X)} t'_i[i_j[l_k[w]]] \tag{4}$$

Let's take an example of itemset <(X, Medium), (Y, Small)> denoted by (X, Medium) as *A* and (Y, Small) as *B*. Fuzzy Itemset transaction weight *FITW* of itemset (A, B) in transaction 1 is calculated as:

$$FITW(A, B) = (0.5 \times 0.7) \times (0.2 \times 0.5) = (0.35) \times (0.1) = 0.035$$

Fuzzy Weighted Support FWS is the aggregated sum of *FITW* of all the transaction's itemsets present divided by the total number of transactions, represented as:

$$FWS(X) = \frac{\sum_{i=1}^{n} \prod_{k=1}^{|A|} {}_{(\forall [i[l[w]]] \in X)} t'_i[i_j[l_k[w]]]}{n} \tag{5}$$

FWS of itemset (A, B) is calculated as: $FWS(A,B) = \dfrac{0.172}{4} = 0.043$

Fuzzy Weighted Confidence FWC is the ratio of sum of votes satisfying both $X \cup Y$ to the sum of votes satisfying X with $Z = X \cup Y$ and given as:

$$FWC(X \to Y) = \frac{FWS(Z)}{FWS(X)} = \sum_{i=1}^{n} \frac{\prod_{k=1}^{|Z|} (\forall [z[w]] \in Z) \, \ell_i^f [z_k[w]]}{\prod_{k=1}^{|X|} (\forall [i[w]] \in X) \, \ell_i^f [x_k[w]]} \qquad (6)$$

FWC of itemset (A, B) is calculated as: $FWC(A,B) = \dfrac{0.043}{0.227} = 0.19$

3.3 Weighted Downward Closure Property (DCP)

In classical ARM algorithm, it is assumed that if the itemset is large, then all its subsets should be large, a principle called downward closure property (DCP). For example, in classical ARM using DCP, it states that if AB and BC are not frequent, then ABC and BCD cannot be frequent, consequently their supersets are of no value.as they will contain non-frequent itemsets. This helps algorithm to generate large itemsets of increasing size by adding items to itemsets that are already large. In the weighted ARM where each item is given a weight, the DCP does not hold in a straightforward manner. Because of the weighted support, an itemset may be large even though some of its subsets are not large and we illustrate this in table 7.

In table 7, all frequent itemsets are generated using 30% support threshold. In column two, itemset {ACD} and {BDE} are frequent with support 30% and 30% respectively. And all of their subsets {AC}, {AD}, {CD} and {BD}, {BE}, {DE} are frequent as well. But in column 3 with weighted settings, itemsets {AC}, {CD} and {DE} are no longer frequent and thus violates the DCP.

We argue that the DCP with binary and quantitative data can be validated using the proposed approach. We prove this by showing that if an itemset is not frequent, then its superset cannot be frequent and $WS(subset) \geq WS(sueprset)$ is always true (see table 7, column 4, Proposed Weighted ARM, only the itemsets are frequent with frequent subsets). A formal proof of the weighted DCP can be found in [14].

4 Frameworks Comparison

In this section, we give a comparative analysis of frequent itemset generation between classical ARM, weighted ARM and the proposed binary and fuzzy ARM frameworks. In table 7 all the possible itemsets are generated using tables 3 and 4 (i.e. 31 itemsets from 5 items), and the frequent itemsets generated using classical ARM (column 2), weighted ARM (column 3) and proposed weighted ARM framework (column 4). Column 1 in table 7 shows itemset's ids.

Table 7. Frequent itemsets comparison

ID	Classical ARM	Classical Weighted ARM	Proposed Weighted ARM
1.	A (50%)	A (30%)	A (0.300)
2.	A→B (30%)	A→B (45%)	A→B (0.162)
3.	A→B→C (20%)	A→B→C (36%)	A→B→C (0.032)
4.	A→B→C→D (20%)	A→B→C→D (38%)	A→B→C→D (0.003)
5.	A→B→C→D→E (10%)	A→B→C→D→E (21%)	A→B→C→D→E (0.000)
6.	A→B→C→E (10%)	A→B→C→E (20%)	A→B→C→E (0.003)
7.	A→B→D (30%)	A→B→D (48%)	A→B→D (0.016)
8.	A→B→D→E (20%)	A→B→D→E (36%)	A→B→D→E (0.002)
9.	A→B→E (20%)	A→B→E (34%)	A→B→E (0.022)
10.	A→C (30%)	A→C (27%)	A→C (0.054)
11.	A→C→D (30%)	A→C→D (30%)	A→C→D (0.005)
12.	A→C→D→E (10%)	A→C→D→E (12%)	A→C→D→E (0.000)
13.	A→C→E (10%)	A→C→E (11%)	A→C→E (0.004)
14.	A→D (50%)	A→D (35%)	A→D (0.030)
15.	A→D→E (20%)	A→D→E (18%)	A→D→E (0.002)
16.	A→E (20%)	A→E (16%)	A→E (0.024)
17.	B (60%)	B (54%)	B (0.540)
18.	B→C (40%)	B→C (48%)	B→C (0.108)
19.	B→C→D (30%)	B→C→D (39%)	B→C→D (0.008)
20.	B→C→D→E (20%)	B→C→D→E (30%)	B→C→D→E (0.001)
21.	B→C→E (30%)	B→C→E (42%)	B→C→E (0.016)
22.	B→D (50%)	B→D (50%)	B→D (0.045)
23.	B→D→E (30%)	B→D→E (36%)	B→D→E (0.005)
24.	B→E (40%)	B→E (44%)	B→E (0.072)
25.	C (60%)	C (18%)	C (0.180)
26.	C→D (40%)	C→D (16%)	C→D (0.012)
27.	C→D→E (20%)	C→D→E (12%)	C→D→E (0.001)
28.	C→E (30%)	C→E (15%)	C→E (0.018)
29.	D (80%)	D (8%)	D (0.080)
30.	D→E (40%)	D→E (12%)	D→E (0.008)
31.	E (50%)	E (10%)	E (0.100)

A support threshold for classical ARM is set to 30% and for classical WARM and proposed Weighted ARM it is set to 0.3 and 0.03 respectively. Itemsets with a highlighted background indicate frequent itemsets. This experiment is conducted in order to illustrate the effect of item's occurrences and their weights on the generated rules.

Frequent itemsets in column 3 are generated using classical weighted ARM pre-processing technique. In this process all the frequent itemsets are generated first with count support and then those frequent itemsets are pruned using their weights. In this case only itemsets are generated from the itemset pool that is already frequent using their count support. Itemsets with shaded background and white text are those that WARM does not consider because they are not frequent using count support. But with weighted settings they may be frequent due to significance associated with them. Also, the generated itemsets do not hold DCP as described in sect. 3.2.

In column 4 frequent itemsets are generated using the proposed weighted ARM framework. It is noted that the itemsets generated are mostly frequent using count support technique and interestingly included fewer rules like {AB→C} that is not frequent, which shows that the non-frequent itemsets can be frequent with weighted settings due to their significance in the data set even if they are not frequent using count support.

In column 4, itemsets {A→B} and {B→C} are frequent due to high weight and support count in transactions. It is interesting to have a rule {B→D} because D has very low weight (0.1) but it has the highest count support i.e. 80% and it appears more with item B than any other item i.e. with 50% support. Another aspect to note is that, B is highly significant (0.9) with high support count (60%). These kinds of rules can be helpful in "Cross-Marketing" and "Loss Leader Analysis" in real life applications.

Also the itemsets generated using our approach holds valid DCP as shown in section 3.3. Table 7 gives a concrete example of our approach and we now perform experiments based on this analysis.

5   Experimental Evaluation

To demonstrate the effectiveness of the approach, we performed several experiments using a real retail data set [13]. The data is a binary transactional database containing 88,163 records and 16,470 unique items. Weights were generated randomly and assigned to all items in the dataset to show their significance.

Experiments were undertaken using four different association rule mining techniques. Four algorithms were used for each approach, namely Binary Weighted ARM (BWARM), Fuzzy Weighted ARM (FWARM), standard ARM as Classical ARM and WARM as post processing weighted ARM algorithm.

The BWARM and FWARM algorithms belongs to *breadth first traversal* family of ARM algorithms, uses tree data structures and works in fashion similar to the Apriori algorithm [9]. Both algorithms consist of several steps. For more details on algorithms and pseudo code, refer to [14].

In this paper, an improvement from [14] is that we have used a real dataset in order to demonstrate performance of the proposed approach. We performed two types of experiments based on quality and performance measures. For quality measures, we compared the number of frequent itemsets and the interesting rules generated using four algorithms described above. In the second experiment, we showed the scalability of the proposed BWARM and FWARM algorithms by comparing the execution time with varying user specified support thresholds.

5.1   Quality Measures

For quality measures, the binary retail dataset described above was used. Each item is assigned a weight range between [0..1] according to their significance in the dataset. For fuzzy attributes we used approach described in [15] to obtain fuzzy dataset. With fuzzy dataset each attribute is divided into five different fuzzy sets.

In figure 1, the x-axis shows support thresholds from 2% to 10% and on the y-axis the number of frequent itemsets. Four algorithms are compared, BWARM (Binary Weighted ARM) algorithm using weighted binary datasets; FWARM (Fuzzy Weighted ARM) algorithm using fuzzy attributes and weighted fuzzy linguistic values; Classical ARM using standard ARM with binary dataset and WARM using

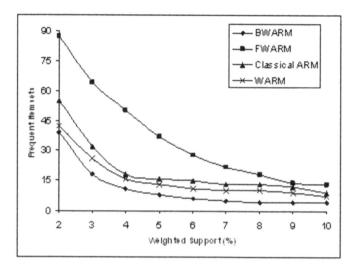

Fig. 1. No. of frequent Itemsets generated using user specified support threshold

weighted binary datasets and applying a post processing approach. Note that the weight of each item in classical ARM is 1 i.e. all items have equal weight.

The results show quite similar behavior of the three algorithms to classical ARM. As expected the number of frequent itemsets increases as the minimum support decreases in all cases. Number of frequent itemsets generated using the WARM algorithm are always less than the number of frequent itemsets generated by classical ARM because WARM uses only generated frequent itemsets in the same manner as classical ARM. This generates less frequent itemsets and misses many potential ones.

We do not use classical ARM approach to first find frequent itemsets and then re-prune them using weighted support measures. Instead all the potential itemsets are considered from the beginning for pruning using Apriori approach [9] in order to validate the DCP. Results of proposed BWARM approach are better than WARM because less arguably better frequent itemsets and rules are generated as we consider both itemset weights and their support count. Moreover, BWARM, classical ARM and WARM utilise binary data. FWARM generates more rules because of the extended fuzzy attributes, and it considers degree of membership instead of attribute presence only (count support) in a transaction.

Figure 2 shows the number of interesting rules generated using confidence measures. In all cases, the number of interesting rules is less because the interestingness measure generates fewer rules.

FWARM produces more rules due to the high number of initially generated frequent itemsets due to the introduction of more fuzzy sets for each quantitative attribute. Given a high confidence, BWARM outperforms classical WARM because the number of interesting rules produced is greater than WARM. This is because BWARM generates rules with items more correlated to each other and consistent at a higher confidence unlike WARM, where rules keep decreasing even at high confidence.

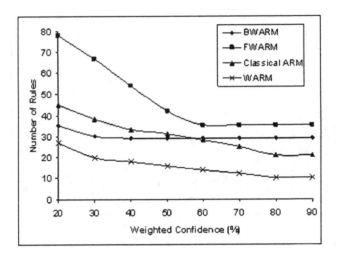

Fig. 2. No. of Interesting Rules generated using user specified confidence

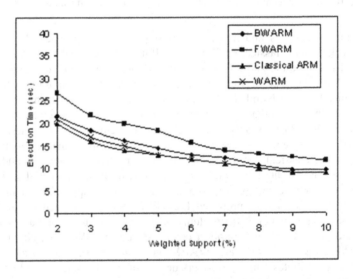

Fig. 3. Performance measures: varying weighted support (WS) threshold

The experiments show that the proposed framework produces better results as it uses all the possible itemsets and generates rules effectively using valid DCP. Further, the novelty is the ability to analyse both binary and fuzzy datasets with weighted settings.

5.2 Performance Measures

Experiment two compares the execution time of BWARM and FWARM algorithms with classical Apriori ARM and WARM algorithms. We investigated the effect on execution time caused by varying the weighted support threshold with fixed data size (number of records). In figure 3, a support threshold from 2% to 10% is used again.

FWARM has comparatively higher execution time due to the fact that it deals with fuzzy data as mentioned earlier. Classical ARM and WARM have almost similar timings as WARM initially uses classical ARM approach and uses already generated frequent sets for post processing. Results show that BWARM has almost similar execution time to WARM. The minor difference is due to the way it generates frequent sets i.e. it considers items weights and their count support. Similarly from figure 3, it can be noted that BWARM and FWARM algorithms scale linearly with increasing weighted support threshold, which is similar behavior to Classical ARM.

6 Conclusion

We have presented a generalised approach for mining weighted association rules from databases with binary and quantitative (fuzzy) attributes. A classical model of binary and fuzzy association rule mining is adopted to address the issue of invalidation of downward closure property (DCP) in weighted association rule mining. The problem of invalidation of the DCP is solved using an improved model. We used classical and weighted ARM examples to compare support and confidence measures and evaluated the effectiveness of the proposed approach experimentally. We have demonstrated the valid DCP with formal comparisons with classical weighted ARM. It is notable that the approach presented here is effective in analysing databases with binary and fuzzy (quantitative) attributes with weighted settings.

References

1. Tao, F., Murtagh, F., Farid, M.: Weighted Association Rule Mining Using Weighted Support and Significance Framework. In: Proceedings of 9th ACM SIGKDD Conference on Knowledge Discovery and Data Mining, Washington, DC, pp. 661–666 (2003)
2. Cai, C.H., Fu, A.W.-C., Cheng, C.H., Kwong, W.W.: Mining Association Rules with Weighted Items. In: Proceedings of Intl. Database Engineering and Applications Symposium (IDEAS 1998), Cardiff, Wales, UK, July 1998, pp. 68–77 (1998)
3. Wang, W., Yang, J., Yu, P.S.: Efficient Mining of Weighted Association Rules (WAR). In: Proceedings of the KDD, Boston, August, pp. 270–274 (2000)
4. Lu, S., Hu, H., Li, F.: Mining Weighted Association Rules. Intelligent data Analysis Journal 5(3), 211–255 (2001)
5. Wang, B.-Y., Zhang, S.-M.: A Mining Algorithm for Fuzzy Weighted Association Rules. In: IEEE Conference on Machine Learning and Cybernetics, vol. 4, pp. 2495–2499 (2003)
6. Gyenesei, A.: Mining Weighted Association Rules for Fuzzy Quantitative Items. In: Proceedings of PKDD Conference, pp. 416–423 (2000)
7. Shu, Y.J., Tsang, E., Yeung, D.S.: Mining Fuzzy Association Rules with Weighted Items. In: IEEE International Conference on Systems, Man, and Cybernetics (2000)
8. Lu, J.-J.: Mining Boolean and General Fuzzy Weighted Association Rules in Databases. Systems Engineering-Theory & Practice 2, 28–32 (2002)
9. Agrawal, R., Srikant, R.: Fast Algorithms for Mining Association Rules. In: 20th VLDB Conference, pp. 487–499 (1994)

10. Bodon, F.: A Fast Apriori implementation. In: ICDM Workshop on Frequent Itemset Mining Implementations, Melbourne, Florida, USA, vol. 90 (2003)
11. Agrawal, R., Imielinski, T., Swami, A.: Mining Association Rules Between Sets of Items in Large Databases. In: 12th ACM SIGMOD on Management of Data, pp. 207–216 (1993)
12. Kuok, C.M., Fu, A., Wong, M.H.: Mining Fuzzy Association Rules in Databases. SIGMOD Record 27(1), 41–46 (1998)
13. Brijs, T., Swinnen, G., Vanhoof, K., Wets, G.: The use of association rules for product assortment decisions: a case study. In: Proceedings of the Fifth International Conference on Knowledge Discovery and Data Mining, San Diego, pp. 254–260 (1999)
14. Sulaiman Khan, M., Muyeba, M., Coenen, F.: Fuzzy Weighted Association Rule Mining with Weighted Support and Confidence Framework. In: Proc. of ALSIP Workshop (PAKDD), Osaka, Japan (to appear, 2008)
15. Sulaiman Khan, M., Muyeba, M., Coenen, F.: On Extraction of Nutritional Patterns (NPS) Using Fuzzy Association Rule Mining. In: Proc. of Intl. Conference on Health Informatics (HEALTHINF 2008), Madeira, Portugal, vol. 1, pp. 34–42. INSTICC press (2008)

A Comparative Impact Study of Attribute Selection Techniques on Naïve Bayes Spam Filters

J.R. Méndez[1], I. Cid[1], D. Glez-Peña[1], M. Rocha[2,*], and F. Fdez-Riverola[1]

[1] Dept. Informática, University of Vigo, Escuela Superior de Ingeniería Informática
Edificio Politécnico, Campus Universitario As Lagoas s/n, 32004, Ourense, Spain
{moncho.mendez,icgomez,dgpena,riverola}@uvigo.es
[2] Dept. Informática, University of Minho, Centro de Ciências e Tecnologias da Computação.
Campus de Gualtar, 4710-057, Braga, Portugal
mrocha@di.uminho.pt

Abstract. The main problem of Internet e-mail service is the massive spam message delivery. Everyday, millions of unwanted and unhelpful messages are received by Internet users annoying their mailboxes. Fortunately, nowadays there are different kinds of filters able to automatically identify and delete most of these messages. In order to reduce the bulk of information to deal with, only distinctive attributes are selected spam and legitimate e-mails. This work presents a comparative study about the performance of five well-known feature selection techniques when they are applied in conjunction with four different types of Naïve Bayes classifier. The results obtained from the experiments carried out show the relevance of choosing an appropriate feature selection technique in order to obtain the most accurate results.

1 Introduction and Motivation

During the last years, Internet has become an extremely important communication and information exchange platform. In this context, relevant authorities have introduced some strategic plans in order to promote the usage of Internet and the newest communication technologies. Included into this plans, we can cite i2010 [1], an European strategic programme designed to boost economical and social evolution based on a new society of knowledge.

For the development of these plans several socioeconomic aspects, communication infrastructures and educational level for the information society have been evaluated. Nevertheless, although the troubles caused by the delivery of spam messages are upsetting Internet users, these institutional programmes have not taken this fact into consideration. We believe that e-mail service and electronic delivery of instant messages are one key matter for the evolution of the information society. In this context, there is a compelling need for increasing the classification accuracy of existing filters in order to automatically detect and drop spam messages.

---

* The work of Miguel Rocha is supported by the Portuguese FCT project PTDC/EIA/ 64541/2006.

P. Perner (Ed.): ICDM 2008, LNAI 5077, pp. 213–227, 2008.

Nowadays, Internet e-mail service is generally used by the vast majority of the Internet users [2]. This fact ensures the efficacy of the advertising messages sent through Internet e-mail [3]. Spam identification is a difficult task because spammers (spam message senders) use tricks in order to avoid spam filters and ensure their deliveries. Moreover, misclassification errors of anti-spam filters present different cost values. False positive errors (FP, legitimate messages classified as spam) are unacceptable for a great amount of Internet users. From a complementary point of view, the presence of a false negative error (FN, spam message classified as legitimate) is less harmful than a false positive misclassification.

Generally speaking, there are two kinds of spam filtering techniques: (i) collaborative systems and (ii) content-based approaches. The former are based on sharing identifying information about spam messages within a filtering community, while the later are based on a deep analysis of the message content in order to identify its class (usually using machine learning techniques). This work is focused in Naïve Bayes filters, a well-known content-based technique able to accurately combine the probability of finding terms in spam and legitimate messages.

Content-based spam filters analyze the words extracted from the available messages (corpus). Although each term could be a putative feature that should be analyzed for spam filtering, in practice this is not possible because of the huge amount of words extracted from the whole corpus. The usage of large feature vectors with machine learning techniques is not advisable because it can cause the loss of efficiency and accuracy in existing filters [4]. Therefore, several feature selection techniques need to be applied as a pre-processing stage previous to the construction of any spam filtering system. These techniques have been designed to perform dimensionality reduction and their goal is to discard attributes that do not provide essential information for the classification task.

This work presents a comparative study about the impact of five feature selection methods when using with four variants of the original Naïve Bayes algorithm working as spam filter. The feature selection techniques analyzed in this study are the following: (i) Information Gain (IG) (ii) Odds ratio (OR), (iii) Document Frequency (DF) (iv) χ^2 statistic, and (v) Mutual Information (MI). Moreover, we have evaluated the following Naïve Bayes alternatives: (i) Multivariate Bernoulli, (ii) Multinomial Naïve Bayes, (iii) Multivariate Gaussian, and (iv) Flexible Bayes.

The remaining of the paper is structured as follows: Section 2 briefly introduces the four successful Naïve Bayes approaches taken into consideration. Section 3 presents the details of several feature selection techniques applied in the spam filtering domain. The experimental protocol and the empirical evaluation results are showed in Section 4. Finally, Section 5 summarizes the main conclusions extracted from this work and outlines future research lines.

2 Naïve Bayes Alternatives for Spam Filtering

Nowadays, the vast majority of commercial anti-spam filtering tools are based on the usage of Naïve Bayes classifiers [5-7]. This section presents a compilation of different ways of applying this technique for spam filtering.

Naïve Bayes classifiers represent each message, d, as a feature vector in the form $\vec{x} = \{x_1, ..., x_m\}$, where each x_i stands for the value of an attribute containing information about a token (term or word) identified in the target e-mail. Keeping in mind the Bayes theorem [8], the probability of a given message belonging to the class c_j (spam| legitimate) can be computed as shown in Expression (1).

$$p(c_j \mid \vec{x}) = \frac{p(c_j) \cdot p(\vec{x} \mid c_j)}{p(\vec{x})} \tag{1}$$

A Naïve Bayes classifier assigns to each message the class maximizing $p(c_j) \cdot p(\vec{x} \mid c_j)$, where $p(c_j)$ represents the probability of class c and $p(\vec{x} \mid c_j)$ is the probability of finding a vector $\vec{x}$ in category c_j. Therefore, in the spam filtering domain, we can classify a message d as spam when Expression (2) becomes true.

$$\frac{p(c_s) \cdot p(\vec{x} \mid c_s)}{p(c_s) \cdot p(\vec{x} \mid c_s) + p(c_l) \cdot p(\vec{x} \mid c_l)} > T \tag{2}$$

where c_s and c_l represent the spam and legitimate classes respectively, and T is a threshold that stands for the required security level belonging to interval [0, 1]. The most common value for T is 0.5 [8].

As we previously stated, there are several variants of the Naïve Bayes algorithm able to cope with the problem of spam filtering. The main difference between them is the way to compute the probability of finding a message that contains the feature vector $\vec{x}$, considering only those e-mails belonging to class c_j, $p(\vec{x} \mid c_j)$. In this context, we can find the following alternatives: (*i*) Multivariate Bernoulli, (*ii*) Multinomial NB (*iii*) Multivariate Gaussian, and (*iv*) Flexible Bayes.

Multivariate Bernoulli is a variant of Naïve Bayes that represents each message as a feature vector $\vec{x} = \{x_1, ..., x_m\}$, where each element x_i is a boolean attribute representing if the word appears on the target message or not. Moreover, this alternative considers that results are independent for each category [9]. Multivariate Bernoulli computes the probability $p(\vec{x} \mid c_j)$ as shown in Expression (3).

$$p(\vec{x} \mid c_j) = \prod_{i=1}^{m} p(t_i \mid c_j)^{x_i} \cdot \left(1 - p(t_i \mid c_j)\right)^{(1-x_i)} \tag{3}$$

where the probability of finding a term t_i in a given message belonging to class c_j, $p(t_i \mid c_j)$, is calculated using a Laplacean prior as shown in Expression (4).

$$p(t_i \mid c_j) = \frac{1 + M_{t_i, c_j}}{2 + M_{c_j}} \tag{4}$$

where M_{t_i,c_j} is the number of messages belonging to class c_j (spam or legitimate) that contains the term t_i and M_{c_j} stands for the amount of messages in class c_j.

Multinomial NB uses a vector $\vec{x} = \{x_1, ..., x_m\}$ in order to describe each message d, where each element x_i is represented by a boolean attribute that indicates the presence or absence of the term t_i in the message [2]. Moreover, there is an alternative for Multinomial NB based on representing each element x_i as the number of appearances of each term t_i in the message d. Nevertheless, this variant has showed to achieve poor results [8]. In this case, $p(\vec{x} \mid c_j)$ is computed as Expression (5) shows.

$$p(\vec{x} \mid c_j) = p(\mid d \mid) \cdot \mid d \mid! \prod_{i=1}^{m} \frac{p(t_i \mid c_j)^{x_i}}{x_i!} \tag{5}$$

where x_i is the value for the term t_i and $p(t_i, c_j)$ stands for the probability of finding documents containing the term t_i when only messages from class c_j are selected. As $\mid d \mid$ is independent from category c_j, there is no need to calculate the sub-expressions $p(\mid d \mid)$ and $\mid d \mid!$ [10]. Moreover, each probability $p(t_i \mid c_j)$ is estimated by means of a Laplacean prior as Expression (6) shows.

$$p(t_i \mid c_j) = \frac{1 + N_{t_i,c_j}}{m + N_{c_j}} \tag{6}$$

where N_{t_i,c_j} is the number of messages belonging to class c_j that contain the term t_i, and N_{c_j} stands for the amount of messages belonging to class c_j.

Multivariate Gaussian uses continuous attributes representing the frequency of the terms in a given message. It assumes, in comparison with Multinomial NB, that the distribution associated to each term is a Gaussian distribution for each class c_j. Moreover, this variant considers that the values of the attributes are independent in each category [8]. The probability $p(\vec{x} \mid c_j)$ is computed as Expression (7) shows.

$$p(\vec{x} \mid c_j) = \prod_{i=1}^{m} g(x_i; \mu_{i,c_j}; \sigma_{i,c_j}) \tag{7}$$

where μ_{i,c_j} and σ_{i,c_j} represent the mean and the standard deviation of the appearance frequency of the term t_i in messages belonging to class c_j.

Flexible Bayes [11] works in a similar way than Multivariate Gaussian, but the distributions of each attribute x_i are estimated by means of L_i Gaussian distributions representing each different value for the attribute in each class. Therefore, the probability $p(\vec{x} \mid c_j)$ is computed as Expression (8) shows.

$$p\left(\vec{x} \mid c_{j}\right)=\prod_{i}^{m} p\left(x_{i} \mid c_{j}\right)=\prod_{i}^{m} \frac{1}{L_{i}} \cdot \sum_{l=1}^{L_{i}} g\left(x_{i} ; \mu_{l, c_{j}}, \sigma_{l, c_{j}}\right) \tag{8}$$

where L_i represents the number of different values for the attribute x_i in the category c_j, μ_{l,c_j} is the l-th value for attribute x_i in category c_j and σ_{l,c_j} represents the standard deviation of the l-th value for the attribute x_i. In order to simplify the calculation, σ_{l,c_j} is estimated as Expression (9) shows [11].

$$\sigma_{l,c_j} = \frac{1}{\sqrt{M_{c_j}}} \tag{9}$$

where M_{c_j} is the number of messages belonging to category c_j [11].

Despite the simplicity of Naïve Bayes algorithms, the execution of feature reduction techniques is advisable in order to drop unhelpful information gathered from e-mails. The application of these techniques can discard irrelevant features from target problem, reducing computational requirements of the training and test stages. Moreover, some confusing attributes can cause a filter effectiveness decrease. The usage of feature selection techniques can reduce the presence of those unhelpful terms increasing the filter performance. Next section introduces some feature selection techniques commonly used on spam filtering domain.

3 Feature Selection Techniques

There are a lot of available techniques for feature selection purposes [12]. In this work, we have considered the utilization of five well-known feature reduction strategies commonly used on the text mining domain. This section describes the application of the following techniques: (*i*) Information Gain (IG), (*ii*) Odds ratio (OR), (*iii*) Document Frequency (DF), (*iv*) χ^2 statistic and (*v*) Mutual Information (MI).

IG is able to measure the number of bits that were acquired for the classification of spam and legitimate messages from using the presence or absence of a term in a message [12]. If $\{c_j\}_{j=1}^n$ represents the set of categories (spam or legitimate), the IG of a term t_i is computed as Expression (10) shows.

$$IG\left(t_{i}\right)=-\sum_{j=1}^{n} p\left(c_{j}\right) \cdot \log p\left(c_{j}\right)$$

$$+p\left(t_{i}\right) \sum_{j=1}^{n} p\left(c_{j}\right) \cdot \log p\left(c_{j} \mid t_{i}\right) \tag{10}$$

$$+p\left(t_{i}\right) \sum_{j=1}^{n} p\left(c_{j}\right) \cdot \log p\left(c_{j} \mid t_{i}\right)$$

where $p(t_i)$ is the probability of finding a message with the term t_i, and $p(c_j \mid t_i)$ stands for the probability of an e-mail belonging to category c_j and containing the term t_i.

OR is able to find terms commonly included in messages belonging to a certain category. The meaning of this measure can be interpreted as follows: words that appear in both spam and legitimate classes are assigned an OR score near to 1, otherwise, terms which are representative of a certain class present an OR value higher than 1 [3]. OR is computed as Expression (11) shows.

$$OR(t_i, c_j) = \frac{p(t_i \mid c_j) \cdot \left[1 - p(t_i \mid \overline{c}_j) \right]}{\left[1 - p(t_i \mid c_j) \right] \cdot p(t_i \mid \overline{c}_j)} \tag{11}$$

where $p(t_i \mid c_j)$ represents the probability of finding the term t_i in messages belonging to category c_j, and $p(t_i \mid \overline{c}_j)$ stands for the probability of finding the term t_i in e-mails from opposite class to c_j.

The DF measure represents the number of messages where a given term appears. For each term t_i, this method computes the number of e-mails from the training corpus containing it. Then, those terms having lower DF values are discarded [12].

χ^2 statistic is able to test the hypothesis of the independency of two different variables. Using a term t_i and a category c_j, Expression (12) shows the way to compute χ^2 statistic.

$$\chi^2(t_i, c_j) = \frac{N \cdot (A \cdot D - C \cdot B)^2}{(A+C) \cdot (B+D) \cdot (A+B) \cdot (C+D)} \tag{12}$$

where A is the amount of messages belonging to class c_j and containing the term t_i, B stands for the number of e-mails containing the term t_i and belonging to the opposite category of c_j, C represents the number of documents from class c_j that do not contain the term t_i, D is the number of messages that do not contain the term t_i and do not belong to category c_j, and finally, N represents the amount of available e-mails. χ^2 statistic reaches a value of 0 if t_i and c_j are independent. This method usually calculates the measure χ^2 from each term and category [12]. The results achieved for the different categories can be computed as Expression (13) shows.

$$\chi^2_{avg}(t_i) = \sum_{j=1}^{n} p(c_j) \cdot \chi^2(t_i, c_j) \tag{13}$$

MI is a technique for statistic modelling of the language. For a given term t_i and a category c_j, MI score can be computed as Expression (14) indicates.

$$MI(t_i, c_j) = \log \frac{p(t_i \wedge c_j)}{p(t_i) \cdot p(c_j)} \approx \log \frac{A \cdot N}{(A+C) \cdot (A+B)} \tag{14}$$

where A, B, C, D and N are defined as previously mentioned, $p(t_i)$ represents the probability of finding a message with the term t_i, and $p(t_i \wedge c_j)$ stands for the probability of finding an e-mail from category c_j containing the term t_i. Similarly to χ^2 statistic, $MI(t_i, c_s)$ and $MI(t_i, c_l)$ (The MI scores for the term t_i in spam and legitimate classes) can easily be combined as shown in Expression (13).

Once the different Naïve Bayes variants and the selected feature selection methods have been introduced, next section presents the experimental protocol designed for the empirical evaluation carried out, as well as the results achieved during the execution of the experiments.

4 Experimental Setup and Results

This section presents the experimental protocol designed for the evaluation of the different feature selection methods when they are used in conjunction with Naïve Bayes filters. Moreover, we analyze the results achieved during the experimentation carried out. The main goal of this work is the development of a useful guide to select the best feature selection technique for the different variants of Naïve Bayes algorithm. Feature selection strategies can be applied for reducing the computational requirements and to improve the performance of spam filters.

Despite privacy issues associated with legitimate messages, recently some e-mail compilations have been published through Internet. From these corpuses, we highlight LingSpam [13] and SpamAssassin [14]. Given its great amount of messages, we have selected the SpamAssassin corpus for the execution of the different experiments. Moreover, we have used only space characters as token separator and a stopword removal pre-processing technique as recommended in [18]. All experiments have been carried out using feature vectors containing 1000 attributes for representing each message. In order to ensure the quality of the achieved results, we have used a 10 stratified fold-cross validation [19]. Finally, we have assigned a value of 0.5 to the T parameter for all Naïve Bayes classifiers during our experiments.

For comparison purposes, we have used 5 well-known accuracy measures: (*i*) percentage of false positives, false negatives and correct classifications, (*ii*) batting average, (*iii*) recall and precision, (*iv*) F-score and balanced F-score using different β values, and (*v*) Total Cost Ratio using different values for λ parameter. We have also executed a ROC analysis [20-22] in order to achieve better conclusions from the effects of feature selection on Naïve Bayes classifiers.

Figure 1 shows, the results for the false positive (FP), false negative (FN) and correct classifications (OK) achieved by the different Naïve Bayes approaches using the proposed feature selection techniques. The amount of FP errors should be carefully studied in order to prevent the elimination of legitimate messages from the inbox of final users [15].

As we can see from Figure 1(a), the worst method for feature selection when using a Multivariate Bernoulli classifier is OR, while DF produces the highest percentage of correct classifications (90.67%) with a lower amount of FP errors. Although, the MI method generates the lowest FP error rate, the number of errors evidenced by using it is unacceptable.

On the other hand Figure 1(b), where a Multinomial Naïve Bayes is used, OR presents a poor performance (achieving the percentages of 65% of FP errors and 9.43%

220   J.R. Méndez et al.

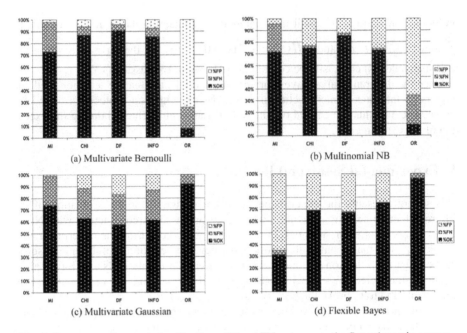

Fig. 1. Percentage of correct classifications, FP and FN errors over the SpamAssassin corpus

of correct classifications). The method with the highest level of correct classifications is DF (85.53%). Nevertheless, this method still exhibits a great amount of FP errors (12%). Finally, although MI presents the lowest FP rate, the percentage of correct classifications is not very good (72%).

In Figure 1(c) results for Mutivariate Gaussian NB are shown, where OR feature selection method behaves better than other techniques achieving a high percentage of correct classifications (92.06%) as well as the lowest amount of FP errors.

Finally, analysing the results from using a Flexible Bayes classifier in Figure 1(d), we can realize that the OR feature selection technique achieves the highest amount of correct classifications (93.43%). Moreover, the use of this method does not introduce FP errors. Therefore, this combination represents a good approach for spam filtering. IG also produces a relatively high performance rate (75.08%) with a low amount of FP errors. Finally, the MI method should be discarded because of its great amount of FP errors (65%).

After a percentage accuracy evaluation we have analyzed precision, recall, F-score and balanced F-score measures for the comparison of the proposed feature selection techniques and anti-spam filters. Recall estimates the filter performance (a high score indicates that the vast majority of spam messages are detected) while precision stands for a secure operation (great score when no FP errors are produced). In order to combine precision and recall measures two new measures have been recently introduced [16]. F-score is equal to 1 when the evaluated classifier does not present errors. Balanced F-score works similarly than F-score but it includes a weighted factor β in order to establish the relevance between precision and recall. It can be computed as Expression (15) shows.

$$f - score_\beta = \frac{(\beta^2 + 1) \cdot precision \cdot recall}{\beta^2 \cdot precision + recall} \tag{15}$$

Following this equation, if β is equal to 1 then F-score gets the same value than balanced F-score. Moreover, if β is greater than 1, precision is considered more important that recall. Otherwise, recall is assumed as the most important criterion [16]. Table 1 presents a precision and recall analysis for the use of different feature selection techniques in combination with Multivariate Bernoulli variant of Naïve Bayes.

A brief look at Table 1 shows that in this case, DF clearly achieves the best scores for recall and precision. Table 2 shows a comparison of the analyzed feature selection techniques with Multinomial Naïve Bayes using the above mentioned measures.

Analyzing Table 2, it is clear that the usage of OR in conjunction with Multinomial NB generates a poor precision (great amount of FP errors). Moreover, the DF method achieves the best value for precision. Finally, χ^2, DF and IG are able to detect a great amount of spam messages (recall is greater).

Table 3 presents the previous measures achieved by using the different feature selection methods with Multivariate Gaussian. Analyzing results from Table 3, it is visible that the OR method presents a high precision rate as well as the highest recall score.

Table 1. Precision, recall, F-score and balanced F-score using Multivariate Bernoulli

	OR	MI	χ^2	DF	IG
Precision	0.098	0.222	0.757	0.824	0.710
Recall	0.303	0.025	0.718	0.807	0.718
F-score	0.109	0.087	0.749	0.821	0.711
balanced F-score ($\beta=2$)	0.143	0.045	0.737	0.815	0.714

Table 2. Precision, recall, F-score and balanced F-score using Multinomial Naïve Bayes

	OR	MI	χ^2	DF	IG
Precision	0.007	0.288	0.503	0.654	0.484
Recall	0.017	0.071	0.924	0.920	0.950
F-score	0.007	0.179	0.554	0.694	0.537
balanced F-score ($\beta=2$)	0.009	0.114	0.652	0.764	0.641

Table 3. Precision, recall, F-score and balanced F-score using Multivariate Gaussian

	OR	MI	χ^2	DF	IG
Precision	1	0.429	0.404	0.006	0.008
Recall	0.689	0.038	0.043	0.004	0.004
F-score	0.917	0.140	0.143	0.006	0.007
balanced F-score ($\beta=2$)	0.816	0.069	0.072	0.005	0.006

Finally, Table 4 shows the evaluation of the same scenario where a Flexible Bayes classifier is used. In this case, OR achieves the highest precision level. This value means that there are no FP errors. Moreover, MI produces a great amount of FP errors. The analysis of F-score measure shows that OR is the most reliable feature selection method for Flexible Bayes.

We have also used batting average and TCR scores for comparison purposes. Batting average measures the ability to detect spam messages (effectiveness) and the capacity of generating a small amount of FP errors (precision). This measure is a pair hit/strike rate, where the former represents the rate of detected spam messages and the later the amount FP error rate [17].

TCR is a performance measure from a cost point of view usually applied to spam filters. It uses a λ parameter that establishes the cost proportion between FP and FN errors. When TCR is computed, a FP is considered λ times more expensive than a FN error [8]. It can be calculated as shown in Expression (16).

$$TCR_{\lambda} = \frac{nspam}{\lambda \cdot fp + fn} \qquad (16)$$

where fp and fn represent the false positive and false negative amount and $nspam$ stands for the total number of spam messages. High TCR scores indicate low global cost while TCR scores under 1 indicate that the usage of the classifier is worse than manually classifying e-mails.

Table 5 presents the TCR and batting average results achieved by using the available feature selection strategies with Multivariate Bernoulli.

As we can see from Table 5, DF, MI and χ^2 methods present the greatest performance in this scenario. Moreover, the DF method obtains the lowest amount of FP errors.

Table 4. Precision, recall, F-score and balanced F-score β=2 using Flexible Bayes

	OR	MI	χ^2	DF	IG
Precision	1	0.251	0.449	0.434	0.504
Recall	0.814	0.857	0.979	0.966	0.979
F-score	0.956	0.293	0.503	0.488	0.558
balanced F-score (β=2)	0.897	0.389	0.616	0.599	0.666

Table 5. TCR and batting average using Multivariate Bernoulli

	OR	MI	χ^2	DF	IG
TCR λ=999	0	0.011	0.004	0.006	0.003
TCR λ=9	0.037	0.565	0.423	0.573	0.341
TCR λ=1	0.276	0.941	1.951	2.736	1.737
Batting average	0.302/1	0.025/0.030	0.718/0.079	0.806/0.058	0.718/0.100

Table 6. TCR and batting average using Multinomial Naïve Bayes

	OR	MI	χ^2	DF	IG
TCR λ=999	0	0.006	0.001	0.002	0
TCR λ=9	0.042	0.397	0.121	0.224	0.110
TCR λ=1	0.282	0.905	1.013	1.763	0.948
Batting average	0.016/0.879	0.071/0.060	0.924/0.312	0.920/0.166	0.953/0.346

Table 7. TCR and batting average using Multivariate Gaussian

	OR	MI	χ^2	DF	IG
TCR λ=999	3.216	0.019	0.002	0.001	0.001
TCR λ=9	3.216	0.706	0.198	0.144	0.178
TCR λ=1	3.216	0.988	0.690	0.604	0.664
Batting average	0.689/0	0.037/0.017	0.004/0.153	0.004/0.225	0.008/0.176

Table 8. TCR and batting average using Flexible Bayes

	OR	MI	χ^2	DF	IG
TCR λ=999	5.368	0	0	0	0.001
TCR λ=9	5.368	0.043	0.092	0.088	0.116
TCR λ=1	5.368	0.371	0.818	0.773	1.026
Batting average	0.813/0	0.857/0.874	0.978/0.411	0.966/0.431	0.983/0.329

Table 6 shows the TCR and batting average scores for the different feature selection methods when Multinomial Naïve Bayes filter is used. As Table 6 shows, DF and χ^2 methods seem to be the most viable feature reduction strategies for use in this scenario.

Table 7 shows the TCR and batting average scores for the analyzed feature selection methods using Multivariate Gaussian classifier. As we can realize from Table 7, OR is clearly the best option for use in this context.

Table 8 shows the results of batting average and TCR for the different feature selection strategies when using Flexible Bayes classifier.

Analyzing the TCR values showed in Table 8, we can see that the best method for this scenario is OR. Moreover, the rest of the methods are very poor from a cost perspective. From another point of view, OR presents a strike rate equal to 0 guaranteeing the great performance of Flexible Bayes working with an OR feature selection.

Finally, we have executed a ROC analysis in order to obtain better conclusions from the experiments carried out. Figure 2 presents four ROC plots comparing the different feature selection methods for each of the analyzed classifiers.

Figure 2(a) shows ROC curve plots for all feature selection methods when Multivariate Bernoulli classifier is used. As we can see from the ROC curves, DF seems to be the best feature selection technique for this classifier. Figure 2(b) represents the

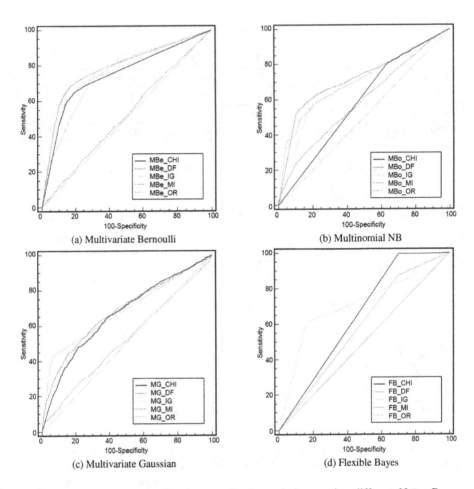

Fig. 2. ROC analysis for the five feature selection techniques using different Naïve Bayes classifiers

ROC analysis for the different attribute selection alternatives when Multinomial NB is used. Although results are clearly poor when combining with those using Multivariate Bernoulli classifier, DF presents again better results than other alternatives. Figure 2(c) includes the ROC curves for Multivariate Gaussian classifier. From these plots we can see that the differences between the areas under the curves are small. Moreover, the best results are found when OR feature selection technique is used. Finally, Figure 2(d) shows the analysis of different feature selection alternatives when using Flexible Bayes classifier. In this scenario, IG seems to be the greatest alternative for spam filtering.

In order to complete the information provided by plots showed in Figure 2, we have computed the area under the ROC curves. Table 9 shows these values for the different analyzed configurations. Results from Table 9 confirm the previously exposed observations from Figure 2.

Table 9. Area under the ROC showed in Figure 2

	OR	MI	χ^2	DF	IG
Multivariate Bernoulli	0.751	0.516	0.747	0.786	0.756
Multinomial NB	0.706	0.591	0.584	0.722	0.704
Multivariate Gaussian	0.685	0.518	0.663	0.680	0.669
Flexible Bayes	0.587	0.500	0.647	0.587	0.713

Analyzing the results shown in Figure 2 and Table 9, we can realize that the usage of all analyzed feature selection techniques works well in conjunction with Multivariate Bernoulli classifier except for MI. In order to use a Multinomial Naïve Bayes classifier, OR, DF and IG are good choices for feature selection. Moreover, IG method achieves better performance when it is used with Flexible Bayes. Finally, Multivariate Gaussian gets the worst results on the ROC analysis.

Some results achieved during our ROC analysis suggest that 0.5 value for T parameter does not improve results obtained by all Naïve Bayes classifiers. For instance, Multivariate Bernoulli and Multinomial Naïve Bayes classifiers present poor performance when they are analyzed using percentage measures, recall, precision and/or TCR scores. Moreover, using the area under the ROC curve (Table 9), we can see that these classifiers can achieve greater performance levels. These classifiers seem to have a greater dependence with training and testing data. Therefore, we discourage their usage. Once the discussion of the results has been concluded, next section presents some significant conclusions achieved from this work. It also includes some promising lines for future research related to this work.

5 Conclusions and Further Work

This paper presents and compares several feature selection techniques commonly used in the context of text mining in conjunction with four different Naïve Bayes classifiers. In order to evaluate the proposals, we have designed an empirical test using several measures commonly used in the domain of spam filtering.

To sum up the main conclusions, generated from our experimentation the OR method presents a high performance level when it is used with Gaussian-based Naïve Bayes algorithms like Flexible Bayes and Multivariate Gaussian. Nevertheless, the results achieved with this technique by using Laplace-based Naïve Bayes variants (Multivariate Bernoulli and Multinomial NB) are very poor.

Experiments have shown that MI method should be discarded for spam filtering. Moreover, OR, χ^2, DF or IG methods should be selected taking into consideration the target classifier.

From another point of view, we have appreciated some differences between Laplace-based and Multivariate Gaussian-based approaches. The usage of former approaches seems to be more dependant from the training and test corpus. This fact discourages their election and aims the usage of Multivariate Gaussian and Flexible Bayes techniques.

OR, IG and DF methods seem to work well in conjunction with Laplace-based approaches. Nevertheless, OR method presents a high dependence with the training

Table 10. Summary of previous conclusions

	Laplace-based NB	Gaussian-based NB
OR	Not very bad	Probably the best decision
MI	Not advisable	Not advisable
χ^2	Quite good combination	Quite good combination
DF	Select only for CPU usage reduction	Select only for CPU usage reduction
IG	Quite good combination	Quite good combination

corpus and parameter T should be adjusted with extreme accuracy. Moreover, IG and DF feature selection techniques seem to work fine using $T=0.5$.

Finally, OR, χ^2, DF or IG methods appear to get better classification accuracy when Gaussian-based Naïve Bayes variants are used. Moreover, experiments has shown the convenience of using OR technique. According with the OR measure computation method (see Expression (12)), when a term is evaluated, both the probability of finding a term in spam messages and in legitimate ones are taken into consideration. Using this strategy, a term that appears frequently in spam messages is discarded from the feature selection unless it appears in the opposite class with lower probabilities. This issue is useful in order to discard noise terms from spam messages, although it can cause the absence of representative words in the feature selection process.

χ^2 and IG methods are conceptually better than OR because they are able to select terms appearing in more than one message and can be used as distinctive attributes between spam and legitimate e-mails. Finally, although DF based feature selection gets poor results than χ^2 and IG methods, it is very interesting in spam filtering domain due to its simplicity for calculation.

Table 10 summarizes the above commented findings and it is useful in order to design filters to avoid spam. As we can see from Table 10, the combination of Gaussian-based filters and OR feature selection method seems to be one of the most reliable configurations for the studied domain. Moreover, χ^2 and IG methods achieve good performance working with any Naïve Bayes variant. Finally, we discourage the usage of the remaining combinations as they are clearly poor.

As future work, a promising direction would be the possibility of using continuous updating strategies in conjunction with Naïve Bayes spam filters. The spam filtering domain is constantly changing and presents the problem of concept drift [3]. Therefore, the usage of lazy learning strategies should work better than the application of eager learning algorithms. Moreover, some dynamical evaluation approaches must be designed for the filters. This kind of assessment should permit the evaluation of the impact of dynamical changes on the environment.

References

1. European Commission: i2010 - A European Information Society for growth and employment (2007), http://ec.europa.eu/i2010/
2. CardCommunications - Sitebrand Corporation: Email Trends Report (2007), http://www.cardcommunications.com/expresso/reports/Q1_07_trends_card.pdf

3. Cunningham, P., Nowlan, N., Delany, S.J., Haahr, M.: A Case-Based Approach to Spam Filtering than Can Track Concept Drift. In: Proceedings of the 5th International Conference on Case Based Reasoning, ICCBR-2003, Workshop of Long-Lived CBR Systems, pp. 115–123 (2003)
4. Jain, A., Zongker, D.: Feature Selection: Evaluation, Application, and Small Sample Performance. IEEE Transactions on Pattern Analysis and Machine Intelligence 19(2), 153–158 (1997)
5. Peters, T., Robinson, G., Hooft, R., Hammond, M., Meyer, T., True, S., Walker, A., Hindle, C., Pickett, N., Stone, T.: SpamBayes Project (2002),
   http://spambayes.sourceforge.net
6. Mozilla Project: Mozilla Spam Filter, http://www.mozilla.org/mailnews/spam.html
7. Burton Computer Corporation: SpamProbe: A Fast Spam Bayesian Filter (2002),
   http://spamprobe.sourceforge.net/
8. Androutsopoulos, I., Metsis, V., Paliouras, G.: Spam Filtering with Naive Bayes – Which Naive Bayes? In: Proceedings of the 3rd Conference on Email and AntiSpam (2006)
9. Androutsopoulos, I., Koustias, J., Chandrinos, K.V., Paliouras, G., Spyropoulos, C.: An Evaluation of Naïve Bayesian Anti-Spam Filtering. In: Proc. of the Workshop on Machine Learning in the New Information Age at 11th European Conference on Machine Learning, pp. 9–17 (2000)
10. Schneider, K.M.: A comparison of event models for Naive Bayes anti-spam e-mail filtering. In: Proceedings of the 10th Conference of the European Chapter of the ACL, Budapest, Hungry (2003)
11. John, G., Langley, P.: Estimating continuous distributions in Bayesian classifiers. In: Proceedings of the 11th Conference on Uncertainty in Artificial Intelligence, pp. 338–345 (1995)
12. Yang, Y., Pedersen, J.: A Comparative Study on Feature Selection in Text Categorization. In: Proceedings of the 14th International Conference on Machine Learning, pp. 412–420 (1997)
13. Androutsopoulos, I.: Ling Spam Corpus (2000),
   http://www.iit.demokritos.gr/~ionandr/lingspam_public.tar.gz
14. Mason, J.: The Apache SpamAssassin Project (2005),
   http://spamassassin.apache.org
15. Fdez-Riverola, F., Iglesias, E.L., Díaz, F., Méndez, J.R., Corchado, J.M.: Applying Lazy Learning Algorithms to Tackle Concept Drift in Spam Filtering. Expert Systems With Applications 33(1), 36–48 (2007)
16. Shaw, W.M., Burgin, R., Howell, P.: Performance standards and evaluations in IR test collections: Cluster-based retrieval models. Information Processing and Management (1997)
17. Graham-Cumming, J.: Understanding Spam Filter Accuracy. JGC spam and anti-spam newsletter (2004)
18. Méndez, J.R., Iglesias, E.L., Fdez-Riverola, F., Díaz, F., Corchado, J.M.: Tokenising, Stemming and Stopword Removal on Anti-Spam Filtering Domain. Lecture notes in artificial intelligence, vol. 4147, pp. 559–558 (2006)
19. Kohavi, R.: A study of cross-validation and bootstrap for accuracy estimation and model selection. In: Proceedings of the 14th International Joint Conference on Artificial Intelligence, pp. 1137–1143 (1995)
20. Egan, J.P.: Signal Detection Theory and ROC Analysis. Academic Press, New York (1975)
21. Metz, C.E.: Basic principles of ROC analysis. Seminars in Nuclear Medicine 8, 283–298 (1978)
22. Zweig, M.H., Campbell, G.: Receiver-operating characteristic (ROC) plots: a fundamental evaluation tool in clinical medicine. Clinical Chemistry 39, 561–577 (1993)

The Impact of Noise in Spam Filtering: A Case Study

I. Cid, L.R. Janeiro, J.R. Méndez, D. Glez-Peña, and F. Fdez-Riverola

Dept. Informática, University of Vigo, Escuela Superior de Ingeniería Informática
Edificio Politécnico, Campus Universitario As Lagoas s/n, 32004, Ourense, Spain
{icgomez,ljaneiro,moncho.mendez,dgpena,riverola}@uvigo.es

Abstract. Unsolicited commercial e-mail (UCE), more commonly known as spam is a growing problem on the Internet. Every day people receive lots of unwanted advertising e-mails that flood their mailboxes. Fortunately, there are several approaches for spam filtering able to detect and automatically delete this kind of messages. However, spammers have adopted some techniques to reduce the effectiveness of these filters by introducing noise in their messages. This work presents a new pre-processing technique for noise identification and reduction, showing preliminary results when it is applied with a Flexible Bayes classifier. The experimental analysis confirms the advantages of using the proposed technique in order to improve spam filters accuracy.

1 Introduction and Motivation

The phenomenon known as 'spamming' consists on sending unwanted advertising e-mails to final users. Spam is junk e-mail that nowadays represents a special problem due to the particular characteristics of the Internet. For example, spam can be sent with almost no charge to the sender. In fact, this cost is paid by both the Internet Server Provider (ISP) and the receiver. Furthermore, it is difficult to enact a law on the Internet that can prevent the receipt of spam e-mails [1].

In this context, some organizations take advantage of electronic mail service to send unsolicited and unwanted messages. Spam sent by well-known companies is sometimes called mainsleaze and is also a medium for fraudsters to scam users to enter personal information on fake web sites. They use forged e-mails to look like banks or other organizations such as PayPal. Spammers often use false names, addresses, phone numbers, and other contact information to set up disposable accounts at various ISPs. This allows them to move quickly from one account to the next when the host ISP discovers and shuts down each one.

On the other hand, some societies and companies fight against spam. The e-mail protocol (SMTP) has no authentication by default, so the spammer can originate messages from any e-mail address. Given these circumstances, some ISPs and domains require the use of SMTP-AUTH extensions, allowing positive identification of the sender account [2]. Furthermore, some associations like Spamhaus [3] provide dependable real-time anti-spam protection for Internet networks and work to identify and pursue spammers. Spamhaus project provides 4 well-known services: (*i*) Spamhaus Block List (SBL) (*ii*) Policy Block List (PBL), (*iii*) Exploits Block List (XBL)

P. Perner (Ed.): ICDM 2008, LNAI 5077, pp. 228–241, 2008.

and (*iv*) Register of Known Spam Operations (ROKSO). SBL is a real-time database of spam-source IP addresses including known spammers, spam gangs, spam operations and spam support services. PBL is a Domain Name Service Block List (DNSBL) database of end-user IP address ranges which should not be delivering unauthenticated SMTP e-mail to any Internet mail server. XBL is a real-time database containing IP addresses of illegal 3rd party exploits and includes open proxies (HTTP, Socks, AnalogX, Wingate, etc), worms/viruses with built-in spam engines, and other types of trojan-horse exploits. Finally, ROKSO is a database that contains information and evidence on known professional spam operations that have been terminated by a minimum of 3 spam offenses.

Nevertheless, spam continues to plague the Internet because a small number of large ISPs sell services to professional spammers or do nothing to prevent spamming operations from their networks. Moreover, some ISPs think that closing the holes used by spammers to access their broadband systems would be too expensive. In this context, some researchers estimate that just fewer than 100 billion spam messages are sent worldwide every twenty-four hours as of June 2007 [4]. This is the culmination of a spam problem that is grown exponentially year on year. February 2007 saw around 90 billion, while June 2006 estimates placed spam rates at 55 billion per day, up from a tiny 25 billion just twelve months before.

Spam filtering is a difficult task because spammers attempt their e-mails look similar to legitimate ones, so they go unnoticed. They use different techniques to avoid spam filters working well. Their tricks include the alteration of some words that could identify a message as spam. The noise included in e-mail terms reduces the effectiveness of spam filtering tools. The usage of these tricks allows the creation of spam messages by using words that look similar for the human eye, but completely different when they are analyzed by spam filters (word obfuscation). Such methods include character level substitutions, repetitions and insertions to reduce the effectiveness of word-based detection techniques [5]. Table 1 exemplifies some of the most common obfuscations used for the word 'Viagra'.

Table 1. Common obfuscations of the term 'Viagra'

V&Igra	VIAGRA	Viiagrra	viagra	visagra
Vi@gra	Viaagrra	Viaggra	Viagraa	Viiaagra
Via-ggra	Viia-gra	V1AAGRRA	Viiagra	Via-gra

Statistical spam filtering tools are not able to detect word obfuscation, so a previous technique for identification should be applied before carrying out the classification. There are several deobfuscation approaches taken from other research fields like computational biology [6] and text classification [7]. Nevertheless, most of them are not suitable for spam message processing, so new noise reduction methods are needed in spam filtering domain.

This work introduces a new pre-processing technique for noise reduction known as Quick Noise Skipping (QNS). In our experimentation we have used Flexible Bayes, a variant of the original Naïve Bayes algorithm [8], working with five feature selection methods (Information Gain, Odds ratio, Document Frequency, χ^2 statistic and Mutual Information) for evaluating the effectiveness of our proposal. We have selected Flexible Bayes because Naïve Bayes algorithms are very popular in spam filtering domain and previous work demonstrate that it is the most reliable Bayesian algorithm for e-mail classification [9].

The remaining of the paper is structured as follows: Section 2 presents an introduction about the problem of noise in spam messages. Our proposed technique is presented in Section 3. The experimental setup is shown in Section 4 while Section 5 gathers the results obtained from the experimentation carried out. Finally, Section 6 summarizes the main conclusions extracted from this work and outlines some feasible alternatives for future research.

2 Noise in Spam Messages

As it was mentioned before, spammers make use of some tricks trying to hide some relevant terms from messages, adding noise to e-mails in order to reduce the effectiveness of spam filters. This section analyzes different techniques used to spawn noise in spam messages including: *(i)* character substitution, *(ii)* fake taps, *(iii)* HTML comments, *(iv)* character encoding tricks and *(v)* hash buster method. Moreover, it also provides a review of different promising alternatives for noise reduction.

Character substitution is the most important way of introducing noise in a message. This technique changes some characters of a word so a human can understand the term, but it goes unnoticed through the filter. There are a lot of possible character substitutions for a given term, so the brute force line of attack is not valid when the number of insertions, deletions and substitutions increases quickly.

Fake taps is a technique that uses an HTML tag to introduce a spam word, so it can not be detected. A new tag is created containing the word the spammer wants to hide. HTML comments are similar to fake taps because it achieves spam words obfuscation by introducing them into an HTML comment. If a spam word is inserted into an HTML comment it is more probably to pass unobserved than if the word is part of the HTML document body.

Character encodings take advantage of different character sets to disguise some words. There are diverse charset encodings (ASCII, UNICODE, ISO-8859-1, ISO-8859-15, etc.) each of them only allowing a restrict set of chars. In this situation, spammers use unknown characters to design spam words. Furthermore, URL-encoding codes could be used to introduce some spam words into Internet addresses by exploiting the fact that each ASCII character has an URL-encode translation.

Hash buster is an obfuscation method that consists on adding seemingly senseless words in order to mislead anti-spam filters [10]. Initially, hash busters methods simply tended to exhibit paragraphs of literally random words, but up to now they present somewhat grammatical. Curiously, many of the examples appearing around the summer of 2006 were distorted to get the desired advertising sites unusable, for example substituting "001" for "www". This may be a good technique for avoiding a filter, but it is inadequate for leading novice-users to websites. Additionally, much of the embedded HTML code as well as any MIME-encoded attachments are scrambled and distorted by the process, thereby decreasing the true effectiveness of the spam [11].

Standard e-mail pre-processing techniques start with the extraction of the words (tokens) that compose each message. In plain texts, this stage consists on identifying groups of characters separated using punctuation marks, white spaces, etc. However, in spam e-mails these actions can drop important information because noise terms would not be well processed (e.g. v!agra or v.agra) [12]. Therefore, a previous noise identification stage is needed when pre-processing is made over e-mails instead of plain text.

There is some previous research work on deobfuscation techniques applied to the spam filtering domain. In [13] a Hidden Markov Model (HMM) approach was introduced. This technique allows the conversion of obfuscated text back into the original one intended by the sender. The method uses a word (or sentence) as input data (observation) and produces the deobfuscated word (or sentence) as output. Thus, the HMM is able to correctly handle misspellings, incorrect segmentations (adding/removing spaces) and other word modifications, such as substitutions and insertions of non-alphabetic characters.

There are also several methods that use simple algorithms composed by some easy steps. The work of Shabbir and Mithun [14] presents an algorithm that is composed by seven straightforward steps that are described below:

1) Remove all non-alpha characters (allowing some exceptions like '/' '\' '|' etc. which can be used together to look like some other characters, such as \/ for 'V').
2) Remove all vowels from the word except for a trailing one.
3) Replace consecutive repeated characters by a single character.
4) Use phonetic algorithms on the resultant string.
5) Give it a numeric score depending on the operations performed over it.
6) Use the previous score to look up a table containing a list of offending words where each entry has a range of acceptable values.
7) Replace the original word with that of the table.

Jonathan Zdziarski, the developer of the open source Dspam filter [15], has developed an innovative technique able to detect the noise that spammers often include in their messages to fool filters. This technique (called Dobly) uses a fixed size sliding window as context window [16]. Words inside a context window are considered as relevant and will be checked for consistency. Furthermore, it introduces the probability for a fixed size sliding window of the e-mail body. Therefore, if the window has a high probability of being spam, then each individual word inside the window should have a high probability of being spam as well. Otherwise, they are considered as inconsistent.

There are several noise reduction methods and some of them are based on concrete experiences. Some researchers have analyzed spammers' tactics for term obfuscation. Others have been developed by keeping in mind the contents of well-known spam messages. For instance, spam e-mails tend to have more sentences in imperative mood, and in those the first word is a verb. So verbs with initial caps have higher spam probabilities than those in lowercase [17].

3 Quick Noise Skipping: A Novel Pre-processing Technique for Noise Reduction

This section presents our QNS technique, a method for noise reduction in spam filtering domain. This technique is proposed as an early pre-processing method able to remove the intrinsic noise in any text (orthographical characters and punctuation marks) as well as identify the noise that spammers put into e-mails in order to avoid spam filters.

Noise is one of the most common tricks for information obfuscation used by spammers to avoid spam filters. Although noise affects the spam message recognition, it also makes difficult the correct classification of legitimate e-mails.

Recently, Naïve Bayes filtering tools have acquired big popularity due to their simplicity and reliability. These tools are able to quickly classify any incoming e-mail and being easy to understand and implement. Generally, they are used in conjunction with some feature selection technique in order to reduce the dimensionality of the problem.

The tokenization process carried out generates a words set that algorithms are able to use as single characteristics. The identification of terms belonging to a message is made by using space, tabs and carriage returns as separators. This method is not able to remove punctuation marks and noise introduced by spammers. Any text composition written in a correct grammatical way contains lots of special characters like punctuation marks, hyphens, question and exclamation marks, etc. These characters can cause the existence of noise in the identified tokens.

Quick Noise Skipping is a technique able to find intrinsic noise of messages due to punctuation marks as well as noise deliberately introduced by spammers. When any kind of these noises are detected it is removed without replacing it for any special character. This policy produces accurate results with an insignificant computational cost, which is the main problem of other noise reduction methods. Table 2 illustrates some situations achieved by using our Quick Noise Skipping approach.

Table 2. Examples of Quick Noise Skipping method for noise removal

Plain word	Deobfuscated word using QNS
Interesting!	Interesting
Interesting.	Interesting
V!agra	Vagra
V1agra	Vagra

As we can see from Table 2, this technique presents an obvious advantage on both kind of noise. It allows words with different punctuation marks or obfuscations converge into the same characteristic. This fact has a direct impact on the feature selection process and the spam filtering classifier operation.

Our QNS reduction technique is able to deobfuscate the noise present in incoming e-mails. In the previous section we have explained some of the main obfuscating methods used by spammers to introduce noise in their messages. From those methods, our QNS technique is able to detect character substitutions, fake taps, HTML comments and character encoding tricks because it is specifically designed for removing special character modifications in single words. Nevertheless, QNS is not able to fight against hash buster because this spam method does not make character substitutions but it adds seemingly senseless words into e-mails.

Next section describes the experimental setup designed for the validation of the proposed technique, while Section 5 presents the results achieved when QNS method is applied over a Bayesian classifier using different selection features approaches.

4 Experimental Setup

This section describes the experimental setup for analyzing the impact of applying our QNS method as pre-processing stage for a Flexible Bayes classifier. We discuss some specific issues about corpus availability, feature selection methods, Naïve Bayes filters, and finally we present the measures we have used for evaluating the results.

Since the earliest spam filtering tools were initially developed, some datasets have been released for benchmarking spam classifiers. Some researchers have created and shared with the scientific community several e-mail corpuses that can be used to make standard tests. Table 3 shows a summary of the main available information.

Table 3. Summary of main available datasets containing spam and legitimate e-mails

Corpus	Legitimate percentage	Spam percentage	Format
Lingspam	83.3 %	16.6 %	tokens
Enron	35.9 %	64.1 %	tokens
SpamAssassin	84.9 %	15.1 %	RFC 822

Lingspam was created by Androutsopoulos *et al.* [8] containing 481 spam messages (16.6%) and 2412 legitimate ones (83.3%) took out from some texts belonging to a moderate linguistic list. This corpus is too small and have not general e-mails but messages linked to a specific field, so it is not a good choice to test our QNS reduction technique.

From another perspective, Enron Creditors Recovery Corporation was an American energy company based in Houston, Texas. Before its bankruptcy in late 2001, Enron employed around 22,000 people and was one of the biggest electricity, natural gas, pulp and paper, and company communications. It achieved infamy at the end of 2001, when it was revealed that it was reported financial condition sustained mostly by institutionalized, systematic, and creatively planned accounting fraud. The Enron corpus was made public during the legal investigation concerning the Enron Corporation. This dataset contains 50895 messages belonging to the company employers [18] and due to its huge size it is not adequate for our preliminary study.

Last corpus, is a project of Apache Group that uses a wide variety of local and network tests to identify spam signatures. This corpus can be downloaded from the project web page [19] and has three different versions (2002, 2003 and 2005). There are five groups identifying the message type and the difficulty of correct classification. These groups are: (*i*) spam: 500 spam messages from non-usually, (*ii*) easy_ham: 2500 ham messages easy to classify, (*iii*) hard_ham: 250 ham e-mails similar to spam ones, (*iv*) easy_ham_2: 1400 legitimate messages currently incorporated into the corpus and (*v*) spam_2: 1397 spam messages recently incorporated into the compilation. Messages in SpamAssassin are encoded in accordance with RFC 822 [20]. This fact makes possible the utilization of different pre-processing techniques. Due to these reasons, it represents a good dataset for testing the new noise reduction technique we are presenting in this work.

In our experimentation with the SpamAssassin corpus we have used five well-known feature selection methods: (*i*) Information Gain (IG), (*ii*) Odds ratio (OR), (*iii*) Document Frequency (DF), (*iv*) χ^2 statistic and (*v*) Mutual Information (MI). Some recent work [9] has gathered together key properties of each one, studying the influence of feature selection over different Naïve Bayes algorithms and concluding that Flexible Bayes is the most reliable Bayesian algorithm for e-mail classification. Based on our previous results we have selected Flexible Bayes classifier in order to carry out the experimental stage. Moreover, we have used a 10 fold-cross stratified validation scheme for improving the quality of the achieved results [21].

In order to determine the effectiveness of our QNS algorithm working with Flexible Bayes we have computed the following percentage measures: (*i*) OK: well-classified mails (*ii*) FP: false positives and (*iii*) FN: false negatives. We have also used *recall* and *precision* measures. Recall is able to compute the ability of classifying spam e-mails (higher values imply more spam detected) while precision shows the competence of low false positive errors (higher values imply lower false positives rate). The equations that govern recall and precision are:

$$ recall = \frac{nspam - fn}{nspam} \qquad precision = \frac{nspam - fn}{nspam - fn + fp} \tag{1} $$

where *fn* and *fp* represent the number of false negatives and false positives respectively, and *nspam* is the number of spam e-mails in the training corpus.

In the work of [22] it is suggested the utilization of *Batting average* in order to show the connection between recall and precision. This measure is built taking into consideration *hitrate/strikerate* ratio, where the former represents the number of detected spam messages and the later the false positive average. It is also a useful measure for comparison purposes, so we have used it.

Another interesting measure is *f*-score, which was originally proposed by Rijsbergen [23] to combine recall and precision. F-score ranges in the interval [0-1] and its value is 1 only if the number of FP and FN errors generated by the filter is 0. Expression (2) shows how to compute this measure.

$$ f - score = \frac{2 \cdot precision \cdot recall}{precision + recall} \tag{2} $$

Recently, a new measure called *balanced f-score* or *f-score$_\beta$* has been introduced [24]. As f-score, balanced f-score combines precision and recall but considering they have different importance. If *β=1* then precision and recall have the same weight, so f-score= f-score$_\beta$. If *β>1* then precision is more important than recall. Otherwise, recall has more weight. f-score$_\beta$ can be computed as Expression (3) shows.

$$ f - score_\beta = \frac{(\beta^2 + 1) \cdot precision \cdot recall}{\beta^2 \cdot precision + recall} \tag{3} $$

In the work of [8] the *Total Cost Ratio* (TCR) measure was introduced. This measure uses λ to ponder false positive and false negative costs. A FP error is λ times more costly than a false negative one and its value represent the problems caused to the end-user. Default values of λ are 1, 9 and 999. Expression (4) shows how to calculate TCR.

$$TCR = \frac{nspam}{\lambda \cdot fp + fn} \tag{4}$$

where *fn* and *fp* represent the number of false negatives and false positives, respectively, and *nspam* is the number of spam e-mails in the training corpus.

From another point of view, we have used ROC (*Receiver Operating Characteristics*) curves [25]. A ROC curve is a graphical plot of the fraction of true positives vs. the fraction of false positives. ROC analysis provides tools to select plausible optimal models and to discard suboptimal ones from a cost-sensitive point of view. ROC analysis is related in a direct and natural way to cost/benefit analysis of diagnostic decision making.

If there are no classification errors, the values of sensitivity and specifity are 1 and the ROC curve is a horizontal line. If the model probability is near 50%, the ROC curve is a line from (0, 0) point to (1, 1) point. An interesting feature for assessing model accuracy is the area under the ROC curve. This area ranges from 1 (perfect test) to 0.5 (pointless test), and it can be understood as the probability of achieving good classification results by using an e-mail classifier.

5   Results and Evaluation

This section presents and explains the results obtained during the experimentation carried out. As we mentioned earlier, we have selected the SpamAssassin corpus following a 10 fold-cross stratified validation schema. Our experimental design is composed of two different scenarios. The two analyzed scenarios consider the utilization of Flexible Bayes with and without our proposed QNS technique. Moreover, in each scenario we analyze the impact of our method over five well-known feature selection algorithms.

Figure 1 shows the connection between the percentage of well-classified messages, false positives and false negatives between the two analyzed scenarios.

As we can realize from Figure 1, the number of well-classified e-mails grows up when our QNS method is used. Furthermore, the amount of false positives is also reduced. Finally, the number of false negative errors is also decreased when MI and IG feature selection methods are used. This percentage increases slightly when using CHI, DF and OR feature selection approaches.

After carrying out an initial percentage evaluation we have used precision, recall, f-score and balanced f-score measures for the comparison of the selected scenarios. Table 4 compares precision and recall results for Flexible Bayes with and without running our QNS technique.

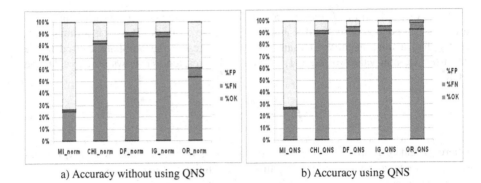

a) Accuracy without using QNS   b) Accuracy using QNS

Fig. 1. Comparison of OK, FN and FP percentages for Flexible Bayes

Table 4. Precision, recall, f-score and balanced f-score using Flexible Bayes

		MI	χ^2	**DF**	**IG**	**OR**
FB w/o QNS	Precision	0.251	0.591	0.703	0.698	0.314
	Recall	0.978	0.938	0.908	0.906	0.701
	F-score	0.295	0.638	0.736	0.732	0.353
	Balanced f-score (β=2)	0.400	0.725	0.792	0.789	0.433
FB with QNS	Precision	0.255	0.735	0.810	0.817	0.914
	Recall	0.991	0.913	0.867	0.872	0.774
	F-score	0.300	0.765	0.821	0.827	0.838
	Balanced f-score (β=2)	0.406	0.814	0.837	0.843	0.882

From Table 4 we can see how DF and IG achieve the best values for precision, f-score and balanced f-score, among all feature selection methods when no noise reduction method is used. When we use our QNS method it can be seen that DF and IG present good values for precision, f-score and balanced f-score improving those initially obtained. Furthermore, OR has a high value of precision and a similar recall to DF as well as it improves original algorithm results. This behaviour is due to QNS application makes terms with noise become similar to their originals, so they are able to belong to the same category. This term convergence lets OR to determinate more clearly if a category is spam or legitimate and consequently the classifier achieves accurate results.

As we suggested before, we have also used the batting average and TCR scores to analyze the filter performance. Table 5 displays a summary of the achieved results by taking into account these measures.

Inspecting table values presented in Table 5, we can realize that DF and IG are the feature selection techniques with best accuracy when we do not use our noise reduction method. In case of QNS is applied, TCR and batting average results become improved. Moreover, OR is able to achieve better values than DF and IG, with a strikerate of 0.025 indicating that the false positives average is close to 0. From

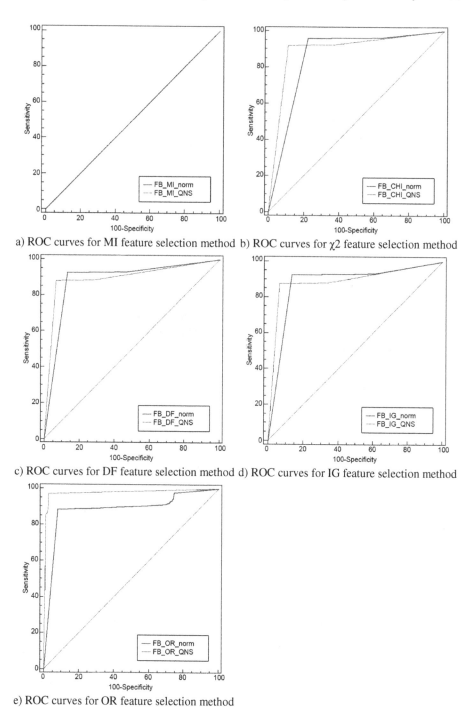

a) ROC curves for MI feature selection method b) ROC curves for χ2 feature selection method

c) ROC curves for DF feature selection method d) ROC curves for IG feature selection method

e) ROC curves for OR feature selection method

Fig. 2. ROC curves for Flexible Bayes using χ^2, DF, IG, MI and OR as feature selection methods with and without our QNS technique

Table 5. Different TCR values and batting average ratio using Flexible Bayes

		MI	χ^2	DF	IG	OR
FB w/o QNS	TCR λ=999	0	0.001	0.002	0.002	0
	TCR λ=9	0.038	0.170	0.284	0.277	0.071
	TCR λ=1	0.341	1.411	2.119	2.071	0.546
	Batting average	0.978/0.996	0.938/0.222	0.908/0.131	0.906/0.133	0.701/0.524
FB with QNS	TCR λ=999	0	0.003	0.005	0.005	0.014
	TCR λ=9	0.038	0.330	0.520	0.547	1.170
	TCR λ=1	0.345	2.422	3.046	3.194	3.388
	Batting average	0.991/0.987	0.913/0.112	0.867/0.069	0.872/0.067	0.774/0.025

another perspective, we can conclude the usage of MI feature selection technique is not advisable for using in conjunction with Flexible Bayes.

Finally, we have plotted several ROC curves in order to check the advisability of the proposed technique using different feature selection methods with Flexible Bayes classifier. Figure 2 represents the ROC curves for the execution of the experiments.

In order to complete the information provided by plotting the ROC curves, we have also computed the area under the curve, the standard error, and the 95% confidence interval for each test. Results are shown in Table 6.

Table 6. ROC analysis for Flexible Bayes using χ^2, DF, IG, MI and OR feature selection methods with and without QNS technique

	Area under curve	Standard error	95% Confidence interval
FB_MI_norm	0.500	0.007	0.490 to 0.510
FB_MI_QNS	0.500	0.007	0.490 to 0.510
FB_CHI_norm	0.861	0.005	0.854 to 0.868
FB_CHI_ QNS	0.895	0.005	0.889 to 0.901
FB_DF_norm	0.885	0.005	0.879 to 0.892
FB_DF_QNS	0.894	0.005	0.887 to 0.900
FB_IG_norm	0.880	0.005	0.873 to 0.887
FB_IG_QNS	0.890	0.005	0.884 to 0.897
FB_OR_norm	0.883	0.005	0.876 to 0.890
FB_OR_QNS	0.976	0.003	0.972 to 0.979

As we can see from the information showed in Table 6 and Figure 2, the utilization of our proposed noise removal technique is able to successfully improve the accuracy of the filtering process in almost all the analyzed scenarios. Moreover, it can be stated that the utilization of MI method should not be used for spam filtering. Results using this feature selection technique are very poor even when the QNS method is applied.

Finally, in order to study the significance level of our findings, we have executed a statistical test assuming the null hypothesis that the areas under the ROC curves are equal. Results from this statistical test are shown in Table 7.

Table 7. Results for the execution of ROC statistical test

	Difference between areas	Standard error	95% Confidence interval	Significance level
FB_MI norm vs. QNS	0.000	0.010	0.019 to 0.019	p = 1.000
FB_CHI norm vs. QNS	0.034	0.004	0.026 to 0.042	p < 0.001
FB_DF norm vs. QNS	0.008	0.003	0.002 to 0.015	P = 0.012
FB_IG norm vs. QNS	0.010	0.003	0.003 to 0.017	p = 0.003
FB_OR norm vs. QNS	0.092	0.005	0.082 to 0.102	p < 0.001

From results shown in Table 7, we can see that p-values achieved in the different scenarios are lower than 0.05 except for the usage of MI feature selection technique. Therefore, from a statistical point of view, the choice of using the QNS technique can significantly improve the accuracy of Flexible Bayes spam filter.

6   Conclusions and Further Work

This work has presented a novel noise reduction technique for spam filtering called Quick Noise Skipping. We have successfully applied this method using a Bayesian algorithm and five different feature selection techniques. Results from our experiments have shown the increment on filter effectiveness for all the feature selection methods used.

Our QNS reduction technique provides more useful information for feature selection methods working together with Flexible Bayes classifier. When removing noise, filters ignore information that adulterates their regular operation. Moreover, some features are taken into account which would not have been considered if QNS or other noise reduction method was not applied. Due to these useful information gain and superfluous data reduction, the number of well classified messages increases while reducing the number of false positives. Therefore, the accuracy of the Flexible Bayes filter is improved by using each analyzed feature selection method. These findings are supported from the results showed in Section 5.

Odds Rate is the best feature selection method to apply with one QNS technique because it utilizes a threshold value with a high variance, so probability values near to 0.5 prevent the odds rate index from correctly classifying a term. When noise is reduced for those terms near to 0.5, in which obfuscated words are included, we get a slight probability variation that implies a big change on their odds rate index. Thus, it clearly implies a better working of the feature selection technique, and consequently, a more accurate operation of the Bayesian filter.

As future research line, we are working on evaluating several regular expressions with our noise reduction technique, in order to improve the convergence of some word variants into the same term. Furthermore, some semantic evaluation approaches can be considered as a previous step to feature selection. This extra level of information should identify different words with the same category by using similarity relations among them. Moreover, we are thinking about the possibility of test our QNS technique with the Enron corpus as well as applying QNS over different Naïve Bayes filters.

References

1. Cunningham, P., Nowlan, N., Delany, S.J., Haahr, M.: A Case-Based Approach to Spam Filtering than Can Track Concept Drift. In: Ashley, K.D., Bridge, D.G. (eds.) ICCBR 2003. LNCS, vol. 2689, pp. 115–123. Springer, Heidelberg (2003)
2. The Spamhaus Project: Working to Protect Internet Networks Worldwide (2007), http://www.spamhaus.org/
3. Spam overview (2007), http://en.wikipedia.org/wiki/E-mail_spam
4. Spam statistics (2007), http://www.spamunit.com/spam-statistics/
5. Wittel, G.L., Wu, S.F.: On attacking statistical spam filters. CEAS: First Conference on E-mail and Anti-Spam (2004)
6. Leslie, C., Kuang, R.: Fast string kernels using inexact matching for protein sequences. Journal of Machine Learning Research, 1435–1455 (2004)
7. Lodhi, H., Saunders, C., Shawe-Taylor, J., Cristianini, N., Watkins, C.: Text classification using string kernels. Journal of Machine Learning Research 2, 419–444 (2002)
8. Androutsopoulos, I., Koustias, J., Chandrinos, K.V., Paliouras, G., Spyropoulos, C.: An Evaluation of Naïve Bayesian Anti-Spam Filtering. In: Proceedings of the 11th European Conference on Machine Learning, Workshop on Machine Learning in the New Information Age, pp. 9–17 (2000)
9. Cid, I., Méndez, J.R., Peña-Glez, D., Fdez-Riverola, F.: A comparative impact study of attribute selection techniques on Naïve Bayes spam filters. In: The 8th Industrial Conference on Data Mining, ICDM 2008 (submitted for publication 2007)
10. Random Act of Spamness (2007), http://www.wired.com/techbiz/it/news/2004/01/61886
11. Hash Buster definition (2007), http://en.wikipedia.org/wiki/Hash_buster
12. Méndez, J.R., Fdez-Riverola, F., Díaz, F., Corchado, J.M.: Sistemas Inteligentes para la Detección y Filtrado de Correo Spam: una Revisión. Inteligencia Artificial, Revista Iberoamericana de Inteligencia Artificial 34, 63–81 (2007)
13. Lee, H., Ng, A.Y.: Spam deobfuscation using a Hidden Markov Model. In: Second Conference on E-mail and Anti-Spam (2005)
14. Shabbir, A., Farzana, M.: Word stemming to enhance spam filtering. In: CEAS: First Conference on E-mail and Anti-Spam (2004)
15. The Dspam project (2007), http://dspam.nuclearelephant.com/
16. SpamAssassin BNR (Bayes Noise Reduction) (2007), http://docs.google.com/View?docid=dfsk849w_13d4zm72
17. Graham, P.: Better bayesian filtering (2003), http://www.paulgraham.com/better.html
18. Klimt, B., Yang, Y.: Introducing the Enron corpus. In: CEAS: First Conference on E-mail and Anti-Spam (2004)
19. The Apache SpamAssassin Public Corpus (2007), http://spamassassin.apache.org/publiccorpus/

20. Crocker, D.: Standard for the Format of ARPA Internet Text Messages. STD 11, RFC 822 (2007), http://www.faqs.org/rfcs/rfc822.html
21. Kohavi, R.: A study of cross-validation and bootstrap for accuracy estimation and model selection. In: Proceedings of the 14th International Joint Conference on Artificial Intelligence, pp. 1137–1143 (1995)
22. Graham-Cumming, J.: Understanding Spam Filter Accuracy. In: jgc spam and anti-spam newsletter (2004) (2007),
   http://www.jgc.org/antispam/
   11162004-baafcd719ec31936296c1fb3d74d2cbd.pdf
23. Rijsbergen, C.J.: Information Retrieval (ed.). Butterworth, London (1979)
24. Shaw, W.M., Burgin, R., Howell, P.: Performance standards and evaluations in IR test collections: Cluster-based retrieval models. Information Processing and Management 33(1), 1–14 (1997)
25. Egan, J.P.: Signal Detection Theory and Roc Analysis (ed.). Academic Press, New York (1975)

Designing Specific Weighted Similarity Measures to Improve Collaborative Filtering Systems

Laurent Candillier, Frank Meyer, and Françoise Fessant

France Telecom R&D Lannion, France
{name.surname}@orange-ftgroup.com

Abstract. The aim of *collaborative filtering* is to help *users* to find *items* that they should appreciate from huge catalogues. In that field, we can distinguish *user-based* from *item-based* approaches. The former is based on the notion of user neighbourhoods while the latter uses item neighbourhoods.

The definition of *similarity* between users and items is a key problem in both approaches. While traditional similarity measures can be used, we will see in this paper that bespoke ones, that are tailored to type of data that is typically available (i.e. very sparse), tend to lead to better results.

Extensive experiments are conducted on two publicly available datasets, called *MovieLens* and *Netflix*. Many similarity measures are compared. And we will show that using weighted similarity measures significantly improves the results of both user- and item-based approaches.

1   Introduction

There has been a growth in interest in *recommender systems* [1] in the last two decades, since the appearance of the first papers on this subject in the mid-1990s [2]. The aim of such systems is to help *users* to find *items* that they should appreciate from huge catalogues.

Items can be of any type, such as movies, music, books, web pages, online news, jokes and restaurants. In this paper, we focus on movie items. The goal, therefore, of recommender systems, in this context, is to help users to find movies of interest, based on some information about their historical preferences.

Three types of recommender systems are commonly proposed:

1. *collaborative filtering*;
2. *content-based filtering*;
3. and *hybrid filtering*.

In the first case, the input to the system is a set of user ratings on items. Users can be compared based upon their shared appreciation of items, creating the notion of user neighbourhoods. Similarly, items can be compared based upon the shared appreciation of users, rendering the notion of item neighbourhoods. The item rating for a given user can then be predicted based upon the ratings given in her user neighbourhood and the item neighbourhood.

P. Perner (Ed.): ICDM 2008, LNAI 5077, pp. 242–255, 2008.

In content-based filtering, however, item content descriptions are used to construct *user thematic profiles*, that contain information about user preferences, such as, in the context of movies, *"like comedy and dislike war"*. The user's predicted appreciation of a given item is then based on the proximity between the item description and the user profile.

Finally, in the case of hybrid filtering, both types of information, collaborative and content-based, are exploited.

Collaborative filtering techniques are more often implemented than the other two and often result in better predictive performance. The main reason is that they do not require these difficult to come by, well-structured item descriptions. Instead, they are based on users' preferences for items, that can carry a more general meaning than is contained in an item description. Indeed, viewers generally select a movie to watch based upon more elements than only its genre, actors and director.

So this paper focuses on collaborative filtering, that can be divided into three sets of approaches:

1. *user-based* approaches [2] associate a set of nearest neighbours with each user, and then predict the user's rating for unscored items using the ratings given by the neighbours on that item;
2. *item-based* approaches [3] associate an item with a set of nearest neighbours, and then predict the user's rating for an item using the ratings given by the user on the nearest neighbours of the target item;
3. and *model-based* approaches and more specifically those based on *clustering* [4], tend to be more scalable, by constructing a set of user groups or item groups, and then predicting a user's rating for an item using the mean rating given by the group members

We concentrate on the first two approaches. The definition of similarity between users and items is a key problem in each approach. Since the data that are typically available in collaborative filtering are very sparse, traditional similarity measures need to be adapted. Generally, when comparing two users or two items, only the set of attributes in common are considered. However, by doing so, many users and items may be compared based only on very few attributes, which can lead to a lack of meaning.

In this paper, we propose to adapt traditional similarity measures to sparse data. Our approach consists in using a new weighting scheme of existing similarity measures, so that users and items that share many attributes are prefered to others that only share a few attributes. We will then show the interest of such new measures, not only in improving the predictive performance of collaborative filtering systems, but also in improving their scalability.

This brings us naturally to the issue of evaluating the performance of a given recommender system [5]. This is typically done by using *cross-validation*. Then the most widely used measures for comparison are:

1. *Mean Absolute Error* (MAE);
2. *Root Mean Squared Error* (RMSE);
3. and *Precision* and *Recall*.

The first two measures evaluate the capability of a method to predict if a user will like or dislike an item, whereas the third set of measures evaluates its capability of providing an ordered list of items that a user should like. These measures thus carry different meanings [6]. In the first two cases, the method needs to be able to predict dislike, but there is no need to order items. In the last case, however, the method only focuses on items that users will like and the order in which these items are ranked is important.

Beyond the importance of the predictive performance of recommender systems, other elements may be taken into consideration in their evaluation. The scalability of the proposed system is for example an important characteristic that needs to be taken into account.

To conduct experiments, we will use two real rating datasets that are publicly available, called *MovieLens* and *Netflix* [7]. The first one contains 1,000,209 ratings collected from 6,040 users on 3,706 movies. The second dataset contains 100,480,507 ratings for 480,189 users and 17,770 movies. Both are completely independent.

To summarise, this paper, on the study of similarity measures for collaborative filtering, will be structured as follows: an overview of the principal approaches for collaborative filtering is presented in section 2, alongside a study of similarity measures for comparing users and items; the results of extensive experiments are reported in section 3 in order to compare various similarity measures for collaborative filtering approaches, using cross-validation on the MovieLens and Netflix datasets; finally, section 4 concludes the paper and topics for future research are suggested.

2   Collaborative Filtering

Let U be a set of N users, I a set of M items, and R a set of ratings r_{ui} of users $u \in U$ on items $i \in I$. $S_u \subseteq I$ stands for the set of items that user u has rated.

Let us set to min_r the minimum rating and Max_r the maximum rating. In the MovieLens and Netflix datasets for example, ratings are integers ranging from 1 (meaning dislike) to 5 (meaning high like).

The goal of collaborative filtering approaches is then to be able to predict the rating p_{ai} of a given user a on a given item i. User a is supposed to be *active*, meaning that she has already rated some items, so $S_a \neq \emptyset$. But the item to be predicted shall not be known by the user, so $i \notin S_a$.

2.1   User-Based Approaches

For user-based approaches [2,8], the prediction of a user rating on an item is based on the ratings, on that item, of the nearest neighbours. So a similarity measure between users needs to be defined. Then a set of nearest neighbours is selected. And finally, a method for combining the ratings of those neighbours on the target item needs to be chosen.

The way the similarity between users is computed will be discussed in subsection 2.3. Let $sim(a, u)$ be that similarity measure between users a and u. The number of neighbours considered is often set by a system parameter that we denote by K. So the set of neighbours of a given user a, denoted by T_a, is made of the K users that maximise their similarity to user a.

A possible way to predict the rating of user a on item i is then to use the weighted sum of the ratings of the nearest neighbours $u \in T_a$ that have already rated item i:

$$p_{ai} = \frac{\sum_{\{u \in T_a | i \in S_u\}} sim(a, u) \times r_{ui}}{\sum_{\{u \in T_a | i \in S_u\}} |sim(a, u)|} \qquad (1)$$

In order to take into account the difference in use of the rating scale by different users, predictions based on deviations from the mean ratings have been proposed. In that case, p_{ai} is computed using the sum of the user mean rating and the weighted sum of deviations from their mean rating of the neighbours that have rated item i:

$$p_{ai} = \overline{r_a} + \frac{\sum_{\{u \in T_a | i \in S_u\}} sim(a, u) \times (r_{ui} - \overline{r_u})}{\sum_{\{u \in T_a | i \in S_u\}} |sim(a, u)|} \qquad (2)$$

$\overline{r_u}$ represents the mean rating of user u:

$$\overline{r_u} = \frac{\sum_{\{i \in S_u\}} r_{ui}}{|S_u|} \qquad (3)$$

Indeed, let us suppose a user rates 4 a movie she likes and 1 a movie she dislikes while another user rates 5 a movie she likes and 2 a movie she dislikes. Then using deviations from their mean rating will reduce the effect of such a difference in use of the rating scale.

The time complexity of user-based approaches is $O(N^2 \times M \times K)$ for the neighbourhood model construction, it is $O(K)$ for one rating prediction, and the space complexity is $O(N \times K)$.

2.2   Item-Based Approaches

Then item-based approaches have known a growing interest [3,9,10,11]. Given a similarity measure between items, such approaches first define item neighbourhoods. Then the rating of a user on an item is predicted by using the ratings of the user on the neighbours of the target item.

The possible choices of the similarity measure $sim(i, j)$ defined between items i and j will be discussed in the next subsection. Then, as for user-based approaches, the item neighbourhood size K is a system parameter that needs to be defined. And given T_i the neighbourhood of item i, two ways for predicting new user ratings can be considered: weighted sum, and weighted sum of deviations from the mean item ratings:

$$p_{ai} = \frac{\sum_{\{j \in S_a \cap T_i\}} sim(i, j) \times r_{aj}}{\sum_{\{j \in S_a \cap T_i\}} |sim(i, j)|} \qquad (4)$$

$$p_{ai} = \overline{r_i} + \frac{\sum_{\{j \in S_a \cap T_i\}} sim(i,j) \times (r_{aj} - \overline{r_j})}{\sum_{\{j \in S_a \cap T_i\}} |sim(i,j)|} \tag{5}$$

$\overline{r_i}$ represents here the mean rating on item i:

$$\overline{r_i} = \frac{\sum_{\{u \in U | i \in S_u\}} r_{ui}}{|\{u \in U | i \in S_u\}|} \tag{6}$$

The time complexity of item-based approaches is $O(M^2 \times N \times K)$ for the neighbourhood model construction, it is $O(K)$ for one rating prediction, and the space complexity is $O(M \times K)$.

2.3   Similarity Measures

The similarity defined between users or items is crucial in collaborative filtering. The first one proposed in [2] is *pearson* correlation. It corresponds to the cosine of deviations from the mean:

$$pearson(a,u) = \frac{\sum_{\{i \in S_a \cap S_u\}} (r_{ai} - \overline{r_a}) \times (r_{ui} - \overline{r_u})}{\sqrt{\sum_{\{i \in S_a \cap S_u\}} (r_{ai} - \overline{r_a})^2 \sum_{\{i \in S_a \cap S_u\}} (r_{ui} - \overline{r_u})^2}} \tag{7}$$

Simple *cosine* can also be used. It has been tested in [3]:

$$cosine(a,u) = \frac{\sum_{\{i \in S_a \cap S_u\}} r_{ai} \times r_{ui}}{\sqrt{\sum_{\{i \in S_a \cap S_u\}} r_{ai}^2 \sum_{\{i \in S_a \cap S_u\}} r_{ui}^2}} \tag{8}$$

Other similarity measures based on cosine have been proposed in [12]. But they have been shown less relevant. We also propose here the simple case of *manhattan* similarity:

$$manhattan(a,u) = 1 - \frac{1}{Max_r - min_r} \times \frac{\sum_{\{i \in S_a \cap S_u\}} |r_{ai} - r_{ui}|}{|\{i \in S_a \cap S_u\}|} \tag{9}$$

For all these similarity measures, only the set of attributes in common between two vectors are considered. Thus two vectors may be completely similar even if they only share one appreciation on one attribute.

Let us illustrate the drawback of such measures on one example. One user is fan of science fiction while another user only watches comedys. They haven't rated any movie in common so their similarity is null. Now they both say they like "*Men In Black*", a comic science fiction movie. So these two users become completely similar according to the previously presented measures.

On the contrary, *jaccard* similarity doesn't suffer from this limitation since it measures the overlap that two vectors share with their attributes:

$$jaccard(a,u) = \frac{|\{S_a \cap S_u\}|}{|\{S_a \cup S_u\}|} \tag{10}$$

On the other hand, such a measure doesn't take into account the difference of ratings between the vectors. In this case, if two users watch the same movies but have completely opposite opinions on them, then they are considered to be similar anyway according to jaccard similarity.

So we propose to combine jaccard with the other similarity measures, in order to benefit from their complementarity. We propose to simply compute the product between the jaccard similarity and the other ones. Jaccard would thus serve as a weighting scheme. So *wpearson* stands for *weighted pearson*, the new similarity measure made by the product of pearson and jaccard. Similarly, we denote by *wcosine* and *wmanhattan* the combination of jaccard with cosine and manhattan respectively.

Cosine-based similarity measures have their values comprised between -1 and 1 while the other similarity values are comprised between 0 and 1.

3   Experiments

3.1   Protocol

We conduct these experiments using the MovieLens and Netflix datasets. We randomly divide them into 2 parts, training the chosen model using 90% of the datasets and testing it on the last 10%. In all experiments, the division of the datasets is always the same, so that all approaches are evaluated under exactly the same conditions.

Given $T = \{(u, i, r)\}$ the set of (user, item, rating) triplets used for test, the Mean Absolute Error Rate (MAE) and Root Mean Squared Error (RMSE) are used to evaluate the performance of the algorithms:

$$MAE = \frac{1}{|T|} \sum_{(u,i,r)\in T} |p_{ui} - r| \qquad (11)$$

$$RMSE = \sqrt{\frac{1}{|T|} \sum_{(u,i,r)\in T} (p_{ui} - r)^2} \qquad (12)$$

The predicted ratings are rounded when the MAE is reported.

We then use precision measures specifically designed for the current context. precision$_5$ stands for the proportion of maximum ratings in the test dataset ($Max_r = 5$) that are retrieved as the best predicted ratings. Similarly, precision$_4$ stands for the proportion of test ratings higher than $Max_r - 1 = 4$ that are considered as the best ratings using the given recommender system.

Finally, we also report the time spent to learn the models, and for predictions. The computer used for these experiments has 32Go RAM and 64 bits 3.40GHz 2-cores CPU.

3.2   Results

Considering only the principal collaborative filtering approaches already leads us to a lot of choices and parameters. When implementing a user- or item-based approach, one may choose:

- a similarity measure: pearson (equation 7), cosine (8), manhattan (9), jaccard (10), or the proposed combinations of jaccard with the others;
- a neighbourhood size K;
- and how to compute predictions: using a weighted sum of rating values (equations 1 and 4), or using a weighted sum of deviations from the mean (2 and 5).

Prediction scheme based on deviations from the mean has been shown to be more effective in [12]. We confirmed this result in our experiments. So in the following, only the results using this scheme are reported.

This paper is about the comparison of various similarity measures. So the following figures 1 to 3 show the Mean Absolute Error Rate obtained using the proposed measures, varying the neighbourhood size K from 10 to the maximum number of possible neighbours, for both user- and item-based approaches, and on both MovieLens and Netflix datasets.

Figure 1 first shows the error rates of item-based approaches depending on the similarity measure used and the neighbourhood size. The optimum is reached with the weighted pearson similarity and 100 neighbours. All similarity measures are improved when they are weighted with jaccard, at least when few neighbours are considered. All these weighted similarity measures reach their optimum when 100 neighbours are selected. On the contrary, non-weighted similarity measures need much more neighbours to reach their optimum. 700 neighbours shall be selected when using simple manhattan similarity, and 1500 when using simple cosine or simple pearson.

Figures 2 and 3 show that the same conclusions hold when using user-based approaches, as well as when the Netflix dataset is used instead of MovieLens. Weighted pearson similarity always leads to the best results. Using our proposed weighting scheme always improves the results. 300 neighbours shall be considered for an user-based approach on MovieLens, and 70 for an item-based approach on Netflix. On the contrary, 2000 to 4000 neighbours need to be selected to reach the minimum error rate of non-weigthed similarity measures.

The results have been presented for the MAE. They are completely similar for the other performance measures: RMSE and precisions. Beyond the improvement of predictive performance when our proposed weighting scheme is used, another important advantage is that fewer neighbours need to be selected, so that the algorithms also gain in scalability.

Tables 1 and 2 summarise the results of the best of each approach, including learning and prediction times, and precisions. The default approach that is reported is Bayes designed to minimise RMSE. More details on this method can be found in [12]. Both user- and item-based approaches reach optimal results when the weighted pearson similarity is used. On MovieLens, 300 neighbours are selected for the best user-based approach, and 100 for the best item-based

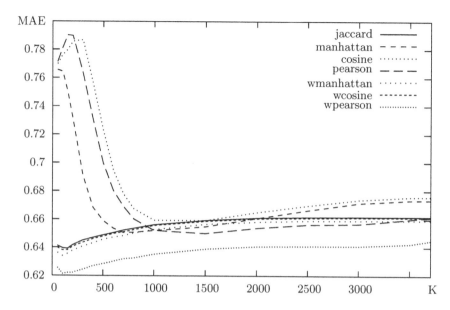

Fig. 1. Comparing MAE on MovieLens when using item-based approaches with different similarity measures and neighbourhood sizes (K)

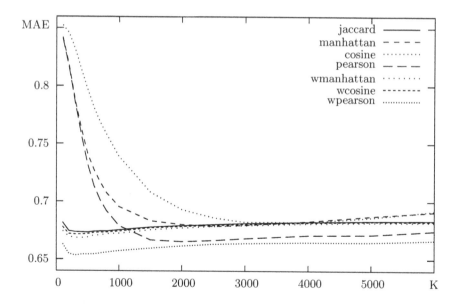

Fig. 2. Comparing MAE on MovieLens when using user-based approaches with different similarity measures and neighbourhood sizes (K)

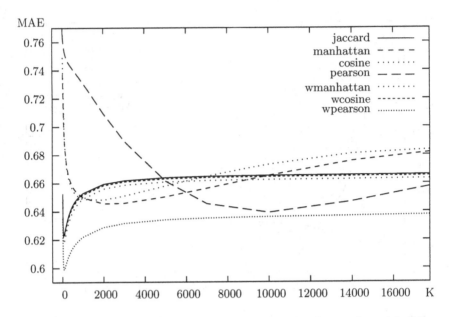

Fig. 3. Comparing MAE on Netflix when using item-based approaches with different similarity measures and neighbourhood sizes (K)

Table 1. Summary of the best results on MovieLens depending on the type of approach

	neighbour number	learning time	prediction time	MAE	RMSE	precision5	precision4
default	/	1 sec.	1 sec.	0.6855	0.9234	0.5451	0.7676
user-based	300	4 min.	3 sec.	0.6533	0.8902	0.5710	0.7810
item-based	**100**	2 min.	1 sec.	**0.6213**	**0.8550**	**0.5864**	**0.7915**

Table 2. Summary of the best results on Netflix depending on the type of approach

	neighbour number	learning time	prediction time	MAE	RMSE	precision5	precision4
default	/	6 sec.	15 sec.	0.6903	0.9236	0.5524	0.7392
user-based	1000	9 h	28 min.	0.6440	0.8811	0.5902	0.7655
item-based	**70**	2 h 30	1 min.	**0.5990**	**0.8436**	**0.6216**	**0.7827**

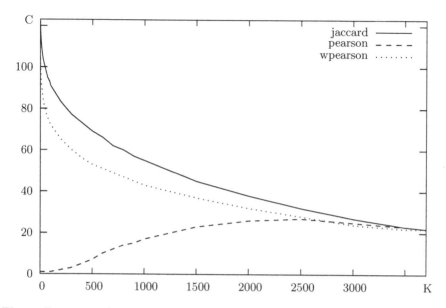

Fig. 4. Comparing the mean number of common attributes (C) between nearest item neighbours on MovieLens when considering different neighbourhood sizes (K)

one. On Netflix, considering 70 neighbours leads to the lowest error rate. User-based approaches, however, face scalability issues. It is too expensive to compute the entire user-user matrix. So instead, we begin by running a clustering, and we then consider as potential neighbours only the users that belong to the same cluster. Considering many neighbours improves the results. But a model based on 1000 neighbours, selected starting from a clustering with 90 clusters, needs 9 hours to learn, 28 minuts to predict, and 10Go RAM. The best overall results are reached using an item-based approach. It needs 2 hours and a half to learn the model on Netflix, and 1 minut to produce 10 millions rating predictions. A precision$_5$ of 0.6216 means that 62.16% of the best rated items are captured by the system and proposed to the users.

Now let us concentrate on item-based approaches and pearson-based similarities. Figure 4 shows the mean number of attributes in common between the nearest item neighbours, depending on the similarity measure used and the number K of nearest neighbours considered. It thus shows that when pearson is used, the nearest neighbours in fact do not share many attributes. They often have only one attribute in common. On the contrary, by using jaccard similarity, the selected neighbours are those that share a maximum number of attributes. Figure 4 shows us that many items share more than 100 attributes on the MovieLens dataset. Jaccard searches to optimise the number of common attributes between vectors, but this may not be the best solution for nearest neighbour selection

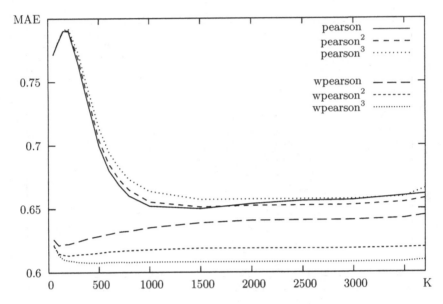

Fig. 5. Comparing MAE on MovieLens when using item-based approaches with different pearson-based similarity measures and neighbourhood sizes (K)

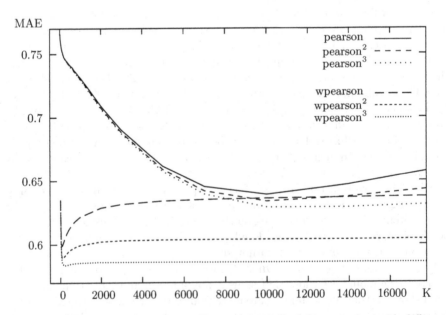

Fig. 6. Comparing MAE on Netflix when using item-based approaches with different pearson-based similarity measures and neighbourhood sizes (K)

since the values of the vectors on the shared attributes may differ. Weighted pearson is then an interesting compromise between pearson and jaccard.

Then figures 5 and 6 show the results obtained when raising the similarities to some different power (but still preserving their sign). By doing that, since the measures are comprised between -1 and 1, the influence of the nearest neighbours increases while the influence of the furthest neighbours decreases. We can thus observe that the results improve when raising weighted pearson to some power. On the contrary, the results do not change very much when simple pearson is raised to some power, and even sometimes results are worse. So this shows that our proposed weighting scheme gives meaning to the notion of neighbourhood, since giving more trust to the nearest neighbours improves the results, and considering more and more neighbours does not decrease the results anymore.

Finally, tables 3 and 4 summarise the results obtained with item-based approaches and pearson-based similarities. They compare the performance depending on whether our proposed weighting scheme is used or not. It thus clearly shows the interest of weighting the traditional similarity measures, since all performance measures are improved. Fewer neighbours need to be selected, learning time and prediction time are lower, and the predictive performance are significantly better.

We have also tested other ensemblist similarity measures than jaccard that can be found in [13]: Simple Matching, Russel and Rao, Sokal and Sneath, Rogers and Tanimoto, Dice, and Yule. But none of these did improve the results.

Table 3. Summary of the best results of item-based approaches on MovieLens depending on the use of our proposed weighting scheme

	neighbour number	raised power	learning time	prediction time	MAE	RMSE	$precision_5$	$precision_4$
non-weighted	1500	1	3 min.	7 sec.	0.6500	0.8798	0.5703	0.7808
weighted	500	3	2 min.	3 sec.	0.6075	0.8404	0.5957	0.7958

Table 4. Summary of the best results of item-based approaches on Netflix depending on the use of our proposed weighting scheme

	neighbour number	raised power	learning time	prediction time	MAE	RMSE	$precision_5$	$precision_4$
non-weighted	10000	3	7 h	38 min.	0.6291	0.8675	0.6042	0.7712
weighted	200	3	2 h 30	1 min.	0.5836	0.8297	0.6352	0.7881

4   Conclusion

We have shown in this paper that weighting traditional similarity measures used for collaborative filtering can substantially improve the results of both user- and item-based approaches. More specifically, we have shown that combining pearson and jaccard similarities leads to the best results. This has been verified on two widely-used and independent movie datasets: MovieLens and Netflix.

We have explain that such weighting schemes are useful because traditional similarity measures, such as pearson, have the drawback of considering as neighbours rating vectors that share too few attributes. On the contrary, by using our proposed weighted measures, we have verified that the vectors considered as neighbours share many attributes.

Moreover, we have observed that using these weighted measures allows us to consider fewer neighbours than when non-weighted measures are used. So learning and prediction times of both user- and item-based collaborative filtering approaches are also improved thanks to our new proposition. Finally, we have also observed that raising the weighted pearson similarity to some power still improves the results.

We have also confirmed that item-based approaches outperform user-based ones. Besides their better results, item-based methods have other advantages. Their learning and prediction times are lower, at least for datasets that contain more users than items. They are able to produce relevant predictions as soon as a user has rated one item. Moreover, such models are also appropriate for the navigation in item catalogues even when no information about the current user is available, since it can also present to a user the nearest neighbours of any item she is currently interested in.

Starting from the new similarity measure we have proposed, many approaches for its optimisation may be considered. Content information on items could be used for its improvement [14]. In particular, when an item is associated to very few ratings, then finding its neighbours based on those ratings has little meaning. Instead, using its content description could lead to find more relevant neighbours. The similarity matrix could also be modified by using stochastic perturbations led by cross-validation, or any other optimisation process [15]. Finally, other schemes for combining jaccard and pearson than simple product may also be tested.

Another way to continue improving item-based approaches is to adapt the neighbourhood size parameter to every item independently. Indeed, instead of considering a global parameter fixing the number of neighbours to be considered by each item, we could associate to each item its own parameter. Such parameters could be optimised using cross-validation.

Finally, combining different approaches could also lead to better results. Many ensemble methods could then be implemented in that field [16,17].

References

1. Adomavicius, G., Tuzhilin, A.: Toward the next generation of recommender systems: A survey of the state-of-the-art and possible extensions. IEEE Transactions on Knowledge and Data Engineering 17, 734–749 (2005)

2. Resnick, P., Iacovou, N., Suchak, M., Bergstrom, P., Riedl, J.: Grouplens: An open architecture for collaborative filtering of netnews. In: Conference on Computer Supported Cooperative Work, pp. 175–186. ACM, New York (1994)
3. Sarwar, B.M., Karypis, G., Konstan, J., Riedl, J.: Item-based collaborative filtering recommendation algorithms. In: 10th International World Wide Web Conference (2001)
4. Ungar, L., Foster, D.: Clustering methods for collaborative filtering. In: Workshop on Recommendation Systems. AAAI Press, Menlo Park (1998)
5. Herlocker, J., Konstan, J., Terveen, L., Riedl, J.: Evaluating collaborative filtering recommender systems. ACM Transactions on Information Systems 22, 5–53 (2004)
6. McNee, S., Riedl, J., Konstan, J.: Being accurate is not enough: How accuracy metrics have hurt recommender systems. In: Extended Abstracts of the 2006 ACM Conference on Human Factors in Computing Systems (2006)
7. NetflixPrize (2006), http://www.netflixprize.com/
8. Shardanand, U., Maes, P.: Social information filtering: Algorithms for automating "word of mouth". In: ACM Conference on Human Factors in Computing Systems, vol. 1, pp. 210–217 (1995)
9. Karypis, G.: Evaluation of item-based top-N recommendation algorithms. In: 10th International Conference on Information and Knowledge Management, pp. 247–254 (2001)
10. Linden, G., Smith, B., York, J.: Amazon.com recommendations: Item-to-item collaborative filtering. IEEE Internet Computing 7, 76–80 (2003)
11. Deshpande, M., Karypis, G.: Item-based top-N recommendation algorithms. ACM Transactions on Information Systems 22, 143–177 (2004)
12. Candillier, L., Meyer, F., Boullé, M.: Comparing state-of-the-art collaborative filtering systems. In: Perner, P. (ed.) MLDM 2007. LNCS (LNAI), vol. 4571, pp. 548–562. Springer, Heidelberg (2007)
13. Janowitz, M.F.: A combinatorial introduction to cluster analysis. Technical report, Classification Society of North America (2002)
14. Vozalis, M., Margaritis, K.G.: Enhancing collaborative filtering with demographic data: The case of item-based filtering. In: 4th International Conference on Intelligent Systems Design and Applications, pp. 361–366 (2004)
15. Bell, R., Koren, Y.: Improved neighborhood-based collaborative filtering. In: ICDM 2007: IEEE International Conference on Data Mining, pp. 7–14. ACM, New York (2007)
16. Polikar, R.: Ensemble systems in decision making. IEEE Circuits & Systems Magazine 6, 21–45 (2006)
17. Bell, R., Koren, Y., Volinsky, C.: Modeling relationships at multiple scales to improve accuracy of large recommender systems. In: KDD 2007: Proceedings of the 13th ACM SIGKDD International Conference on Knowledge Discovery and Data Mining, pp. 95–104. ACM, New York (2007)

Browsing Assistance Service for Intranet Information Systems

Peter Géczy, Noriaki Izumi, Shotaro Akaho, and Kôiti Hasida

National Institute of Advanced Industrial Science and Technology (AIST)
Tokyo and Tsukuba, Japan

Abstract. Improved usability and efficiency of organizational information systems brings economical benefits to the organization and time
benefits to the users. We present a browsing assistance service suitable
for the organizational intranet environments. It helps users to shorten
their browsing interactions and achieve their goals faster. These benefits
are accomplished by providing relevant suggestions on the potential navigation targets of interest to the users. The system design employs the
analytics of user browsing behavior and its appropriate segmentation.
It efficiently utilizes the initial and the terminal navigation points for
providing recommendations. The performance of the system has been
evaluated on the real world data of a large scale intranet portal.

1 Introduction

Commercial organizations have been devoting substantial resources to improving
usability of their electronic environments oriented toward the customers. They
have been collecting significant amounts of behavioral data in order to advance
their operations and services [1]-[3]. Unfortunately, the progress in the internal
improvement, aiming at the usability enhancements of their own information
systems—benefiting the employees, remains distantly behind [4].

Effective way to alleviate the usability of the electronic environments is to provide the browsing recommendations to the users implemented via service oriented
architecture [5],[6]. However, the recommender systems should more efficiently
utilize browsing analytics [7]. Researches have been exploring several approaches
to analyzing human behavior in electronic environments—ranging from behavioral studies [8], and their automation [9], to click-stream analysis [10]. Significant
attention has been paid to modeling browsing behavior for predictive purposes
by utilizing: Markov models [11], cluster analysis methods [12], adaptive learning
strategies [13], and frequent pattern mining [14]. Unfortunately, these methods
are computationally intensive and non-scalable.

This work presents a novel approach to developing browsing assistance systems based on the segmentation of human web behavior and identification of
the essential navigation points [15]. These analytics are effectively utilized for
designing scalable algorithms economically implementable into large scale organizational information systems.

P. Perner (Ed.): ICDM 2008, LNAI 5077, pp. 256–267, 2008.

2   Concept Formalization

We present and formalize the essential conceptual elements and introduce the corresponding terminology. Along with the formal definitions are provided intuitive and illustrative explanations that highlight the important aspects and facilitate further comprehension.

The concept of segmenting human behavior in electronic environments has been presented in [4]. We recall relevant constructs. Human undertakings in electronic spaces are recorded as click-stream sequences. The click-streams of page views can be divided into sessions, and sessions can further be divided into subsequences. The segmentation of long click-streams into browsing sessions and subsequences is carried out according to the user activity—primarily accounting for the temporal characteristics of the activity bursts.

The click-stream sequence of page transitions is recorded as a sequence of pairs $\{(p_i, t_i)\}_i$, where p_i denotes the visited page URL_i at the time t_i. For analytic convenience this sequence is converted into the form: $\{(p_i, d_i)\}_i$ where $d_i = t_{i+1} - t_i$ represents a delay between the consecutive page views $p_i \rightarrow p_{i+1}$ (see Figure 1). Human behavior in web environments displays periods of activity followed by a longer inactivity periods. Delays d_i between the page transitions serve a suitable segmentation indicator.

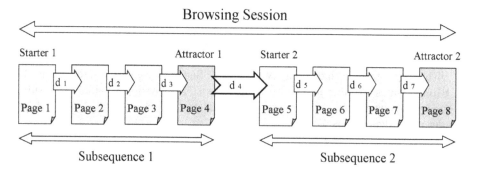

Fig. 1. Depiction of click-stream segmentation into browsing session and subsequences. The first element of the subsequence is the starter. The last element is the attractor.

Definition 1. *(Browsing Session, Subsequence)*
Let $\{(p_i, d_i)\}_i$ be a sequence of pages p_i associated with delays d_i between consecutive transitions $p_i \rightarrow p_{i+1}$.

Browsing session is a sequence $B = \{(p_i, d_i)\}_i$ where each $d_i \leq T_B$. The length of the browsing session is $|B|$.

Subsequence of an individual browsing session B is a sequence $Q = \{(p_i, d_i)\}_i$ where each delay $d_i \leq T_Q$, and $\{(p_i, d_i)\}_i \subset B$. The subsequence length is $|Q|$.

Behavioral segments delineate tasks of various complexities users undertake in electronic environments. Sessions represent more complex tasks accomplished via several subtasks—subsequences. Example: during a browsing session a user may target a recent company policy announcement by navigating from the initial page on a corporate portal - subsequence 1; and then locate and download the related document templates - subsequence 2. The user's task of complying with the latest company policy has been divided into two subtasks represented by subsequences: **1**—familiarizing himself/herself with the new policy, and **2**—obtaining the proper document templates.

Pertinent issues in segmenting the users' browsing behavior into sessions and subsequences are the proper values of T_B and T_Q. Analysis of students' web behavior revealed that their browsing sessions last on average 25.5 minutes [16]. Knowledge workers' browsing sessions on a corporate intranet portal have been reported to last 48.5 minutes on average [4]. The study utilized empirically set $T_B = 1$ *hour*, and dynamically calculated value of T_Q as an average delay in the browsing session bounded from below by 30 seconds.

The ambition to understand the human browsing behavior in more detail calls for observing where the users initiate their browsing actions and which resources do they target. This leads to the identification of the starting and ending points of subsequences, as well as the single user actions.

Definition 2. *(Starter, Attractor, Singleton)*
Let $B = \{(Q_i, d_i)\}_i^M$ be a browsing session, and $Q_i = \{(p_{ik}, d_{ik})\}_k^N$ be its subsequence.

Starter is the point p_1 of the first pair element of subsequence Q_i or session B with length greater that 1. Set of starters is denoted as S.

Attractor is the point p_l of the last pair element of subsequence Q_i or session B with length greater that 1; $l \equiv N$ or $l \equiv M$. Set of attractors is denoted as A.

Singleton is a point p such that there exist a browsing session B or subsequence Q_i where $|B| = 1$ or $|Q_i| = 1$, and $(p, d) \in B$ or $(p, d) \in Q_i$. Set of singletons is denoted as Z.

As illustrated in Figure 1, the starters correspond to the initial navigation points of users' actions; e.g. the initial intranet portal page from our former example. The attractors refer to the users' goals of subtasks (example: the company policy announcement page) as well as the complete browsing sessions (example: document templates download page). The singletons underline the single user actions such as use of hotlists (e.g. history or bookmarks) [17]. Note that a single navigation point can be starter, attractor, and singleton.

It is equally relevant to observe the multitudes of navigational pathways from the starting points together with the transitions to the following subsequences.

Definition 3. *(Starter and Attractor Mappings)*
Let $B = \{(Q_i, d_i)\}_i$ be a browsing session with consecutive subsequences Q_i and Q_{i+1}.

Starter-attractor mapping $\omega : S \to A$ is a mapping where for each starter $s \in S$, $\omega(s)$ is a set of attractors of the subsequences Q_i having starter s.

Attractor-starter mapping $\psi : A \to S$ is a mapping where for each attractor $a \in A$ of the subsequences Q_i, $\psi(a)$ is a set of starters of the existing consecutive subsequences Q_{i+1}.

The starter-attractor mapping indicates the spectrum of targets the users accessed when initiating their browsing actions from the given starter. Note that it does not relate to the number of links on the starter page; it rather indicates the range of detected abstract browsing patterns: starter $\to$ set of attractors. It is an important 'long-range' access pattern indicator, since there may be several intermediate pages in the subsequence. On the other hand, the attractor-starter mapping outlines an important 'close-range' access pattern indicator: attractor $\to$ set of starters. It relates more closely to the spectrum of links exposed on the attractor page (static or dynamic) and/or utilization of hotlists.

Definition 4. *(Top Sets)*
Let ω be a starter-attractor mapping and ψ be an attractor-starter mapping. Top-n sets $\omega^{(n)}(s) \subseteq \omega(s)$ and $\psi^{(n)}(a) \subseteq \psi(a)$ are the ordered sets of the first n points $p \in \omega^{(n)}(s)$ and $p \in \psi^{(n)}(a)$ selected with respect to an ordering defined by a function $f : \Phi \to \Re$; where Φ is either/or $S \cup A$, $S \times A$, $A \times S$.

The top sets outline the sampling from the mappings with respect to an ordering function. Consider for example a starter s with $\omega(s) = \{a_1, \ldots, a_x\}$, $x \in N$. Top-n set $\omega^{(n)}(s) = \{a_1, \ldots, a_n\}$, $n \leq x$, can be the selection of the highest ranking attractor points according to a ranking function f defining ordering on the set $S \cup A$. The ordering function can be for instance a relative frequency of occurrences of points a_i obtained during the behavioral analysis of given user population. Numerous other ordering functions may be utilized.

3 System Conceptualization and Design

The system concept draws from a valuable analysis of knowledge worker behavior on a large corporate intranet portal. We highlight several important characteristics of the intranet usability and browsing behavior of its users before introducing the assistance system. This provides an empirical base for the decisions and choices made during the system design.

Exploratory analysis of knowledge worker behavior in the intranet environment revealed numerous relevant usability and behavioral aspects [4]. Although the intranet portal traffic was substantial, the available resources were generally underutilized. Knowledge workers had a tendency to form browsing and behavioral patterns. However, the patterns were largely diversified. The concise list of the exposed browsing features relevant to our study follows.

- Browsing tasks were divided into three subtasks on average.
- General browsing strategy: familiarity with the starting navigation point and knowledge of the traversal pathway to the target.
- Habituation of browsing behavior.

- Rapid page transitions toward the target resource - within seconds.
- Short attention span - approximately seven minutes on average.
- Utilization of a small spectrum of starting navigation points and aiming at small number of resources.
- Focused browsing interests and minuscule exploratory behavior.

The knowledge workers' browsing behavior analysis uncovered important aspects of their navigation and utilization of the intranet portal. They should be taken into consideration when developing an efficient assistance system. Accounting for these observations enables us to determine the essential strategic design requirements and characteristics.

Focus on starters and attractors. Appropriate assistance to users should be offered on the starter and attractor pages. Since the common browsing strategy highly utilized familiarity with the frequent starters, they are the natural points for assistance services. The intermediate points between the starter and attractor were passed in rapid transitions. Users did not pay sufficient attention to the information displayed on these pages and basically proceeded directly to the known link leading to the next page.

Aim at attractors and consecutive starters. Targets of the browsing assistance services should be the attractors and consecutive starters. These are the desired points in the intranet space that users want to access. The pages between the starter and attractor are essentially transitional. The navigation assistance service should be offering a list of possible desired attractors and starters. Numerous models utilized in browsing assistance systems are focused on attempting to predict what would be the next general page the users are going to visit. Our approach fundamentally differs from these techniques by focusing on the essential navigation points rather than just on any next page in the click-stream sequence.

Sufficient prediction depth: ≤ 3. Browsing sessions of knowledge workers contained on average three subsequences. Thus the more complex intranet tasks were divided into three subtasks. Each subtask has its starter s_i and attractor a_i. A general session may be described as:

$$s_1 \xrightarrow{1} a_1, s_2 \xrightarrow{2} a_2, s_3 \xrightarrow{3} a_3 ,$$

where the numbers above the right arrows denote the depth. In providing assistance on attractors (and possibly subsequent starters) it is generally sufficient to limit the depth level to less than or equal to three. This observation may lead to computationally more efficient algorithms.

Fast and scalable system design. Significantly shortened attention span of knowledge workers requires that assistance systems should be fast and computationally inexpensive, in order to provide on-the-fly recommendations. Extended waiting times, due to the computationally intensive algorithms, may result in negative browsing experiences. These requirements naturally depend on the complexness of the portal, number of users, resources, traffic, and available computing power.

3.1 System Design

Accounting for the presented requirements and characteristics we derive a browsing assistance system providing suggestions on the closest attractors and/or starters. The system is computationally efficient for providing on-the-fly page recommendations and scalable.

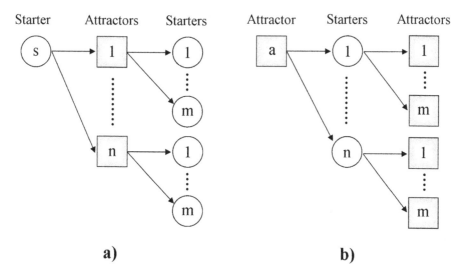

Fig. 2. Depiction of system's formation of page recommendation sets depending on the user's browsing state: **a)** - the starter page; **b)** - the attractor page

Intuitive illustration of how the assistance system extracts the initial sets of recommendation pages is shown in Figure 2. Assume the user has reached a starter page s (Figure 2-a). The system maps the starter s to a set of attractors $\omega(s)$. From the set $\omega(s)$, only top-n attractors $\omega^{(n)}(s)$ are selected with respect to the suitable ranking/ordering function. The top-n attractors in $\omega^{(n)}(s)$ are the seeds for the second stage attractor-starter mapping. For each attractor $a_i \in \omega^{(n)}(s)$ a set of corresponding top-m starters $\psi^{(m)}(a_i)$ is chosen according to the ordering function. The process is outlined as follows:

$$s \longrightarrow \omega^{(n)}(s) \longrightarrow \bigcup_{a_i \in \omega^{(n)}(s)} \psi^{(m)}(a_i) \; .$$

This leads to the initial recommendation set $r(s)$ having $n(1 + m)$ elements:

$$r(s) = \omega^{(n)}(s) \cup \left(\bigcup_{a_i \in \omega^{(n)}(s)} \psi^{(m)}(a_i) \right) \; .$$

The initial recommendation set $r(s)$ undergoes further selection. A subset of w most suitable elements is chosen—again according to the proper ordering function f:

$$r^{(w)}(s) = \left[\omega^{(n)}(s) \cup \left(\bigcup_{a_i \in \omega^{(n)}(s)} \psi^{(m)}(a_i)\right)\right]_{\triangleright f}. \tag{1}$$

Analogously, the recommendations are provided also when the user reaches an attractor page a (Figure 2-b). From the attractor-starter mapping ψ, the top-n seed set $\psi^{(n)}(a)$ is selected with respect to the given ordering. Then for each starter $s_i \in \psi^{(n)}(a)$ a corresponding set of top-m attractors $\omega^{(m)}(s_i)$ is obtained. This processing:

$$a \longrightarrow \psi^{(n)}(a) \longrightarrow \bigcup_{s_i \in \psi^{(n)}(a)} \omega^{(m)}(s_i) \,,$$

forms the initial recommendation set $r(a)$ with $n(1+m)$ elements:

$$r(a) = \psi^{(n)}(a) \cup \left(\bigcup_{s_i \in \psi^{(n)}(a)} \omega^{(m)}(s_i)\right) \,,$$

out of which only the top w elements are selected (with respect to the ordering function f):

$$r^{(w)}(a) = \left[\psi^{(n)}(a) \cup \left(\bigcup_{s_i \in \psi^{(n)}(a)} \omega^{(m)}(s_i)\right)\right]_{\triangleright f}. \tag{2}$$

Among the requirements for the assistance system design is to maintain the computational efficiency in order to facilitate scalability and on-the-fly processing. Pertinent issue here is the appropriate choice of the ordering function f. The computational complexity and ranking efficiency of the ordering function may significantly impact the system's performance both computationally and qualitatively. It is crucial to reach the right balance between the system's computational complexity and quality of recommendations.

To preserve the simplicity and consistency, the relative frequency of use of the essential navigation points detected during the knowledge worker behavioral analysis was chosen as the ordering function. It facilitates the reuse of the analytic data, is efficiently implementable, and permits easy extension to various domains of definition.

Recall that a navigation point p can be either a common point and/or starter, attractor, and singleton (Definition 2). The sets of starters, attractors, and singletons are not necessarily disjunct. Hence the question rises: how to compute the relative frequency of a point that has been detected to be starter, attractor, and singleton (or any of their combinations)? We resolved this issue by taking the average of the applicable relative frequencies.

$$f(p) = avrg(f_S(p) + f_A(p) + f_Z(p)) \,; \quad f_S(p) \neq 0,\; f_A(p) \neq 0,\; f_Z(p) \neq 0, \tag{3}$$

where f_S denotes the starter relative frequency, f_A stands for the attractor relative frequency, and f_Z indicates the singleton relative frequency.

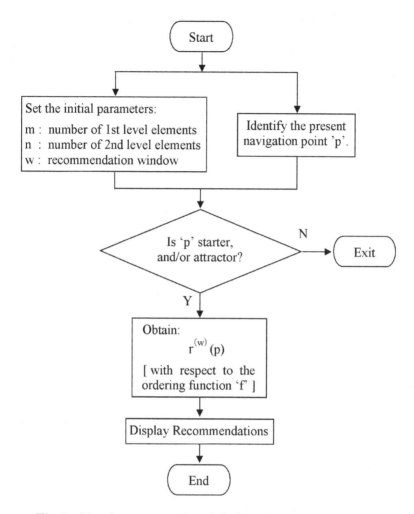

Fig. 3. Flowchart presentation of the browsing assistance system

The browsing assistance system functionality can be graphically presented by the flowchart depicted in Figure 3. Initially, the parameters of the first and second level expansions, and the size of the recommendation window w are set. In parallel, the user's current browsing state p is identified. If the browsing state p is identified to be the starter and/or attractor, the system calculates the appropriate recommendation set $r^{(w)}(p)$, as described by equations (1) and (2), with respect to the relative frequency ordering function f expressed in (3). The recommendations are then presented to the user on-the-fly at the given page p.

4   Practical Evaluation of the Assistance System

Practical evaluation of the assistance system has been performed on the real-world data of a large scale intranet portal with significant knowledge worker user base. The voluminous data incorporates browsing and behavioral characteristics of wide ranging user assemblage. It serves as a suitable testbed for the evaluation of the presented system. First, we briefly introduce the case study portal, plus related data, and then present the essential evaluation of the introduced system.

4.1   Intranet and Data

The present study draws from the web log data analysis of The National Institute of Advanced Industrial Science and Technology (Table 1). Intranet portal of the institute is significantly large and complex. The web-core consists of six servers connected to the high-speed backbone in a load balanced configuration. Its accessibility ranges from high-speed optical to wireless connectivity.

The institute has a number of branches throughout the country. Services and resources are decentralized. Intranet portal incorporates a rich set of resources including documents (in various formats), multimedia, software, etc. Wide spectrum of services facilitate implementation of institutional business processes, management of cooperation with industry, academia, and other institutes, localization of internal resources, etc. They also feature blogging and networking services. Visible web space is in the excess of 1 GB, and deep web space is considerably larger. However, it is difficult to estimate its size due to the distributed architecture and alternating back-end data.

Vast intranet traffic produces a considerably large set of web log data. The traffic is both human and machine generated. Data preprocessing, elimination of

Table 1. Information about the data used in the study

Data Volume	~60 GB
Average Daily Volume	~54 MB
Number of Servers	6
Number of Log Files	6814
Average File Size	~9 MB
Time Period	3/2005 — 4/2006
Log Records	315 005 952
Resources	3 015 848
Sessions	3 454 243
Unique Sessions	2 704 067
Subsequences	7 335 577
Unique Subsequences	3 547 170
Valid Subsequences	3 156 310
Unique Valid Subsequences	1 644 848
Users	~10 000

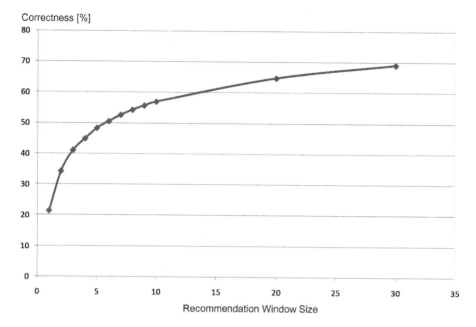

Correctness [%]

Recommendation Window Size

Fig. 4. Recommendation correctness of the browsing assistance system with respect to the size of the recommendation window w between 1 and 30. The steepest performance increase is noticeable in the range $w \in <1,5>$.

the machine generated traffic, and segmentation of the detected human behavior into sessions and subsequences has been presented in [4] and is not detailed here. The resulting working data and basic portal statistics are described in Table 1. The large scale data contains records of significant and behaviorally diverse user population.

4.2 System Evaluation

The processed data of the studied intranet portal were utilized for the practical evaluation of the introduced browsing assistance system. Required evaluation should verify the correctness of the system's recommendations given the actual pages users have accessed during their browsing experiences. We start with the concise description of the test set selection and then present the obtained correctness values for the various sizes of recommendation sets.

Individual users can be identified by the distinct IP addresses. Unfortunately, this holds only for the registered static IP addresses. The available data contained both statically and dynamically assigned IP addresses. Smaller portion of the distinct IP addresses were static and larger portion of the addresses were dynamic. It should be noted that the exact identification of individual users is generally not possible for dynamically assigned IP addresses. However, the detected IP address space proportionally reflected the number of users.

We identified IP addresses having more than fifty sessions originating from them. This, on average, approximately relates to at least once per working week browsing activity. There were 8739 such IPs. The activity logs of these addresses were randomly sampled for subsequences. Ten subsequences were selected from each of these IP addresses. The sequences were then sampled for the test navigation points and their actual corresponding attractors and/or starters. Thus the testing set consisted of the pairs (p, y): point $p \rightarrow$ target y. The total number of the testing pairs was 87390.

Given a navigation point p_i in the testing set $\{(p_i, y_i)\}_i$ the introduced browsing assistance system generated the recommendation set $r^{(w)}(p_i)$. The generated recommendation set $r^{(w)}(p_i)$ was then checked whether it contains the corresponding target element y_i. The correctness of the recommendation was measured as a simple indicator function of y_i on $r^{(w)}(p_i)$. If the actual corresponding point y_i was present in the recommendation set $r^{(w)}(p_i)$, the recommendation was considered correct, otherwise it was considered incorrect. The assistance system indicated satisfactory performance. The results of the evaluation for varying recommendation window size $w \in\ <1, 30>$ are displayed in Figure 4. The correctness performance rises sharply until the window size five, and starts flattening from around twenty. In practice the window size ten may be the most suitable.

5   Conclusions and Future Work

A novel effective browsing assistance service for organizational information systems has been introduced. Its design stands on the solid analysis of the knowledge worker browsing behavior. The case study platform was a large corporate intranet portal. The assistance system is computationally efficient and scalable.

The design benefits from the identification of the essential navigation points where users initiate and terminate their browsing tasks—starters and attractors. These are the desired points where users pay the most attention, and thus are the most appropriate for providing assistance services. Given the identified browsing state of a user the presented system offers recommendations on the potentially desirable set of pages. The recommendation set is comprised of the suitably selected attractors and starters. The selection accounts for observed behavioral analytics and average relative frequency of occurrences. Practical verification of the system's correctness indicates satisfactory results.

The future work aims at enhancing the assistance system by employing profiling. Individual user profiles contain more accurate information about user browsing characteristics. Deriving the recommendations based on the data in user profiles may further improve correctness.

Acknowledgment

The authors would like to thank Tsukuba Advanced Computing Center (TACC) for providing raw web log data.

References

1. Wei, Y.Z., Moreau, L., Jennings, N.R.: A market-based approach to recommender systems. ACM Transactions on Information Systems 23, 227–266 (2005)
2. Moe, W.W.: Buying, searching, or browsing: Differentiating between online shoppers using in-store navigational clickstream. Journal of Consumer Psychology 13, 29–39 (2003)
3. Ahuya, M., Gupta, B., Raman, P.: An empirical investigation of online consumer purchasing behavior. Communications of the ACM 46, 145–151 (2003)
4. Géczy, P., Akaho, S., Izumi, N., Hasida, K.: Knowledge worker intranet behaviour and usability. Int. J. Business Intelligence and Data Mining 2, 447–470 (2007)
5. Zhang, J., Chang, C.K., Zhang, L.J., Hung, P.C.K.: Toward a service-oriented development through a case study. IEEE Transactions on Systems, Man and Cybernetics, Part A 37, 955–969 (2007)
6. Papazoglou, M.P., Georgakopoulos, D.: Service-oriented computing: Introduction. Communications of the ACM 46, 24–28 (2003)
7. Adomavicius, G., Tuzhilin, A.: Toward the next generation of recommender systems: A survey of the state-of-the-art and possible extensions. IEEE Transactions on Knowledge and Data Engineering 17, 734–749 (2005)
8. Benbunan-Fich, R.: Using protocol analysis to evaluate the usability of a commercial web site. Information and Management 39, 151–163 (2001)
9. Norman, K.L., Panizzi, E.: Levels of automation and user participation in usability testing. Interacting with Computers 18, 246–264 (2006)
10. Bucklin, R.E., Sismeiro, C.: A model of web site browsing behavior estimated on clickstream data. Journal of Marketing Research 40, 249–267 (2003)
11. Deshpande, M., Karypis, G.: Selective markov models for predicting web page accesses. ACM Transactions on Internet Technology 4, 163–184 (2004)
12. Wu, H., Gordon, M., DeMaagd, K., Fan, W.: Mining web navigaitons for intelligence. Decision Support Systems 41, 574–591 (2006)
13. Zukerman, I., Albrecht, D.W.: Predictive statistical models for user modeling. User Modeling and User-Adapted Interaction 11, 5–18 (2001)
14. Jozefowska, J., Lawrynowicz, A., Lukaszewski, T.: Faster frequent pattern mining from the semantic web. Intelligent Information Processing and Web Mining, Advances in Soft Computing, 121–130 (2006)
15. Géczy, P., Akaho, S., Izumi, N., Hasida, K.: Human web behavior mining. In: Proceedings of WWW/Internet, Vila Real, Portugal, pp. 163–170 (2007)
16. Catledge, L., Pitkow, J.: Characterizing browsing strategies in the world wide web. Computer Networks and ISDN Systems 27, 1065–1073 (1995)
17. Thakor, M.V., Borsuk, W., Kalamas, M.: Hotlists and web browsing behavior–an empirical investigation. Journal of Business Research 57, 776–786 (2004)

WebAngels Filter: A Violent Web Filtering Engine Using Textual and Structural Content-Based Analysis

Radhouane Guermazi[1], Mohamed Hammami[2], and Abdelmajid Ben Hamadou[1]

[1] MIRACL-ISIMS, Road Tunis Km 10 BP 242, 3021 Sfax Tunisia
[2] MIRACL-FSS, Road Sokra Km 3 BP 802, 3018 Sfax Tunisia
rguermazi@laposte.net
mohamed.hammami@fss.rnu.tn
abdelmajid.benhamadou@isimsf.rnu.tn
http://www.miracl.rnu.tn/

Abstract. The development of the Web has been paralleled by the proliferation of harmful Web pages content. Using Violent Web page as a case study, we review some existing solutions, then we propose a violent Web content detection and filtering system called "WebAngels filter" which uses textual and structural analysis. "WebAngels filter" has the advantage of combining several data mining algorithms for Web site classification. We present a comparative study of different data mining techniques to block violent contentWeb pages. Also, we discuss how the combination learning based methods can improve filtering performances. Our results show that it can detect and filter violent content effectively.

Keywords: Web Filtering Engine, Web classification and categorization, data-mining, violent Web site filtering.

1 Introduction

Recently, research has shown that media violence has increased in quantity and has also become much more graphic and sadistic. The Internet is adding an entirely new dimension to the issue of media violence. Kids are exposed to continuous violence on the Internet, ranging from sites with sophomoric cruel humour to disturbing depiction of torture and sadism. The emergence of violent content on the Web involved the necessity of providing filtering systems designed to secure the internet access. A significant number of these products concentrates on IPbased black list filtering, and their classification of Web sites is mostly manual, that is to say no truly automatic classification process exists. But, as we know, the Web is a highly dynamic information source. The ever-changing nature of the Web calls for new techniques designed to classify and filter Web sites and URLs automatically.

In this paper, we investigate this problem and describe "WebAngels filter", our automatic machine learning based violent Web sites classification and filtering system. We place the emphasis on the comparison of different data mining techniques for automatic violent website classification and filtering, and we

P. Perner (Ed.): ICDM 2008, LNAI 5077, pp. 268–282, 2008.
© Springer-Verlag Berlin Heidelberg 2008

demonstrate that a combination of classifiers can be applied to improve the filtering efficiency of violent web pages.

The remainder of this article is organized as follows. In section 2, we start out with over viewing related work according to Web filtering. Following that the "WebAngels filter" architecture and principle are presented in section 3. The violent Web sites classification is reviewed in section 4. An intensive experimental evaluation and comparison results are discussed in section 5. In section 6, we show the efficiency of combining data mining techniques to improve the filtering accuracy rate. An experimental evaluation and comparison of "WebAngels filter" with other products are presented in section 7. Finally section 8 summarizes the "WebAngels filter" approach and presents some concluding remarks and future work directions.

2   Related Research Work

Several litigious Web sites filtering approaches were proposed. Among these approaches, we can quote :

- The Platform for Internet Content Selection (PICS)[1] is a set of specification for content-rating systems which is supported by Microsoft Internet Explorer, Netscape Navigator and other several Web filtering systems. PICS can only be used as supplementary means for Web content filtering because it is a voluntary self-labelling system freely rated to content provider.
- The exclusion filtering approach, which allows all access except to sites on a manually constructed black list.
- The inclusion filtering approach, which only allows access to sites on a manually constructed white list.

The main problem with these two latest approaches is that because of the continuous emergence of new sites, it is hard to construct and maintain complete and up-to-date lists.

- The automated content filtering approach, where the acceptability of content is dynamically assessed in real-time, based on Web page content, removing the need to manually maintain a black list or a white list. We can distinguish two types:

  Keyword Blocking: In this approach a list of prohibited words is used to identify undesirable Web pages. If a page contains a certain number of forbidden keywords, it is considered undesirable. The problem with this method is the phenomenon of "overblocking" which blocks access to inoffensive Web sites for instance Web pages which fight against violence.

  Intelligent content Web filtering: Takes part of a more general problem of automatic Web sites categorization and needs to rely on machine learning. At least, three categories of intelligent content Web Filtering can be distinguished :

---

[1] http://www.w3.org/PICS

1. Textual content Web filtering: includes the work based only on a textual content analysis. Caulkins et al. presented a general and flexible classification method based on statistical techniques applied to text material for managing access to Web pages [1]. Su et al. used a hybrid model consisted of two classifiers: the first is a key word matching and the second is a text classifier [2]. Du et al. used a text classification for Web filtering [3]. Gao et al. proposed a web filtering scheme based on hypertext classification and present a novel classifier combined SVM and K- nearest neighbor to filter pornographic information on the WWW [4]. Polpinij et al. [5] presented a web filtering system based on content and investigate a text classifier to filter the pornography information on the WWW. They focused on Thai-language and English-language web sites and they aimed to investigate whether Support Vector Machine and Naive Bayes algorithms are suitable for web sites classification.

2. Structural content Web filtering: includes the work based on an analysis of structural content, such as, Lee et al. presented an implementation of a Web content filtering system that combines the use of an artificial neural network and the knowledge gained in the analysis of pornographic Web pages [6]. Wai and Paul used a content filtering of pornographic Web pages test based on structural and statistical analysis with Bayesian classification [7]. Zhang et al. described an URL-based objectionable content categorization approach and its application to Web filtering. They broke the URL into a sequence of n-grams with a range of n's and then a machine learning algorithm is applied to the n-gram representation of URLs to learn a classifier of pornographic Web sites [8]. Agrawal et al. proposed a density threshold model and a density based SVM model to classify objectionable content. They utilized multiple information sources and show that treating them independently for classification improves accuracy [9].

3. Visual content Web filtering: includes the work based on an analysis of visual content. Arentz and Olstad proposed a method of detection of pornographic images to discriminate the adult Web pages [10]. Wang et al. proposed a system named IBCOW (Image - based Classification of Objectionable Websites) which is capable of classifying a website as objectionable or benign based on image content . The system uses WIPETM (Wavelet Image Pornography Elimination) and statistics to provide robust classification of on-line objectionable World Wide Web sites [11]. Sibunruang et al offered a content-based web pornography filtering and image processing system to detect and filter inappropriate web sites. Their system consists of three main processes: filtering by normalized R/G ratio, histogram analysis, and filtering by matrix composition based on skin detection [12].

Other Web filtering solutions are based on an analysis of textual, structural, and visual contents, of a Web page [14,15,16].

Litigious Web pages classification and filtering are diversified. However, the majority of these works treat only adult character. We propose in the following section our tool for the violent Web page filtering which is useful for both kids and adults.

3   "WebAngels filter" Tool

Given the dynamic nature of Web and its huge amount of documents, we decide to build an automatic violent content detection engine, named "WebAngels filter".

The most important step for machine learning is the selection of the appropriate features, according to our a-priori knowledge of the domain, which best discriminates the different classes of the application. Informed by our previous study on the state of the art solutions, we decide that the analysis of Web pages for classification should not only rely on textual content but also on its structural one.

In order to speed up navigation, we opt to use a black list which creation and update is automatic thanks to the machine learning based classification engine. We also turn to use a keyword dictionary as the occurrence of violent explicit terms. This dictionary is an important clue for textual content and its use in the current commercial products.

"WebAngels filter" offers to users the following functionalities :

- Profiles managing: the access to the Web is restraint only for the users that have an account. Each profile is characterized by:
  - User and password for identification
  - Prohibited protocols list
  - Prohibited file extensions list
  - Authorized schedule navigation
  - White list used only for the very young profiles. The system only allows access to recorded site in this list.
- White list managing: The administrator has the possibility to add, remove or empty all the URLs of the white list.
- Black list managing: the black list gathers URLs considered as violent by the system. Its update is automatic thanks to the machine learning based classification engine. Nevertheless, the administrator can also update this list manually by adding or removing an URL from the list.
- Dictionary managing: the administrator can update the words of the dictionary.
- Navigation history: the system saves the navigation history including all visited URLs, schedules of the visits and decision of the system.

All those functionalities that we have already talked about, are just secondary options offered to specific categories. White list can be used for children in case we want them to navigate in some designed sites, however, black list is just an option which accelerates analysis's processus of Web pages.

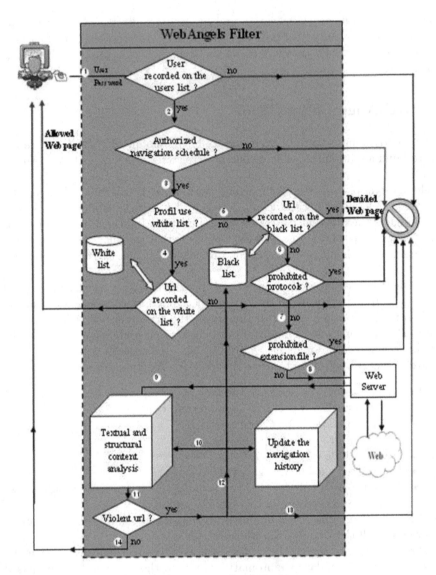

Fig. 1. "WebAngels filter" Architecture

The main thing that has to be stressed on is the textual and structural content based analysis of the Web pages by our intelligent content Web filtering tool.

Figure 1 depicts the general architecture of WebAngels. As we can see, when a URL is launched, "WebAngels filter" carries out the following actions:

(1) Make sure that the user exists in profiles list. If the user is not recorded on the users list then the page is blocked.
(2) Verify the navigation schedule. If the user is not authorized to navigate in the current system time then the page is blocked.

(3) Check if the profile uses the white list. (4) The system blocks/allows the access to Web page if it is/is'nt on the White list.

(5) Verify whether the URL is recorded in the black list. If the URL is recorded on the black list then the system blocks the page.

(6) Control if the URL protocol is prohibited, and in this case, the system blocks the page.

(7) Check, if the URL file extension is a prohibited file extension, and in this case, the system blocks the page.

(8) Load the HTML code source of the Web page.

(9) Analyze the textual and structural information code by detecting the HTML tags of the code and extracting the text itself. One can then calculate a certain number of criteria (percentage of words identified as violent, etc.) by analyzing the text and the tags. Rules make it possible to say, by knowing the value of the criteria, if the page is or not authorized.

(10) Update the navigation history.

(11) If the URL is violent then (12) Update black list and (13) Block the page, If not (14) Allow the page.

4   Violent Web Sites Classification

We propose to build an automatic violent content detection solution based on a machine learning approach using a set of manually classified sites, in order to produce a prediction model, which makes it possible to know which URLs are suspect and which are not.

To classify the sites into two classes, we are based on the KDD process for extracting useful knowledge from volumes data [17]. The general principle of the approach of classification is the following: Let S be the population of samples to be classified. To each sample s of S one can associate a particular attribute, namely, its class label C. C takes its value in the class of labels (0 for violent, 1 for non violent)

$$C : S \rightarrow \Gamma = \{Violent, nonViolent\}$$
$$s \in S \mapsto C(s) \in \Gamma$$

Our study consists in building a means to predict the attribute class of each Web site. To do it three major steps are necessary: (a) extraction of textual and structural content based features; (b) supervised learning; (c) evaluation and validation of the predict model of the learnt model on the training and test data set.

4.1   Data Preparation

In this stage we identify exploitable information and check their quality and their effectiveness in order to build a two-dimensional table from our training corpus. Each table row represents a web page and each column represents a feature. in the last column, we save the web page class (0 or 1).

Learning and test data set. The data mining process for violent Web sites classification requires a representative learning data set consisting of a significant set of manually classified Web sites. Considering the diversity and the enormous number of Web sites on Internet, this phase constitutes a considerable work.

In the collection of violent Web sites, we try to have a diversified base in terms of content, treated languages and structure.

In the selection of non violent sites, we include those which can make confusion, in particular sites which fight against the violence and the sites of law, etc. The rest of sites are randomly selected, we find source code sites, comic sites, educational sites, children sites, etc.

Our training data set is composed of 700 sites of which 350 are violent, and 350 are non violent.

Independently of the learning data set, we manually collect a test data set, consisting of 300 Web sites, half of them being violent while the other half being inoffensive.

Textual and structural contents analysis. The selection of features used in a machine learning process is a key step which directly affects the performance of a classifier. Our study of the state of the art and manual collection of our test data set helped us a lot to gain intuition on violent website characteristics and to understand discriminating features between violent web pages and inoffensive ones. These intuitions and understandings have enabled us to select both textual and structural content-based features for better discrimination purpose. A semi-automatic collected violent words vocabulary (or dictionary) has been used to calculate these features. Indeed, we take into our consideration the construction of this dictionary and, unlike a lot of commercial filtering products, we build a semi-automatic multilingual dictionary based on a statistical method which uses n-gram as a means of representing texts and statistics of χ^2 as a means of selection of the relevant terms [18].

The features used to classify the Web pages are n_v_words_page (total number of violent words in the current Web page), %v_words_page (frequency of the violent words of the page), n_v_words_url (total number of violent words in the URL), %v_words_url (frequency of violent words in the URL), n_v_words_title (total number of violent words which appear in the title tag), %v_words_title (frequency of violent words which appear in the title tag), n_v_words_body (number of violent words which appear in the body tag), %v_words_body (frequency of violent words which appear in the body tag), n_v_words_meta (number of violent words which appear in the meta tag), %v_words_meta (frequency of violent words which appear in the meta tag), n_links (total number of links), n_links_v (total number of links containing at least one violent word), %links_v (frequency of links containing at least a violent word), n_img (total number of images), n_img_v (total number of images whose name contains at least a violent word), n_v_img_src (total number of violent words in the attribute src of the img tag), n_v_img_alt (total number of violent words in the attribute alt of the img tag) and %img_v (frequency of the images containing a violent word).

4.2   Supervised Learning

In the literature, there are several techniques of supervised learning, each having its advantages and disadvantages. In our approach, we study the use of the support vector machines [19], the graphs of decision tree [20] and the neural networks [25].

Support Vector Machines (SVM). is defined over a vector space where categorization is achieved by linear or non-linear separating surfaces in the input space of the original data set [19].

$$\min_{w,b,\xi} \frac{1}{2} w^T w + C \sum_{i=1}^{N} \xi_i \tag{1}$$

$$Subject\ to \begin{cases} y_i(w^T \Phi(x_i) + b) \geq +1 - \xi_i, & i = 1, ..., N \\ \xi_i \geq 0, & i = 1, ..., N \end{cases}$$

where ξ_i's are slack variables needed to allow misclassifications in the set of inequalities, and $C \in R^+$ is a tuning hyper parameter, weighting the importance of classification errors to the margin width.

Here training vectors x_i are mapped into a higher (maybe infinite) dimensional space by the function ϕ . Then SVM finds a linear separating hyperplane with the maximal margin in this higher dimensional space. Furthermore, $K(x_i, x_j) = \phi(x_i)^T \phi(x_j)$ is called the kernel function. In our work,we use a SVM using a Radial Basis Function as kernel defined by the following formula:

$$K(x_i, x_j) = \exp(-\gamma \|x_i - x_j\|^2) \tag{2}$$

Actually, in order to use the RBF kernel, appropriate values for the kernel parameters C (penalty) and γ (kernel width) need to be determined. Intensive experiments were used to determine the optimal values for these parameters. ($C = 1, \gamma = 10$) give the maximum prediction accuracy on learning and test data set.

Decision tree algorithms. The decision tree algorithms are a well-known machine learning approach to automatic induction of classification trees based on training data. In a decision tree, we begin with a learning data set and look for the particular attribute which will produce the best partitioning by maximizing the variation of uncertainty $\Im_\lambda$ between the current partition and the previous one. As $I_\lambda(S_i)$ is a measure of entropy for partition S_i and $I_\lambda(S_{i+1})$ is the measure of entropy of the following partition S_{i+1}. The variation of uncertainty is:

$$\Im_\lambda = I_\lambda(S_i) - I_\lambda(S_{i+1}) \tag{3}$$

For $I_\lambda(S_i)$ we can make use of the quadratic entropy (4) or Shannon entropy (5) according to the selected method:

$$I_\lambda(S_i) = \sum_{j=1}^{K} \frac{n_j}{n} (-\sum_{i=1}^{m} \frac{n_{ij} + \lambda}{n_i + m\lambda} (1 - \frac{n_{ij} + \lambda}{n_i + m\lambda})) \tag{4}$$

$$I_\lambda(S_i) = \sum_{j=1}^{K} \frac{n_j}{n} \left(-\sum_{i=1}^{m} \frac{n_{ij} + \lambda}{n_i + m\lambda} \log_2 \frac{n_{ij} + \lambda}{n_i + m\lambda} \right) \tag{5}$$

Where n_{ij} is the number of elements of class i at the node S_j with $i \in \{c1, c2\}$; n_i is the total number of elements of the class i, $n_i = \sum_{j=1}^{K} n_{nj}$; n_j is the number of elements of the node S_j, $n_j = \sum_{i=1}^{2} n_{ij}$; n is the total number of elements, $n = \sum_{i=1}^{2} n_i$; $m = 2$ is the number of classes (c_1, c_2). λ is a variable controlling effectiveness of graph construction, it penalizes the nodes with insufficient effective. In our work, four data mining algorithms including ID3 [21], C4.5 [22], IMPROVED C4.5 [23], SIPINA [24] have been experimented.

Multilayer Perceptron (MLP) is a network of simple neurons called perceptrons. The perceptron computes a single output from multiple real-valued inputs by forming a linear combination according to its input weights and then possibly putting the output through some nonlinear activation function [25]. Mathematically this can be written as:

$$y = \varphi\left(\sum_{i=1}^{n} \omega_i x_i + b\right) = \varphi(\omega^T + b) \tag{6}$$

where ω denotes the vector of weights, x is the vector of inputs, b is the bias and φ is the activation function.

In order to train a model based on an Artificial Neural Networks, we use a multilayer perceptron with the backpropagation learning algorithm, composed of an input layer, an output layer, and one hidden layer with 17 neurons. The activation function for the neurons in the hidden layer and in the output layer are sigmoid function:

$$f(x) = \frac{1}{1 - \exp(-x)} \tag{7}$$

In a backpropagation network, a supervised learning algorithm controls the training phase. Then, the input and output (desired) data need to be provided, thus permitting the calculation of the error of the network as the difference between the calculated output and the desired vector. The network's weights adjustment is conducted by backpropagating such an error to the network. The weight change rule is a development of the perceptron learning rule. Weights are changed by an amount proportional to the error at that unit times the output of the unit feeding into the weight. Equation 8 shows the general weight correction for the delta rule.

$$\Delta W_{ij} = \eta \delta_j y_i \tag{8}$$

δ_j is the local gradient, y_i is the input signal of neuron j, and η is the learning rate parameter that controls the strength of change.

5   Experimental Results

Actually, we carried out a series of experiments, focusing on textual and structural content-based features, we decide to use several data mining techniques

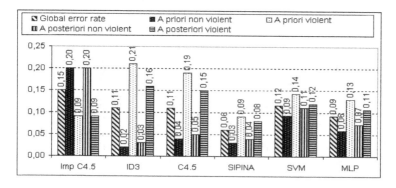

Fig. 2. Experimental results by the six algorithms on learning data set

to derive learnt model, from a learning data set. The quality of this model is evaluated using random error rate technique. The stability of the learnt model is further validated using cross-validation and bootstrapping techniques. Indeed, three measures have been used: *global error rate, a priori error rate* and *a posteriori error rate*. As global error rate is the complement of classification accuracy rate, while *a priori error rate* (respectively, *a posteriori error rate*) is the complement of the classical *recall rate* (respectively, *precision rate*).

During this series of experiments, six data mining algorithms have been studied, including ID3, C4.5, Improved C4.5, SIPINA, MLP and SVM. The following figure shows the individual error rates of these algorithms on learning data set.

As we can see in figure 2, the SIPINA and MLP algorithm echoed very similar performance on the feature vector, displaying a global error rate of less than 9% and only 6% for the best one (SIPINA). The a priori error rates show that different algorithms are not as efficient as SIPINA concerning a priori error rate on violent Web sites. This rate is important for evaluation because it quantifies the efficiency of the classification of violent Web sites. Also, the same algorithm shows the best a priori non violent error rate. We can conclude that SIPINA produce the best learnt model for our problem. A good predict model obtained by a data mining algorithm from the learning data set should not only produce good classification performance on data already seen but also on unseen data as well. In order to ensure the performance stability of our learnt model from learning data set and validated by random error rate technique, we have also tested the learnt model on our test data set. The experimental results are depicted in figure 3.

All the six data mining algorithms echoed very similar performance, displaying a global error rate of less than 13% and only 9% for the best one (C4.5). There is clearly a tradeoff between a priori error rate on violent Web sites and non violent ones, and the much the same between a posteriori error rate on violent Web sites and non violent ones. For instance, when Improved C4.5 displayed the best performance on violent Web sites with a priori rate of 4%, it recorded on the other hand a priori error rate of 18% on non violent Web sites which is

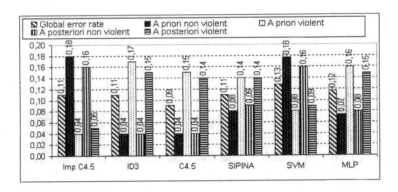

Fig. 3. Experimental results by the six algorithms on test data set

the first worst performance. The same tradeoff was also observed on a posteriori error rate side, when ID3 achieved, with 4%, the best a posteriori error rate on non violent Web sites, it recorded on the other hand a 15% of a posteriori error rate on violent Web sites which is the second worst performance among the six algorithms.

It seems that the best average behavior has been achieved by C4.5 and SIP-INA, which records the best compromise between violent and non violent error rate. Taking into account both obtained results (learning and test), we choose it as the best algorithms.

6   Classifier Combination

As shown in figure 2 and figure 3, the six data mining algorithms (ID3, C4.5, SIPINA, Improved C4.5,SVM and MLP) displayed different performances on the various error rate. Although SIPINA and C4.5 give the best results, combining algorithms may provide better results [26]. In our work, we have experimented three combining classifiers methods:

1. Majority voting: The page is considered violent if the majority of classifiers (SIPINA, C4.5, Improved C4.5, ID3, MLP and SVM) considered it as violent. In case of conflict votes, we consider the decision of SIPINA.
2. Scores combination: we thus affect a weight associated to the score classification decision of each data mining algorithm according to the following formula:

$$Score_{page} = \sum_{i=1}^{4} \alpha_i S_i \qquad (9)$$

Where α_i presents the belief value we accord to the algorithm i.

$$\alpha_i = \frac{\beta_i}{\sum_{j=1}^{K} \beta_i} \ with \ \beta_i = (1 - (\varepsilon_i - \delta))^n \qquad (10)$$

where

- ε_i: the global error rate of the i-th algorithm ;
- N : the number of algorithms used for classification. In our case N=4 ;
- n : the power in order to emphasize the difference in weight ;
- δ : a threshold value that we take away from the error rate again to emphasize the difference in weight.

To calculate the different weights, we are based on the global error rate in order to have the best balanced behavior of our filter, on both violent and non violent classes. After several experiments, we have fixed n=5 and δ=0.085 giving the best results on test data set.

S_i presents the score to be a violent Web page according to i-th data mining algorithm $(i = 1 \ldots 4)$. This combining method can be used only if all algorithms have same type. In our case we used four algorithms based on decision tree to know ID3, C4.5, SIPINA, Improved C4.5. The score is calculated for each algorithm as follows: having the decision tree, we seek for each page the node that corresponds to the values of its characteristics vector. Once found the score is calculated as the ratio of pages judged as violent from the current node and its previous.

The page is considered as violent if its score is greater than a threshold equal to 0.496 which is determinate automatically.

3. Learning the parameters of the combination function: The task of combining classifiers can be considered as a simple classification problem. the input vector contains the score of each individual decision tree, and the output is the forecast class. Indeed, it is possible to use classification algorithms to learn the parameters of the combination function. In our case, we have applied SIPINA to solve this combination problem.

Figure 4 presents a comparison between majority voting model, the scores combination model and the learning combination models on the test data set.

The obtained results confirm the well interest to use the scores combination method. Indeed, it provides better results than Majority Voting and SIPINA combination function. One clearly observes the reduction in the global error rate of 0.13 for Majority Voting and 0.11 by the SPINA combination method to

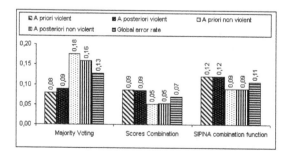

Fig. 4. Comparison combination approaches on test data set

0.07 by the scores combination method. So, to build our violent Web filtering tool, we are based on the predict model of the scores combination method.

7   Comparison of "WebAngels filter" with Others Products

Encouraged by the previous validation results, we then compare the WebAngels to other Web-based violent content detection and filtering systems. The comparison chart is shown in Figure 5. The selected systems are control kids[2] , content protect[3] , k9-webprotection[4] and Cyber patrol[5]. these tools are parametric to filter violent Web Content.

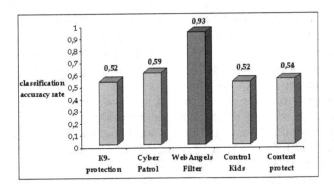

Fig. 5. Classification accuracy rate of "WebAngels Filter" compared to some products

Figure 5 further highlights the performance of "WebAngels Filter" compared to other violent content detection and filtering systems. The comparison was conducted on 300 Web sites, including 150 violent Web sites and 150 non violent Web sites. The least effective results come from control kids and K9protection with 52% success rates while our system is the best with a 93% success rate. Other systems give success rates between 54% (content-protect) and 59% (Cyber Patrol). Thanks to our textual and structural content-based features, the results from these experiments show that "WebAngels Filter" outperform the existing commercial products in the market.

8   Conclusion

In this paper, we have proposed, a machine learning-based system for detecting and filtering violent Web pages named "WebAngels filter", which combines

---

[2] http://www.controlkids.com/fr
[3] http://www.contentwatch.com
[4] http://www.k9webprotection.com
[5] http://www.cyberpatrol.com

textual and structural content-based analysis. Our solution has shown its effectiveness, scoring a 93% classification accuracy rate on our test data set.

We can thus summarize our major contribution by : first, the study and comparison of several techniques of data mining, knowing, Support Vector Machines, Artificial Neural Networks and Decision tree to build a violent Web classifier based on a textual and structural content-based analysis for improving Web filtering. Second, combining classifiers to improve the filtering accuracy rate.

Our experimental evaluation shows the effectiveness of our approach, however, many future work directions can be considered. Actually, we analyze only the Web page, and an additional analysis of the neighbor's Web page can be one of the directions of our future work. It is a fact that Web has become more and more multimedia, including music, images and videos. The work described in this paper suggests that Web document classification can benefit from visual content-based analysis; so, we must think to integrate the treatment of the visual content in the predict models. Future research is underway to develop effective filtering tools for other types of harmful Web pages,such as nazi, racist, etc.

References

1. Caulkins, J.P., Ding, W., Duncan, G., Krishnan, R., Nyberg, E.: A method for managing access to web pages: Filtering by statistical classification (fsc) applied to text. Decision Support Systems 42, 144–161 (2006)
2. SU, G.-y., LI, J.-h., MA, Y.-h., LI, S.-h.: Improving the precision of the keyword-matching pornographic text filtering method using a hybrid model. Journal of Zhejiang University Science 5(9), 1106–1113 (2004)
3. Du, R., Naini, R.S., Susilon, W.: Web filtering using text classification. In: IEEE International Conference on Networks, pp. 325–330 (2003)
4. Gao, Z., Lu, G., Dong, H., Wang, S., Wang, H., Wei, X.: Applying a novel combined classifier for hypertext classification in pornographic web filtering. In: 30th International Conference on Software Engineering (to appear, 2008)
5. Polpinij, J., Chotthanom, A., Sibunruang, C., Chamchong, R., Puangpronpitag, S.: Content-based text classifiers for pornographic web filtering. In: IEEE International Conference on Systems, Man and Cybernetics, pp. 1481–1485 (2006)
6. Lee, P.Y., Hui, S.C., Fong, A.C.M.: Neural networks for web content filtering. IEEE Intelligent Systems 17(5), 48–57 (2002)
7. Ho, W.H., Watters, P.A.: Statistical and structural approaches to filtering internet pornography. In: IEEE Conference on Systems, Man and Cybernetics, pp. 4792–4798 (2004)
8. Zhang, J., Qin, J., Yan, Q.: The role of urls in objectionable web content categorization. In: Proceedings of the 2006 IEEE/WIC/ACM International Conference on Web Intelligence, pp. 277–283 (2006)
9. Agarwal, N., Liu, H., Zhang, J.: Blocking objectionable web content by leveraging multiple information sources. SIGKDD Explorations 8(1), 17–26 (2006)
10. Arentz, W.A., Olstad, B.: Classifying offensive sites based on image content. In: Computer Vision and Image Understanding, pp. 295–310 (2004)
11. Wang, J.Z., Li, J., Wiederhold, G., Firschein, O.: Classifying objectionable websites based on image content. In: Plagemann, T., Goebel, V. (eds.) IDMS 1998. LNCS, vol. 1483, pp. 113–124. Springer, Heidelberg (1998)

12. Sibunruang, C., Polpinij, J., Chamchong, R., Chotthanom, A., Puangpronpitag, S.: A pornographic web patrol system based-on hierarchical image filtering techniques. In: JCIS (2006)
13. Dreyfus, G.: Neural Networks methodology and applications. Springer, Heidelberg (2005)
14. Denoyer, L., Vittaut, J.N., Gallinari, P., Brunessaux, S.: Structured multimedia document classification. In: ACM Symposium on Document Engineering, pp. 153–160 (2003)
15. Hammami, M., Chahir, Y., Chen, L.: A web filtering engine combining textual, structural, and visual content-based analysis. IEEE Transactions on Knowledge and Data Engineering 18(2), 272–284 (2006)
16. Hu, W., Wu, O., Chen, Z., Fu, Z., Maybank, S.J.: Recognition of pornographic web pages by classifying texts and images. IEEE Trans. Pattern Anal. Mach. Intell. 29(6), 1019–1034 (2007)
17. Fayyad, U., Piatetsky-Shapirop, G., Smyth, P.: The kdd process for extracting useful knowledge from volumes data. Communication of the ACM 39, 27–34 (1996)
18. Guermazi, R., Hammami, M., Ben-Hamadou, A.: Using a Semi-Automatic Keyword Dictionary for Improving Violent Web Site Filtering. In: International Conference on Signal-Image Technology & Internet based Systems (2007)
19. Boser, B., Guyon, E.I., Vapnik, V.: A training algorithm for optimal margin classifiers. In: Proceedings of the Fifth Annual Workshop on Computational Learning Theory, pp. 144–152 (1992)
20. Zighed, D.A., Rakotomalala, R.: Graphes d'Induction - Apprentissage et Data Mining. Hermes (2000)
21. Quinlan, J.R.: Induction of decision trees. Machine Learning 1, 81–106 (1986)
22. Quinlan, J.R.: C4.5: Programs for Machine Learning. Morgan Kaufmann, San Mateo (1993)
23. Rakotomalala, R., Lallich, S.: Handling noise with generalized entropy of type beta in induction graphs algorithm. In: International Conference on Computer Science and Informatics, pp. 25–27 (1998)
24. Zighed, D.A., Rakotomalala, R.: A method for non arborescent induction graphs. Technical report, Loboratory ERIC, University of Lyon 2 (1996)
25. Mitchell, T.: Machine learning. Mc Graw-Hill International edn., pp. 86–126 (1997)
26. Guermazi, R., Hammami, M., Ben-Hamadou, A.: Combining classifiers for web violent content detection and filtering. In: International Conference on Computational Science, pp. 772–779 (2007)

Mining Unexpected Web Usage Behaviors

Dong (Haoyuan) Li[1], Anne Laurent[2], and Pascal Poncelet[1]

[1] LGI2P - École des Mines d'Alès, Parc Scientifique G. Besse, 30035 Nîmes, France
{Haoyuan.Li,Pascal.Poncelet}@ema.fr
[2] LIRMM - Université Montpellier II, 161 rue Ada, 34392 Montpellier, France
laurent@lirmm.fr

Abstract. Recently, the applications of Web usage mining are more and more concentrated on finding valuable user behaviors from Web navigation record data, where the sequential pattern model has been well adapted. However with the growth of the explored user behaviors, the decision makers will be more and more interested in unexpected behaviors, but not only in those already confirmed. In this paper, we present our approach USER, that finds unexpected sequences and implication rules from sequential data with user defined beliefs, for mining unexpected behaviors from Web access logs. Our experiments with the belief bases constructed from explored user behaviors show that our approach is useful to extract unexpected behaviors for improving the Web site structures and user experiences.

1   Introduction

Recently, the applications of Web usage mining are more and more concentrated on finding valuable user behaviors from Web navigation record data (also known as Web access logs). A great deal of research work has been performed on porting data mining technologies to the Web usage analysis, in order to improve the personalization, the recommendation, and even the effectiveness of Web sites [1,2,3,4,5,6,7,8,9,10] by exploring the question: *what resources are frequently visited by whom during which periods?*

Among existing technologies, *sequential pattern* mining [11] has been well adapted to answer the above question [4,6,7,8,9]. All those sequential patterns extracted from Web access logs are typically the relationships like "on the Web site of customer support forum, 40% of users visited the TopicList page, then the Search page, then the Login page, and then the PostTopic page", or like "in the online store, 10% of customers visited the notebook cases page after having added a notebook computer to the shopping cart". This kind of relationships reflect the most general and reasonable user behaviors during Web navigations, however it become less important once we interpreted them as domain knowledge. When we regularly perform sequential pattern based Web usage mining on access logs, with the growth of the explored user behaviors, the decision makers will be more and more interested in exploring unexpected user behaviors that contradict existing knowledge, but not only in those the already confirmed.

P. Perner (Ed.): ICDM 2008, LNAI 5077, pp. 283–297, 2008.
© Springer-Verlag Berlin Heidelberg 2008

In this paper, we focus on finding *unexpected behaviors* (that contradict the explored user behaviors) from Web access logs within the context of domain knowledge (that corresponds to the explored user behaviors). To illustrate our goal, let us consider an online news Web site, where the latest news are listed on the static home page index.html by categories. The latest previous news can be visited from static category index pages like cat1.html, cat2.html, etc., and all news can be visited from server side script page listnews.php by specifying the category, like listnews.php?cat=1&page=3. The server side script page readnews.php provides the detail of a specified news identified by news, like readnews.php?news=20080114-002. Assume that (1) 60% of users visit index.html, then various readnews.php, then cat1.html, then various readnews.php, then listnews.php, then various readnews.php, and then other categories and various readnews.php, etc.; (2) 10% of users visit index.html, then cat5.html, then various readnews.php; (3) 8% of users visit readnews.php only once; (4) 0.005% of users visit a large number of readnews.php only. From traditional sequential pattern mining approaches, we may find the most general user behaviors described in (1) with a suitable minimum support threshold, but it is quite hard to find the behaviors described in (2), (3) and (4) because:

1. Most existing sequential pattern mining approaches do not consider the missing elements, neither the semantic contradictions between elements (e.g. between cat1.html and cat5.html) in a sequence. The constraint based approaches like SPIRIT [12] may find the sequences of (2) and (3), but the main drawback is that we cannot find all sequences like the one described in (2) by saying "categories contradicting cat1.html", but will have to indicate cat2.html, cat3.html, etc. exactly, since the constraint **not** cat1.html implies all pages different to cat1.html.
2. According to the model of sequential patterns, the sequences representing (2), (3) and (4) are *contained* in the sequences representing (1). Existing approaches that distinguish the support value of each frequent sequence (instead of maximal frequent sequence), like the *closed sequential pattern* [13], may find the existence of (2), (3) and (4) by computing and comparing the support values, but it is also difficult to indicate them.

The rest of this paper is organized as follows. Section 2 presents the application of our approach USER (Mining Unexpected SEquential Rules) for finding unexpected behaviors from Web access log files. In Sect. 3 we show our experimental results. We introduce the related work in Sect. 4. The conclusion is listed in Sect. 5.

2   Finding Unexpected Behaviors from Web Access Logs

In this section, we present the application of our approach USER for finding unexpected Web usage behaviors. We first propose a formal definition of the session sequence contained in Web access logs, then we detail our approach USER, that finds unexpected sequences and implication rules with user defined beliefs within the context of session sequences. We also briefly introduce the main algorithm of the approach USER.

2.1 Session Sequences for Web Access Log Analysis

The process of Web usage mining contains three phases, including the data preparation, pattern discovery and pattern analysis [3]. The first phase is necessary for cleaning the uninterested information contained in access logs, and also for converting the log content. Then at the second phase we can apply data mining algorithms to find interesting (sequential) patterns. Finally the last phase helps the user to further analyze the results with visualization and report tools. The principle of these three phases is not different from a general data mining process.

We consider the server side access log files in the NCSA Common Logfile Format (CLF) [14], which is supported by most mainstream Web servers including the Apache HTTP Server and the Microsoft Internet Information Services. The Common Logfile Format is as follows:

```
remotehost rfc931 authuser [date] "request" status bytes
```

A log file is a ASCII text-based file, each line contains a CLF log entry that represents a request from a remote client machine to the Web server. Figure 1(a) shows the CLF log entries contained in a log file of an Apache HTTP Server. Additional fields can be combined into the CLF log entries, such as the *referer* field and the *user agent* field, shown in Fig. 1(b).

```
146.19.33.138 - - [11/Jan/2008:17:40:00 +0100] "GET /~li/ HTTP/1.1" 200 5480
146.19.33.138 - - [11/Jan/2008:17:40:00 +0100] "GET /~li/deepred.css HTTP/1.1" 304 -
146.19.33.138 - - [11/Jan/2008:17:40:27 +0100] "GET /~li/TPBD/TP07.html HTTP/1.1" 200 2599
146.19.33.138 - - [11/Jan/2008:17:40:32 +0100] "GET /~li/TPBD/create.sql HTTP/1.1" 200 1376
146.19.33.138 - - [11/Jan/2008:17:49:21 +0100] "GET /~li/TPBD/TP07.pdf HTTP/1.1" 200 111134
```

(a)

```
146.19.33.138 - - [11/Jan/2008:18:27:35 +0100] "GET /~li/ HTTP/1.1" 200 1436 "-" "Mozilla/5
.0 (Macintosh; U; Intel Mac OS X; fr-fr) AppleWebKit/523.10.6 (KHTML, like Gecko) Version/3
.0.4 Safari/523.10.6"
146.19.33.138 - - [11/Jan/2008:18:29:38 +0100] "GET /~li/doc/ HTTP/1.1" 200 854 "http://www
.lgi2p.ema.fr/~li/" "Mozilla/5.0 (Macintosh; U; Intel Mac OS X; fr-fr) AppleWebKit/523.10.6
(KHTML, like Gecko) Version/3.0.4 Safari/523.10.6"
```

(b)

Fig. 1. (a) Common CLF log file entries. (b) Combined CLF log file entries.

According to the definitions of *item*, *itemset* and *sequence* introduced in [11], an attribute is an item; an itemset $\mathcal{I} = (i_1, i_2, \ldots, i_m)$ is an unordered collection of items; a sequence is an ordered list $s = \langle \mathcal{I}_1 \mathcal{I}_2 \ldots \mathcal{I}_k \rangle$ of itemsets. This model is generally represented as a "Customer-Transaction-Items" relation where each sequence stands for all transactions of a customer identified by "CID" and each itemset stands a transaction identified by "TID". We propose the *session sequences* representation of Web access log entries, shown in Definition 1.

Definition 1 (Session Sequence). *Let $\mathcal{L}$ be a set of Web server access log entries and $l \in \mathcal{L}$ be a log entry. A session sequence $s \sqsubseteq \mathcal{L}$ is a sequence*

$$s = \langle (ip_s, S_0^s)(l_1^s.url, S_1^s) \ldots (l_n^s.url, S_n^s) \rangle,$$

such that for $1 \leq i \leq n$, $l_i^s.url$ is the URL requested from IP address ip_s, and for all $1 \leq i < j \leq n$, $l_i^s.time < l_j^s.time$, where $l_i^s.time$ and $l_j^s.time$ denote the request time for log entries l_i^s and l_j^s. S_0^s is a set of items that contains all optional information of the session s. $S_1^s \ldots S_n^s$ are sets of items that contain optional information for each log entry $l_1^s \ldots l_n^s$.

The set S_0^s can be empty or contain IP, date, time, user agent, and etc., for reducing the repetition of items. The sets $S_1^s \ldots S_n^s$ can be empty or contain HTTP query parameters of each access log entry. So that with session sequences, the log entries can be represented as shown in Fig. 2.

Session No.		IP/URL	Optional Information		CID	TID	Items
1	0	146.19.33.*	17h		1	1	11, 15
1	1	/~li/			1	2	21
1	2	/~li/deepred.css			1	3	22
1	3	/~li/TPBD/TP07.html			1	4	35
1	4	/~li/TPBD/create.sql		$\Rightarrow$	1	5	51
1	5	/~li/TPBD/TP07.pdf			1	6	52
2	0	146.19.33.*	17h		2	1	11, 15
2	1	/~li/TPBD/TP07.html			2	2	35
2	2	/~li/TPBD/TP07.pdf			2	3	52
2	3	/index.php	page=2		2	4	25, 59

Legend
11: 146.19.3.*   15: 17h   21: /~li/
22: /~li/deepred.css   25: /index.php   35: /~li/TPBD/TP07.html
51: /~li/TPBD/create.sql 52: /~li/TPBD/TP07.pdf 59: page=2

Fig. 2. Session sequence mapped CLF log entries

The sequential pattern mining finds all maximal frequent sequences with a user defined minimum support value, where the *support* of a sequence is defined as the fraction of the total number of sequences in the database that contain the sequence. So with the minimum support value 0.5, the session sequences shown in Fig. 2 contain a sequential pattern $\langle (11, 15)(35)(52) \rangle$, that is, $\langle (146.19.33.*, 17h)(TP07.html)(TP07.pdf) \rangle$.

2.2   Belief and Unexpectedness on Session Sequences

Before formalizing the belief and unexpectedness on session sequences, we first introduce several additional notions, then propose the occurrence relation and implication rules between sequences.

The *length* of a sequence s is the number of itemsets contained in the sequence, denoted as $|s|$. The *concatenation* of sequences is denoted as the form $s_1 \cdot s_2$, so that we have $|s_1 \cdot s_2| = |s_1| + |s_2|$. The notation $s \sqsubseteq^c s'$ denotes that the sequence s is a *contiguous subsequence* of the sequence s', for example $\langle (a)(b)(c) \rangle \sqsubseteq^c \langle (b)(\underline{a})(a, \underline{b})(\underline{c})(d) \rangle$. We denote the first itemset in a sequence s as $s^\top$ and the last itemset as $s_\bot$. We therefore note $s \sqsubseteq^\top s'$ if $s^\top \sqsubseteq s'^\top$, note $s \sqsubseteq_\bot s'$ if $s_\bot \sqsubseteq s'_\bot$, and note $s \sqsubseteq_\bot^\top s'$ if $s^\top \sqsubseteq s'^\top$ and $s_\bot \sqsubseteq s'_\bot$.

Given a sequence s such that $s_1 \cdot s_2 \sqsubseteq s$, the *occurrence relation* $\mapsto^{\langle \mathbf{op}, n \rangle}$ is a constraint on the occurrences of s_1 and s_2 in s, where $\mathbf{op} \in \{\neq, =, \leq, \geq\}$ and $n \in \mathbb{N}$. Let $|s'| \models \langle \mathbf{op}, n \rangle$ denote that the length of sequence s' satisfies the constraint $\langle \mathbf{op}, n \rangle$, then the relation $s_1 \mapsto^{\langle \mathbf{op}, n \rangle} s_2$ depicts $s_1 \cdot s' \cdot s_2 \sqsubseteq^c s$ where $|s'| \models \langle \mathbf{op}, n \rangle$. In addition, we have $\langle \leq, 0 \rangle$ implies $\langle =, 0 \rangle$. We also note $s_1 \mapsto^{\langle \geq, 0 \rangle} s_2$ as $s_1 \mapsto^* s_2$, and note $s_1 \mapsto^{\langle =, 0 \rangle} s_2$ as $s_1 \mapsto s_2$. For example, we have $\langle (a)(b)(c) \rangle \models \langle \geq, 2 \rangle$, $\langle (a)(b) \rangle \not\models \langle >, 2 \rangle$, $\langle (a)(b)(c)(a)(b)(c) \rangle$ satisfies $\langle (a)(b) \rangle \mapsto \langle (c) \rangle$ and $\langle (a)(b) \rangle \mapsto^{\langle \leq, 3 \rangle} \langle (c) \rangle$.

We also propose the *implication rule* on sequences, of the form $s_\alpha \Rightarrow s_\beta$ where s_α and s_β are two sequences. The rule $s_\alpha \Rightarrow s_\beta$ means that the occurrence of s_α in a sequence s, that is, $s_\alpha \sqsubseteq s$, implies $s_\alpha \cdot s_\beta \sqsubseteq^c s$. We constrain such a rule $s_\alpha \Rightarrow s_\beta$ with the occurrence relation $s_\alpha \mapsto^{\langle \mathbf{op}, n \rangle} s_\beta$ between the sequences s_α and s_β, then the constrained rule, denoted as $s_\alpha \Rightarrow^{\langle \mathbf{op}, n \rangle} s_\beta$, means therefore that $s_\alpha \sqsubseteq s$ implies $s_\alpha \cdot s' \cdot s_\beta \sqsubseteq^c s$ where $|s'| \models \langle \mathbf{op}, n \rangle$. Nevertheless, we further consider a semantics constraint on the rule. If the sequence s_γ semantically contradicts to the sequence s_β, denoted as $s_\beta \not\sim s_\gamma$, then $s_\alpha \sqsubseteq s$ implies $s_\alpha \cdot s_\gamma \not\sqsubseteq^c s$, or implies $s_\alpha \cdot s' \cdot s_\gamma \not\sqsubseteq^c s$ with the occurrence relation constraint, where $|s'| \models \langle \mathbf{op}, n \rangle$. With such occurrence relation and semantic contradiction constrained implication rules, we therefore define the belief on sequences as follows.

Definition 2 (Belief). *A belief on sequences consists of a rule $s_\alpha \Rightarrow s_\beta$, an occurrence relation constraint $\tau = \langle \mathbf{op}, n \rangle$, and a semantic contradiction $s_\beta \not\sim s_\gamma$, denoted as $[s_\alpha; s_\beta; s_\gamma; \tau]$. The rule $s_\alpha \Rightarrow s_\beta$ and the occurrence relation constraint $\tau = \langle \mathbf{op}, n \rangle$ depict that given a sequence s, $s_\alpha \sqsubseteq s$ implies $s_\alpha \cdot s' \cdot s_\beta \sqsubseteq^c s$, where $|s'| \models \tau$. The semantic contradiction $s_\beta \not\sim s_\gamma$ further depicts that $s_\alpha \sqsubseteq s$ implies $s_\alpha \cdot s' \cdot s_\gamma \not\sqsubseteq^c s$, where $|s'| \models \tau$.*

Example 1. Let us consider the online news site illustrated in Sect. 1, assume that most users visit the page `index.html` and then at least 3 news by the page `readnews.php`, and then the page `cat1.html`. This fact can therefore be stated by a belief with the implication rule $\langle (\texttt{index.html}) \rangle \Rightarrow (\texttt{cat1.html})$ and the occurrence relation constraint $\langle \geq, 3 \rangle$. If we know (by the Web site layout strategies or the previous result of sequential pattern mining) that the page `cat5.html` is not considered being visited two early, then we can further add the semantics constraint $\langle (\texttt{cat1.html}) \rangle \not\sim \langle (\texttt{cat5.html}) \rangle$. So that finally we have the belief $[\langle\langle\texttt{index.html}\rangle\rangle; \langle\langle\texttt{cat1.html}\rangle\rangle; \langle\langle\texttt{cat5.html}\rangle\rangle; \langle\geq, 3\rangle]$ for describing such Web usage behaviors.

According to different beliefs, we therefore propose three forms of unexpectedness on sequences: α-unexpectedness, β-unexpectedness and γ-unexpectedness.

Definition 3 (α−unexpectedness). *Given a belief $b = [s_\alpha; s_\beta; s_\gamma; *]$ and a sequence s, if $s_\alpha \sqsubseteq s$ and there does not exist s_β, s_γ such that $s_\alpha \mapsto^* s_\beta \sqsubseteq s$ or $s_\alpha \mapsto^* s_\gamma \sqsubseteq s$, then s contains the α−unexpectedness with respect to the belief b, and s is so called an α−unexpected sequence.*

A belief with the occurrence relation constraint $\tau = *$ states that s_β should occur after the occurrence of s_α, so that a sequence s violates $\tau = *$ if and only if no s_β occurs in s after s_α. We also require that s_γ should not occur after s_α in an α-unexpected sequence, since the occurrence of s_γ with respect to any constraint τ will be categorized to the γ-unexpectedness, see Definition 5. For example, with the belief $[\langle\langle\texttt{index.html}\rangle\langle\texttt{readnews.php}\rangle\rangle; \langle\langle\texttt{index.html}\rangle\rangle; \emptyset; *]$, we can find the users who went never back to the home page $\texttt{index.html}$ after reading news.

Definition 4 (β−unexpectedness). *Given a belief $b = [s_\alpha; s_\beta; s_\gamma; \tau]$ ($\tau \neq *$) and a sequence s, if $s_\alpha \mapsto^* s_\beta \sqsubseteq s$ and there does not exist s' such that $|s'| \models \tau$ and $s_\alpha \mapsto s' \mapsto s_\beta \sqsubseteq^c s$, then s contains the β−unexpectedness with respect to belief b, and s is so called a β−unexpected sequence.*

A β-unexpectedness reflects that the implication rule is broken because the occurrence of s_β violates the constraint τ. For instance, as illustrated in Example 1, even as we expected that most users will visit the category index page $\texttt{cat1.html}$ after reading at least 3 news from the home page $\texttt{index.html}$, there exist users who read less than 3 news before leaving the home page. With analyzing the sequences containing such a β-unexpectedness stated by the belief $[\langle\langle\texttt{index.html}\rangle\rangle; \langle\langle\texttt{cat1.html}\rangle\rangle; \langle\langle\texttt{cat5.html}\rangle\rangle; \langle\geq, 3\rangle]$, we might further find that, for example, this unexpected behavior mostly happens at the moments when the site is usually less updated. So that new site promotion strategies can be positioned for these periods.

Definition 5 (γ−unexpectedness). *Given a belief $b = [s_\alpha; s_\beta; s_\gamma; \tau]$ and a sequence s, if $s_\alpha \mapsto^* s_\gamma \sqsubseteq s$ and there exists s' such that $|s'| \models \tau$ and $s_\alpha \mapsto s' \mapsto s_\gamma \sqsubseteq^c s$, then s contains the γ−unexpectedness with respect to belief b, and s is so called an γ−unexpected sequence.*

The γ-unexpectedness is concentrated on semantics: the occurrence of s_β is replaced by its semantic contradiction s_γ with respect to the constraint τ. Considering again the above example, we know that the news listed on the page $\texttt{cat1.html}$ are semantically different to those listed on the page $\texttt{cat5.html}$ (e.g. "All latest news" vs. "All old news", or "Politics" vs. "Entertainments"). If a lot of users visit the news listed on $\texttt{index.html}$ then those listed on $\texttt{cat1.html}$, and only a few users visit $\texttt{cat5.html}$ instead of $\texttt{cat1.html}$, then it may valuable to explore such an unexpected behavior. For example, assume that (1) from

08h to 23h, 60% of users confirm the explored behavior $\langle(\texttt{index.html})\rangle \mapsto^{\langle\geq,3\rangle}$ $\langle(\texttt{cat1.html})\rangle$; (2) from 23h to 08h of the second day, 80% of users confirm the unexpected behavior $\langle(\texttt{index.html})\rangle \mapsto^{\langle\geq,3\rangle} \langle(\texttt{cat5.html})\rangle$, then it is not difficult to see that the frequency of the sequence describing the behavior (1) can be much more higher than that of those describing the behavior (2). For this reason, frequency based sequence mining approaches are difficult to extract the behavior (2), but such a behavior is valuable to decision makers.

2.3 Unexpected Sequences and Implication Rules

The fact that a sequence s violates a given belief b is denoted as $s \not\models b$. For better describing the structure of unexpectedness, we propose the notions of the *bordered unexpected sequence*, the *antecedent sequence* and the *consequent sequence* within an unexpected sequence.

Definition 6 (Unexpected Sequence). *A sequence s that violates a belief $b = [s_\alpha; s_\beta; s_\gamma; \tau]$ is an unexpected sequence. The bordered unexpected sequence s_u is the maximum contiguous subsequence of s, (1) if s is α-unexpected, we have $s_a \cdot s_u = s$ ($|s_a| \geq 0$) such that $s_\alpha \sqsubseteq^\top s_u$; (2) if s is β-unexpected, we have $s_a \cdot s_u \cdot s_c = s$ ($|s_a|, |s_c| \geq 0$) such that $s_\alpha \sqsubseteq^\top s_u$ and $s_\beta \sqsubseteq_\perp s_u$; (3) if s is γ-unexpected, we have $s_a \cdot s_u \cdot s_c = s$ ($|s_a|, |s_c| \geq 0$) such that $s_\alpha \sqsubseteq^\top s_u$ and $s_\gamma \sqsubseteq_\perp s_u$. The subsequence s_a is the antecedent sequence. The subsequence s_c is the consequent sequence.*

Given a belief b and a sequence database $\mathcal{D}$, let $\mathcal{D}_U$ be the set of bordered unexpected sequences of each $s \in \mathcal{D}$ that $s \not\models b$, $\mathcal{D}_A$ be the set of antecedent sequences of each $s \in \mathcal{D}$ that $s \not\models b$, and $\mathcal{D}_C$ be the set of consequent sequences of each $s \in \mathcal{D}$ that $s \not\models b$. We therefore define the *unexpected sequential patterns* and the *unexpected implication rules* (including the *antecedent rules* and the *consequent rules*) as follows.

Definition 7 (Unexpected Sequential Pattern). *An unexpected sequential pattern s_U is a maximal frequent sequence in $\mathcal{D}_U$.*

The support for an unexpected sequential pattern is defined as the fraction of total number of sequences in $\mathcal{D}_U$ that support this unexpected sequential pattern:

$$supp(s_U) = \frac{|\{s \mid s_U \sqsubseteq s, s \in \mathcal{D}_U\}|}{|\mathcal{D}_U|}. \tag{1}$$

Definition 8 (Unexpected Implication Rules). *Let u denote the unexpectedness stated by b, an antecedent rule is a rule $s_A \Rightarrow u$ where s_A is a maximal frequent sequence in $\mathcal{D}_A$, and a consequent rule is a rule $u \Rightarrow s_C$ where s_C is a maximal frequent sequence in $\mathcal{D}_C$.*

The support for an antecedent rule $s_A \Rightarrow u$ (or for a consequent rule $u \Rightarrow s_C$) is defined as the fraction of total sequences in $\mathcal{D}_A$ (or in $\mathcal{D}_C$) that support the sequence s_A (or the sequence s_C):

$$supp(s_A \Rightarrow u) = \frac{|\{s \mid s_A \sqsubseteq s, s \in \mathcal{D}_A\}|}{|\mathcal{D}_A|} \tag{2}$$

or

$$supp(u \Rightarrow s_C) = \frac{|\{s \mid s_C \sqsubseteq s, s \in \mathcal{D}_C\}|}{|\mathcal{D}_C|}. \tag{3}$$

In order to describe the implication relations between the antecedent/consequent sequences and the unexpectedness, we measure the confidence of the antecedent rule within the sequence database $\mathcal{D}$, and that of the consequent rule within the consequent sequence set $\mathcal{D}_C$:

$$conf(s_A \Rightarrow u) = \frac{|\{s \mid s_A \sqsubseteq s, s \in \mathcal{D}_A\}|}{|\{s \mid s_A \sqsubseteq s, s \in \mathcal{D}\}|} \tag{4}$$

and

$$conf(u \Rightarrow s_C) = \frac{|\{s \mid s_C \sqsubseteq s, s \in \mathcal{D}_C\}|}{|\{\mathcal{D}_C\}|}. \tag{5}$$

The unexpected sequential patterns reflect the internal structure of the unexpectedness, and the unexpected implication rules reflect the implications and influences of the unexpectedness. For instance, as illustrated in the precedent example for describing Definition 5, assume that from 23h to 08h of the second day, 80% of users confirm the unexpected behavior $\langle(\text{index.html})\rangle \mapsto^{\langle 2,3\rangle}$ $\langle(\text{cat5.html})\rangle$, then an antecedent rule can be:

$$\langle(\text{23h-08h})\rangle \Rightarrow \langle(\text{index.html})\rangle \not\mapsto^{\langle 2,3\rangle} \langle(\text{cat5.html})\rangle,$$

and the confidence of such a rule is 0.8. Assume that 60% of users who violate the behavior $\langle(\text{index.html})\rangle \mapsto^{\langle 2,3\rangle} \langle(\text{cat5.html})\rangle$ like to click visit the advertisement displayed by ads3.php, then a consequent rule can be:

$$\langle(\text{index.html})\rangle \not\mapsto^{\langle 2,3\rangle} \langle(\text{cat5.html})\rangle \Rightarrow \langle(\text{ads3.php})\rangle.$$

If we further assume that a unexpected sequential pattern within the unexpectedness $\langle(\text{index.html})\rangle \not\mapsto^{\langle 2,3\rangle} \langle(\text{cat5.html})\rangle$ is explored with support value 0.9, for example, the sequential pattern $\langle(\text{index.html})(\text{readnews.php})(\text{cat3.html})\rangle$, then all those facts can interpreted as following:

"Between 23h and 8h, 80% of users do not follow the explored behavior, 90% of those users read news listed on the home page, and then visit the category 3 instead of visiting the category 5, and then 60% of them follow the same kind of online advertisements."

Such unexpected user behaviors can be enough important to decision makers for pushing new Web site design or collaboration strategies.

2.4   The Algorithm USER

Figure 3 briefly shows the algorithm USER. The algorithm accepts a sequence database $\mathcal{D}$, a user defined belief base $\mathcal{B}$, a minimum support threshold min_supp and a minimum confidence threshold min_conf as inputs. It finds all unexpected sequences contained in the sequence database $\mathcal{D}$ with respect to each unexpectedness u stated by each belief $b \in \mathcal{B}$. If a sequence s is unexpected to b, then

s will be partitioned to the antecedent sequence s_a ($|s_a| \geq 0$), the unexpected bordered sequence s_u and the consequent sequence s_c ($|s_c| \geq 0$). When all unexpected sequences have been extracted, the algorithm starts finding unexpected sequential patterns from each group of unexpected bordered sequences with the minimum support threshold min_supp. Finally the algorithm generates the antecedent/consequent rules from each group of antecedent/consequent sequences with the minimum confidence threshold min_conf.

A detailed description of the algorithm USER is listed in [15].

Input : a sequence database $\mathcal{D}$, a user defined belief base $\mathcal{B}$, a minimum
           support threshold min_supp and a minimum confidence threshold
           min_conf
Output: all unexpected sequential patterns and implication rules for each
           unexpectedness

1 for each sequence $s \in \mathcal{D}$ do
2   for each belief $b \in \mathcal{B}$ do
3   if s is $\alpha/\beta/\gamma$-unexpected to b
4   partition s to s_a, s_u and s_c
5   save s_a, s_u and s_c
6 for each unexpectedness u stated by each $b \in \mathcal{B}$ do
7   find sequential patterns from each $\mathcal{D}_U$ with min_supp
8   generate rules $s_A \Rightarrow u$ from each $\mathcal{D}_A$ with min_conf
9   generate rules $u \Rightarrow s_C$ from each $\mathcal{D}_C$ with min_conf

---

Fig. 3. Sketch of the algorithm USER

3 Experiments

To evaluate of our approach, we performed a number of experiments on two large access log files containing the access records of two Web servers during a period of 3 months. The first log file, labeled as LOGBBS, corresponds to a PHP based discussion forum Web site of an online game provider; the second log file, labeled as LOGWWW, corresponds to a laboratory Web site that also hosts the personal home pages of researchers and teaching staffs.

In our experiments, we split each log file into three 1-month period files, i.e., LOGBBS-{1,2,3} and LOGWWW-{1,2,3}. We generate the session sequences with the information of day (Monday to Sunday) and hour (0h to 23h) of the first log entry of a session. If the interval time of two log entries with the same remote client IP address is greater than 30 minutes, then the last log entry starts a new session sequence. Because the CLF log entry does not contain the "remoteport" information, at this moment we do not identify the accesses from remote clients hidden behind the proxy servers and NAT gateways, so that long session sequences with the same remote address will be cut into multiple sequences with a length no more longer than 50.

For each session sequence, the remote client IP address is considered as a block, such that the IP 146.19.33.138 will be converted to 146.19.33.*. We map only significant HTTP query parameters to items. For LOGBBS, the number of PHP page are very limited and the parameter can stand for an access request. For example, the request

/forumdisplay.php?f=2&sid=f2efeb85fcfd94ecbc2dba0f97b678a1

can be considered as an itemset (f=2), and the request

/viewtopic.php?t=57&sid=f2efeb85fcfd94ecbc2dba0f97b678a1

can be replaced by the itemset (t=57). We focus on the accesses of static HTML pages, server side script pages and JavaScript scripts, hence all other unconcerned files like cascading style sheets, images and data files are ignored (PDF and PS files are kept for LOGWWW). Table 1 details the number of session sequences and distinct items, and the average length of the session sequences contained in the Web access logs.

Table 1. Web access logs in our experiments

Access Log	Sessions	Distinct Items	Average Length
LOGBBS-1	27,294	38,678	12.8934
LOGBBS-2	47,868	42,052	20.3905
LOGBBS-3	28,146	33,890	8.5762
LOGWWW-1	6,534	8,436	6.3276
LOGWWW-2	11,304	49,242	7.3905
LOGWWW-3	28,400	50,312	9.5762

In order to compare our approach with the sequential pattern mining, we first apply the sequential pattern mining algorithm to find the frequent behaviors from LOGBBS-{1,2,3} and LOGWWW-{1,2,3} with different minimum support thresholds, shown in Fig. 4 and Fig. 5. In an acceptable range, the number of sequential patterns discovered are similar in all of the 3 periods, that increases the difficulty in analyzing new user behaviors. The post analysis shows that, for instance, within the frequent behaviors discovered with the minimum support 0.04 from LOGBBS-{1,2,3}, 149 sequential patterns discovered from all of LOGBBS-{1,2,3} are similar (contained in each other), i.e. > 40% of LOGBBS-1; 197 sequential patterns discovered from LOGBBS-{2,3} are similar, i.e., the similarity of accesses is > 70% in these two periods.

We then construct the belief bases from the workflow and those frequent behaviors discovered by sequential pattern mining. For LOGBBS, we first generate 5 beliefs from the workflow considered on this forum site, and then generate 5 beliefs from a set of selected sequential patterns discovered from LOGBBS-1. The following belief corresponds to an "expected" forum browsing order:

$$[\langle\langle(/)\rangle\rangle; \langle\langle(t=2)(t=5)\rangle\rangle; \langle\langle(t=5)(t=2)\rangle\rangle; \langle=, 0\rangle].$$

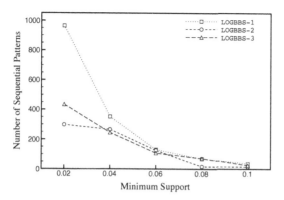

Fig. 4. Frequent behaviors discovered by the sequential pattern mining algorithm from LOGBBS

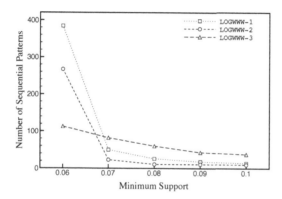

Fig. 5. Frequent behaviors discovered by the sequential pattern mining algorithm from LOGWWW

For LOGWWW we create 10 beliefs corresponding to the most frequent behaviors discovered from LOGWWW-1, for example, according to the navigation menu on the home page of this Web server, we have:

$$[\langle(0018\text{-}04.\texttt{html})\rangle; \langle(0019\text{-}27.\texttt{html})\rangle; \langle((0018\text{-}04.\texttt{html})\rangle; \langle\leq, 5\rangle],$$

where 0018-04.html corresponds to the index page of the section "Research", 0019-27.html corresponds to a subsection in "Research" and 0018-04.html corresponds to a subsection in the section "Publications".

Figure 6 and Fig. 7 show the number of unexpected sequential rules discovered by our approach USER. With comparison between the quantities of frequent user behaviors, our approach generates less than the sequential pattern mining approaches. The analysis of similarity shows that, with the minimum confidence 0.2, only 3 rules are similar from LOGBBS-{1,2,3}, 4 similar rules from LOGBBS-{1,3}, and from LOGWWW-{2,3} we find only 1 similar rule.

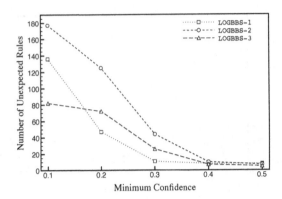

Fig. 6. Unexpected implication rules discovered by the approach USER from LOGBBS

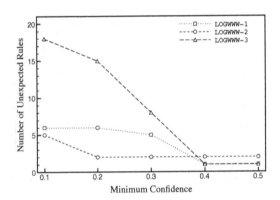

Fig. 7. Unexpected implication rules discovered by the approach USER from LOGWWW

We add 10 unexpected rules discovered from LOGBBS-1/LOGWWW-1 into the belief base for LOGBBS-2/LOGWWW-2, and add 10 (or all of them if the number is less than 10) unexpected rules discovered from LOGBBS-{1,2}/LOGWWW-{1,2} into the belief base for LOGBBS-3/LOGWWW-3, where these latest discovered unexpected rules are considered as explored behaviors. Table 2 [1] shows the results.

The principle of our approach does not imply that to add latest-discovered unexpected behaviors into the belief base will regularly affect the number of unexpected rules discovered from the next-period data, however, Table 2 also shows that the number of unexpected implication rules increased between LOGWWW-1 and LOGWWW-{2,3}, that is because the Web server corresponding to LOGWWW contains a great deal of non-profilable (e.g. personal home pages) Web content and the accesses are highly depended on the period, so that the beliefs could not be coherent to all data contained in LOGWWW.

---

[1] Number of beliefs: Number of unexpected implication rules.

Table 2. Unexpected rules with the increment of explored behaviors

min_conf	LOGBBS-1	LOGBBS-2	LOGBBS-3	LOGWWW-1	LOGWWW-2	LOGWWW-3
0.1	10:136	20:24	30:127	10:6	16:42	26:34
0.2	10:47	20:18	30:19	10:6	16:18	26:25
0.3	10:11	20:16	30:12	10:5	15:9	24:12
0.4	10:8	18:9	27:4	10:1	11:9	20:10
0.5	10:8	18:6	24:4	10:1	11:8	19:10

4 Related Work

In application domain, a great deal of Web analyzing tools, like the Webalizer [16], offer statistics based Web access analysis. In research domain, many Web usage mining approaches focus on the personalization and recommendation of Web sites by finding the most frequent user behaviors. To the best of our knowledge, we propose the first approach to unexpected Web usage mining, in considering constraints on both of the occurrence and semantics. Our approach considers a knowledge based *subjective interestingness measure* on sequence mining. As being summarized in [17], the interestingness measures for data mining can be classified as objective measures and subjective measures. Objective measures typically depend on the structure of extracted patterns and the criteria based on the approaches of probability and statistics (e.g. support and confidence); subjective measures are generally user and knowledge oriented, the criteria can be actionability, unexpectedness etc.. The belief driven unexpectedness is first introduced by [18] as a subjective measure where beliefs are categorized to hard beliefs and soft beliefs. A hard belief is a constraint that cannot be changed with new evidences, and any contradiction of a hard belief implies the error in gathering new evidence. A soft belief is a constraint that can be changed with new evidences by updating the degree of belief, and the interestingness of new evidence is measured by the changes of degree of belief.

Based on the proposition of [18], in the most recent approach to unexpected association rule mining presented by [19]. The mining process is done by the *APriori* based algorithms that find the minimal set of unexpected association rules with respect to a set of user defined beliefs. On sequence mining, [20] proposed a framework for finding unexpected sequential rules based on the frequency. In this approach, the author defines the unexpectedness from the constraints on sequential rules that depict the frequency of the content contained in a discovered sequential pattern, and the goal is to find all the sequences that do not satisfy given statistical frequency constraints.

5 Conclusion

In this paper we present the application of our general purposed approach USER for mining unexpected Web usage behaviors. We propose the formal definition of session sequence contained in Web access log files, for mining user behaviors. We introduce the notions of belief and unexpectedness within the context of session sequences, and propose the unexpected sequential patterns and implication rules. With finding unexpected implication rules, unexpected user behaviors can be studied in order to improve Web site structures and user experiences.

We also present our experiments on the access log files form different kinds of Web sites, the preprocess of such access logs and the management of beliefs are also introduced. Our experimental results show that the effects of unexpected Web usage mining highly depend on explored user behaviors, purpose, and structure of a Web site.

We are interested in applying similarity and fuzzy related approaches to unexpected Web usage mining, such as "many users like to visit the pages similar to **some.page** at **about** 6h p.m.". We are also interested in mining unexpected user behaviors with hierarchical data, for example, with the categories of Web pages, or with the path information contained in the URLs, in order to find more pertinent rules.

References

1. Büchner, A.G., Mulvenna, M.D.: Discovering internet marketing intelligence through online analytical web usage mining. SIGMOD Record 27(4), 54–61 (1998)
2. Spiliopoulou, M., Pohle, C., Faulstich, L.: Improving the effectiveness of a web site with web usage mining. In: WEBKDD, pp. 142–162 (1999)
3. Srivastava, J., Cooley, R., Deshpande, M., Tan, P.-N.: Web usage mining: Discovery and applications of usage patterns from web data. SIGKDD Explorations 1(2), 12–23 (2000)
4. Mobasher, B., Dai, H., Luo, T., Nakagawa, M.: Using sequential and non-sequential patterns in predictive web usage mining tasks. In: ICDM, pp. 669–672 (2002)
5. Eirinaki, M., Vazirgiannis, M.: Web mining for web personalization. ACM Trans. Internet Techn. 3(1), 1–27 (2003)
6. Masseglia, F., Teisseire, M., Poncelet, P.: HDM: A client/server/engine architecture for real-time web usage mining. Knowl. Inf. Syst. 5(4), 439–465 (2003)
7. Huang, Y.-M., Kuo, Y.-H., Chen, J.-N., Jeng, Y.-L.: NP-miner: A real-time recommendation algorithm by using web usage mining. Knowl.-Based Syst. 19(4), 272–286 (2006)
8. Missaoui, R., Valtchev, P., Djeraba, C., Adda, M.: Toward recommendation based on ontology-powered web-usage mining. IEEE Internet Computing 11(4), 45–52 (2007)
9. Masseglia, F., Poncelet, P., Teisseire, M., Marascu, A.: Web usage mining: Extracting unexpected periods from web logs. In: DMKD (2007)
10. Mobasher, B.: Data mining for web personalization. In: The Adaptive Web, pp. 90–135 (2007)
11. Agrawal, R., Srikant, R.: Mining sequential patterns. In: ICDE, pp. 3–14 (1995)
12. Garofalakis, M.N., Rastogi, R., Shim, K.: SPIRIT: Sequential pattern mining with regular expression constraints. In: VLDB, pp. 223–234 (1999)
13. Yan, X., Han, J., Afshar, R.: CloSpan: Mining closed sequential patterns in large databases. In: SDM (2003)

14. NCSA HTTPd Development Team: NCSA HTTPd Online Document: TransferLog Directive (1995), http://hoohoo.ncsa.uiuc.edu/docs/setup/httpd/TransferLog.html
15. Li, D.H., Laurent, A., Poncelet, P.: Mining unexpected sequential patterns and rules. Technical Report RR-07027 (2007), Laboratoire d'Informatique de Robotique et de Microélectronique de Montpellier (2007)
16. Barrett, B.L.: Webalizer (1997-2006), http://www.mrunix.net/webalizer/
17. McGarry, K.: A survey of interestingness measures for knowledge discovery. Knowl. Eng. Rev. 20(1), 39–61 (2005)
18. Silberschatz, A., Tuzhilin, A.: On subjective measures of interestingness in knowledge discovery. In: KDD, pp. 275–281 (1995)
19. Padmanabhan, B., Tuzhilin, A.: On characterization and discovery of minimal unexpected patterns in rule discovery. IEEE Trans. Knowl. Data Eng. 18(2), 202–216 (2006)
20. Spiliopoulou, M.: Managing interesting rules in sequence mining. In: Żytkow, J.M., Rauch, J. (eds.) PKDD 1999. LNCS (LNAI), vol. 1704, pp. 554–560. Springer, Heidelberg (1999)

Generalized Graph Matching for Data Mining and Information Retrieval

Alexandra Brügger[1], Horst Bunke[1],
Peter Dickinson[2], and Kaspar Riesen[1]

[1] Institute of Computer Science and Applied Mathematics, University of Bern,
Neubrückstrasse 10, CH-3012 Bern, Switzerland
{bruegger,bunke,riesen}@iam.unibe.ch
[2] C3I Division, DSTO, PO Box 1500, Edinburgh SA 5111, Australia
peter.dickinson@dsto.defence.gov.au

Abstract. Graph based data representation offers a convenient possibility to represent entities, their attributes, and their relationships to other entities. Consequently, the use of graph based representation for data mining has become a promising approach to extracting novel and useful knowledge from relational data. In order to check whether a certain graph occurs, as a substructure, within a larger database graph, the widely studied concept of subgraph isomorphism can be used. However, this conventional approach is rather limited. In the present paper the concept of subgraph isomorphism is substantially extended such that it can cope with don't care symbols, variables, and constraints. Our novel approach leads to a powerful graph matching methodology which can be used for advanced graph based data mining.

1 Introduction

In recent years powerful methods for knowledge mining and information retrieval have become available [1,2,3,4]. The vast majority of these mining and retrieval methods rely on data represented as a set of independent entities and their attributes. However, this approach does not consider a relevant part of the information in the underlying data, viz. the relationships between different entities. Graphs, which are in fact one of the most general forms of data representation, are able to represent not only the values of an entity, but can be used to explicitly model structural relations that may exist between different parts of an object [5,6]. This crucial benefit of graphs recently led to an emerging interest in graph based data mining [7].

In the current paper, mining of graph data refers to the process of extracting useful knowledge from the underlying data represented as a large database graph. Typically, the extracted knowledge mined from the database graph is also a graph, which may be, for instance, a subgraph of the underlying database graph [7]. In the present paper the concept of subgraph isomorphism is employed for information retrieval from the database graph. Subgraph isomorphism, which

P. Perner (Ed.): ICDM 2008, LNAI 5077, pp. 298–312, 2008.

can be seen as a formal concept for checking subgraph equality, intuitively indicates that a smaller graph is contained in a larger graph. Let us assume that
we represent a query by means of an attributed graph q, termed query graph.
Given q and the database graph G, we can check whether the query graph q is
contained in the underlying database G. That is, the knowledge mining system is
able to come to a binary decision (**yes** if q is a subgraph of G, and **no** otherwise).

Yet, the use of conventional subgraph isomorphism in graph based data mining implicates some severe limitations. First of all, the underlying database graph
often includes a rather large number of attributes, some of which might be irrelevant for a particular query. The second restriction arises from the limited answer
format to a given query graph q. That is, conventional subgraph isomorphism is
only able to check whether or not a query graph is embedded in a larger database
graph and can thus answer only **yes** or **no**. Thirdly, subgraph isomorphism in its
original mode does not allow constraints that may be imposed on the attributes
of a query graph q to model restrictions or dependencies.

The generalized subgraph isomorphism methodology described in the present
paper overcomes these three restrictions. First, the novel approach offers the
possibility to mask out attributes in query graphs. To this end, *don't care* values
are introduced for attributes that are irrelevant for a particular query. Secondly,
for the retrieval of more specific information from the database graph than just
a binary decision **yes** or **no**, *variables* are used. By means of these variables
we are able to retrieve values of predefined attributes in our database graph.
Thirdly, through the concept of *constrained variables*, for example variables that
can assume only values from a certain interval, we become able to define more
specific queries.

Our work is somewhat related to the work presented in [8,9]. In [8] a visual
language for querying and updating graph databases is introduced. In [9] another
generalization of subgraph isomorphism is introduced. However, our approach is
more rigorously embedded in a graph-theoretical context. The novelty and main
contribution of the present paper is a generalization of subgraph isomorphism
that leads to a powerful and flexible graph matching framework suitable for
general graph based data mining.

2   Graphs and Exact Graph Matching

The basic structures we are dealing with in this paper are attributed graphs, or
graphs, for short.

Definition 1 (Graph). *A graph is a 4-tuple $g = (V, E, \mu, \nu)$, where*

- *V is the finite set of nodes*
- *$E \subseteq V \times V$ is the set of edges*
- *$\mu : V \rightarrow \{(t, \mathbf{x}(t)) | t \in T_{nodes}, \mathbf{x}(t) \in (D_1(t) \times \ldots \times D_{n_t}(t))\}$ is the node attribute function*
- *$\nu : V \times V \rightarrow \mathcal{P}(\{(t, \mathbf{x}(t)) | t \in T_{edges}, \mathbf{x}(t) \in (D_1(t) \times \ldots \times D_{n_t}(t))\}) \setminus \emptyset$ is the edge attribute function.*

Through the node attribute function μ, each node in a graph is labeled by a type and a number of attributes. Formally, function μ assigns a (type,attribute)-pair $(t, \mathbf{x}(t))$ to each node $u \in V$. The first component of a (type,attribute)-pair, t, denotes the type of node u. The type t is an element of a finite set of node types, T_{nodes}. The second component is an attribute vector, i.e. $\mathbf{x}(t) = (x_1, \ldots, x_{n_t})$, where each attribute, x_i, belongs to some domain, $D_i(t)$. The dimension of vector $\mathbf{x}(t)$, i.e. the number n_t of its attributes, as well as each individual attribute domain, $D_i(t)$, is dependent on the type t of the node.

Edges are pairs of nodes, $(u, v) \in V \times V$. According to Def. 1, there exists at most one edge (u, v) from node u to node v. In some applications, it may be necessary to include more than one edge between the same two nodes, because of the existence of multiple relations. In the formal graph model provided in Def. 1, this can be accomplished by assigning several (type,attribute)-pairs, $(t_1, \mathbf{x}(t_1)), \ldots, (t_n, \mathbf{x}(t_n))$, to an edge (u, v) by means of edge attribute function ν, i.e. $\nu(u, v) = \{(t_1, \mathbf{x}(t_1)), \ldots, (t_n, \mathbf{x}(t_n))\}$. Note that the range of function ν is the power set of all (type,attribute)-pairs $(t, \mathbf{x}(t))$, where t is an edge type, i.e. an element of the finite set T_{edges}. Assigning $(t_1, \mathbf{x}(t_1)), \ldots, (t_n, \mathbf{x}(t_n))$ to edge (u, v) by means of ν is equivalent to providing n individual edges from node u to node v. The meaning of a pair $(t, \mathbf{x}(t))$, which is assigned to an edge by means of function ν is the same as for the nodes.

The identity of two graphs g_1 and g_2 is commonly established by defining a bijective function, termed graph isomorphism, mapping the nodes of g_1 to the nodes of g_2 such that the edge structure is preserved and the node and edge labels are consistent. In Fig. 1 (a) and (b) two isomorphic graphs are shown.

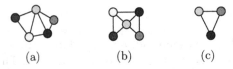

(a)   (b)   (c)

Fig. 1. Graph (b) is isomorphic to (a), and graph (c) is isomorphic to a subgraph of (a). Node attributes are indicated by color.

Definition 2 (Graph Isomorphism). *Assume that two graphs $g_1 = (V_1, E_1, \mu_1, \nu_1)$ and $g_2 = (V_2, E_2, \mu_2, \nu_2)$ are given. A graph isomorphism is a bijective function $f : V_1 \rightarrow V_2$ satisfying*

1. *$\mu_1(u) = \mu_2(f(u))$ for all nodes $u \in V_1$*
2. *for each edge $e_1 = (u, v) \in E_1$, there exists an edge $e_2 = (f(u), f(v)) \in E_2$ such that $\nu_1(e_1) = \nu_2(e_2)$*
3. *for each edge $e_2 = (u, v) \in E_2$, there exists an edge $e_1 = (f^{-1}(u), f^{-1}(v)) \in E_1$ such that $\nu_1(e_1) = \nu_2(e_2)$*

Two graphs are called isomorphic if there exists an isomorphism between them.

In contrast with the components of a feature vector, the nodes and edges cannot be ordered in general. Therefore, the problem of graph isomorphism is computationally very demanding. Standard procedures for testing graphs for isomorphism are based on tree search techniques with backtracking. The basic idea is that a partial node matching, which assigns nodes from the two graphs to each other, is iteratively expanded by adding new node-to-node correspondences. This is repeated until either the edge structure is violated or node or edge labels are inconsistent. In this case a backtracking procedure is initiated, i.e. the last node mappings are undone until a partial node matching is found for which an alternative extension is possible. Obviously, if there is no further possibility for expanding the partial node matching without violating the constraints, the algorithm terminates indicating that there is no isomorphism between the considered graphs. Conversely, finding a complete node-to-node correspondence without violating both structure and label constraints proves that the investigated graphs are isomorphic.

A popular algorithm implementing the idea of a tree search for graph isomorphism is described in [10]. More recent algorithms for graph isomorphism also based on the idea of tree search can be found in [11,12].

Closely related to graph isomorphism is subgraph isomorphism, which can be seen as a concept describing subgraph equality. A subgraph isomorphism is a weaker form of matching in terms of requiring only that an isomorphism holds between a graph g_1 and a subgraph of g_2. Intuitively, subgraph isomorphism is the problem to detect if a smaller graph is present in a larger graph. In Fig. 1 (a) and (c), an example of subgraph isomorphism is given.

Definition 3 (Subgraph Isomorphism). *Let* $g_1 = (V_1, E_1, \mu_1, \nu_1)$ *and* $g_2 = (V_2, E_2, \mu_2, \nu_2)$ *be graphs. An injective function* $f : V_1 \to V_2$ *from* g_1 *to* g_2 *is a subgraph isomorphism if there exists a subgraph* $g \subseteq g_2$ *such that* f *is a graph isomorphism between* g_1 *and* g.

Obviously, the tree search based algorithms for graph isomorphism [10,11,12] described above can be also applied to the subgraph isomorphism problem.

3 Generalized Subgraph Isomorphism for Information Retrieval

Relational databases offer a popular possibility to represent relational structured data. Another approach recently emerged is that of representing the underlying data by means of graph based representation. In fact, the graph representation supports all aspects of the relational data mining process [7]. The approach to knowledge mining and information retrieval proposed in this paper is based on the idea of specifying a query by means of a query graph, possibly augmented by some constraints.

Definition 4 (Query Graph). *A query graph is a 4-tuple,* $q = (V, E, \mu, \nu)$, *where* V, E, μ, *and* ν *are the same as in an attributed graph (see Def. 1), except* for the domains $D_i(t)$ of all node and edge attributes, x_i. These domains include the don't care symbol, $-$, and variables, X, from a finite set of variables, Σ. Any variable $X \in \Sigma$ can not occur more than once in a query graph.

Query graphs are more general than graphs following Def. 1 in the sense that the don't care symbol and variables may occur as the values of attributes on the nodes or edges. The purpose of the variables is to define those attributes whose values are to be returned as an answer to a query (we will come back to this point later). Furthermore variables may occur in a query because they may be used to express constraints on one or several attribute values.

Definition 5 (Constraint). Let $q = (V, E, \mu, \nu)$ be a query graph and Σ the set of all variables occurring in q. A constraint on set Σ is a condition on one or several variables from Σ that evaluates to **true** or **false** if we assign a concrete attribute value to each variable occurring in Σ.

Once the query graph has been construced by the user, it is matched against a database graph. The process of matching a query graph to a database graph essentially means that we want to find out whether there exists a subgraph isomorphism from the query to the database graph. Obviously, as our query graph may include don't care symbols and variables, we need a more general notion of subgraph isomorphism. According to the next definition, we call such a generalized subgraph isomorphism a *match* between a query and a database graph.

Definition 6 (Match). Let $q = (V_1, E_1, \mu_1, \nu_1)$ be a query graph, Σ the set of all variables occurring in q, C be a set of constraints on Σ, and $G = (V_2, E_2, \mu_2, \nu_2)$ be an attributed graph, called the database graph. Query graph q matches database graph G if there exists an injective mapping $f : V_1 \to V_2$ such that the following conditions hold:

1. For each edge $(u, v) \in E_1$ there exists an edge $(f(u), f(v)) \in E_2$
2. For each node $u \in V_1$, let $\mu_1(u) = (t, \mathbf{x}(t)) = (t, (x_1, \ldots, x_n))$ and $\mu_2(f(u)) = (t', \mathbf{x}'(t')) = (t', (x'_1, \ldots, x'_m))$; then
   (a) $t = t'$ and $n = m$
   (b) if $x_i \notin \Sigma$ and $x_i \neq -$ then $x_i = x'_i$
   (c) if $x_i \in \Sigma$ and there are one or several constraints from C imposed on x_i, then all these constraints are satisfied if the value of x'_i is substituted for x_i
3. For each edge $(u, v) \in E_1$ let $\nu_1(u, v) = \{(t_1, \mathbf{x}(t_1)), \ldots, (t_k, \mathbf{x}(t_k))\}$ and $\nu_2(f(u), f(v)) = \{(t'_1, \mathbf{x}'(t'_1)), \ldots, (t'_s, \mathbf{x}'(t'_s))\}$; then for each $(t, \mathbf{x}(t)) = (t, (x_1, \ldots, x_n)) \in \nu_1(u, v)$ there exists $(t', \mathbf{x}'(t')) = (t', (x'_1, \ldots, x'_m))$ such that
   (a) $t = t'$ and $n = m$
   (b) if $x_i \notin \Sigma$ and $x_i \neq -$ then $x_i = x'_i$
   (c) if $x_i \in \Sigma$ and there are one or several constraints from C imposed on x_i, then all these constraints are satisfied if the value of x'_i is substituted for x_i

Conditions 2 and 3 ensure that the (type,attribute)-pair $(t', \mathbf{x}'(t'))$ matches $(t, \mathbf{x}(t))$ under f. If a query graph q matches a database graph G, we call the injective function f a match between q and G. Note that for given q and G and a given set of constraints C over Σ, there can be zero, one, or more than one matches.

For a match we require each edge of the query graph being included in the database graph (condition 1 in Def. 6). A node, u, can be mapped, via injective function f, only to a node of the same type (condition 2.a)[1]. If the (type,attribute)-pair of a node u of the query graph includes an attribute value x_i then it is required that the same value occur at the corresponding position in the (type,attribute)-pair of the node $f(u)$ in the database graph (condition 2.b). Don't care symbols occuring in the (type, attribute)-pair of a node u will match any attribute value at the corresponding position in the (type,attribute)-pair of node $f(u)$. Similarly, unconstrained variables match any attribute value at their corresponding position in $f(u)$. In case there exist constraints on a variable in the query graph, the attribute values at the corresponding positions in $f(u)$ must satisfy these constraints (condition 2.c). The conditions 3.a to 3.c imposed on the (type, attribute)-pairs of edges are similar to 2.a to 2.c.

To query a given database graph G we not only need a query graph q and a set of constraints C over the variables occurring in q, but also a set of answer variables $A \subseteq \Sigma$. An answer variable is a variable occurring in the query graph. By means of answer variables we indicate which attribute values are to be returned by our knowledge mining system as an answer to a query. Therefore, the answer to a query can be **no**, if there is no such structure as the query graph contained as a substructure in the database graph, or **yes** if the query graph exists (at least once) as a substructure in the database graph and the query graph does not contain any answer variables. In the case where answer variables are defined in the query graph and one or several matches are found, for each match, f_j, an individual answer is generated. An answer is of the form $X_1 = x'_1, \ldots, X_n = x'_n$ where $X_1, \ldots, X_n$ are the answer variables occurring in set A and x'_i are the values of the attributes in the database graph that correspond to the variables X_i under match f_j.

In Fig. 2 an example of a query graph (Fig. 2 (a)) and database graph (Fig. 2 (c)) are illustrated. In this illustration nodes are of the type *person* and labeled with the person's first and second name, and e-mail address. Edges are of the type *e-mail* and labeled with the e-mail's subject, the date, and the size[2]. We can easily verify that there exists a subgraph isomorphism from the query graph (Fig. 2 (a)) to the database graph (Fig. 2 (c)).

In Fig. 2 (b) an example of a query graph with variables (X, Y) and don't care symbols $(-)$ is given. In contrast with the query graph given in Fig. 2 (a) we are now particularly interested in the subject (X) and the date (Y) of the e-mail sent from John Arnold to Ina Rangel (if there exists such an e-mail).

---

[1] Note that the condition $n = m$ actually follows from $t = t'$.

[2] Note that in general there may occur nodes as well as edges of different type in the same graph.

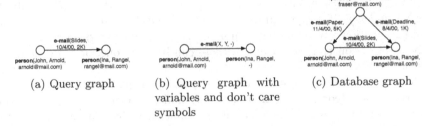

(a) Query graph   (b) Query graph with   (c) Database graph
            variables and don't care
            symbols

Fig. 2. A subgraph isomorphism from the query to the database graph

As we do not care about the size of the e-mail and we do not know the e-mail address of Ina Rangel, two don't care symbols are used. Obviously, there is a match between this query graph and the database graph in Fig. 2 (c). Hence, the variables are linked by $X = $ Slides and $Y = 10/4/00$. If the query in Fig. 2 (b) is augmented by the constraint that the e-mail has to be sent between October 1 and October 3 (formally $9/31/00 < Y < 10/4/00$) no match can be established as the linkage of variable Y violates the constraint imposed on the query.

Next we describe an algorithm for finding matches between a query and a database graph. The procedure given in Algorithm 1 checks two given graphs, q and G, whether there exists a match from q to G by constructing all possible mappings $f \colon V_1 \to V_2$ and checking the conditions of Def. 6.

The algorithm uses a set OPEN to store a set of partial matches. A partial match is a set of pairs, $\{(u_1, v_{i_1}), \ldots, (u_k, v_{i_k})\}$, where $f(u_1) = v_{i_1}, \ldots, f(u_k) = v_{i_k}$. If a partial match is included in OPEN, it has been verified before that all its pairs fulfil the conditions of Def. 6. Initially, OPEN is empty. In lines 2 to 6, the first node of V_1, u_1, is matched against all nodes, w, of V_2 and any pair (u_1, w) that satisfies the Boolean predicate *feasible* is added to OPEN as a partial match[3]. In the main loop of the algorithm, from lines 7 to 18, any partial match, $\{(u_1, v_{i_1}), \ldots, (u_k, v_{i_k})\}$, retrieved from OPEN, is first checked for completeness in line 9. If the considered partial match is complete, the algorithm outputs match $\{(u_1, v_{i_1}), \ldots, (u_k, v_{i_k})\}$ as the solution and terminates. If partial match $\{(u_1, v_{i_1}), \ldots, (u_k, v_{i_k})\}$ is not complete, i.e. $k < n$, then it is extended by one additional pair (u_{k+1}, w). This extension is done for all nodes, w, in V_2 that are not yet included in $\{(u_1, v_{i_1}), \ldots, (u_k, v_{i_k})\}$. If the extended set of pairs, $\{(u_1, v_{i_1}), \ldots, (u_k, v_{i_k})\} \cup \{(u_{k+1}, w)\}$ satisfies Boolean predicate *feasible*, it is added to open. Otherwise it is discarded.

The Boolean predicate *feasible* is used to verify that a partial match satisfies all conditions stated in Def. 6. In line 3 of the algorithm, as there are no edges included in the partial match under consideration, only condition 2 needs to be checked. In line 14, partial match $\{(u_1, v_{i_1}), \ldots, (u_k, v_{i_k})\}$ has already been checked for feasibility. Therefore, we only need to check condition 2 for nodes

---

[3] For the purpose of computational efficiency, it is advisable to process nodes and edges with more constraints and fewer variables and don't care symbols first. Fewer nodes and edges of this type will survive the feasibility test, which results in a search with fewer alternatives.

u_{k+1} and $f(u_{k+1}) = w$, and conditions 1 and 3 for all edges that start or end at u_{k+1} or w.

Algorithm 1 terminates as soon as it has found the first match from q to G. However, if we explicitly consider the situation of multiple matches, $f_1, \ldots, f_n$, it takes only a small modification to change the algorithm in such a way that it finds and outputs all matches from q to G. We only need to replace the **terminate** statement in line 11 by a statement to re-enter the main loop starting in line 7.

The matching algorithm given in Algorithm 1 is exponential. However, as the underlying query graphs are often limited in size and due to the fact that the attributes and constraints limit the potential search space for a match, the computational complexity of our algorithm is expected to be still manageable, as shown in the experiments reported in Section 4.

Algorithm 1. Algorithm for finding a match between a query graph q to a database graph G

Input: $q = (V_1, E_1, \mu_1, \nu_1)$, where $V_1 = \{u_1, \ldots, u_n\}$,
             $G = (V_2, E_2, \mu_2, \nu_2)$, where $V_2 = \{v_1, \ldots, v_m\}$,
             a set of constraints, C, over the variables occurring in q

Output: injective mapping $f : V_1 \rightarrow V_2$, given as
             $f = \{(u_1, v_{i_1}), (u_2, v_{i_2}), \ldots, (u_n, v_{i_n})\}$

```
 1: initialise OPEN to be the empty set
 2: for each node w ∈ V₂ do
 3:    if feasible[{(u₁, w)}] then
 4:       add partial match {(u₁, w)} to OPEN
 5:    end if
 6: end for
 7: while OPEN is not empty do
 8:    remove the first element, {(u₁, v_{i₁}), ..., (u_k, v_{i_k})} from OPEN
 9:    if k = n then
10:       output{(u₁, v_{i₁}), ..., (u_k, v_{i_k})} as the solution and
11:       terminate
12:    end if
13:    for each w ∈ V₂ \ {v_{i₁}, ..., v_{i_k}} do
14:       if feasible[{(u₁, v_{i₁}), ..., (u_k, v_{i_k})} ∪ {(u_{k+1}, w)}] then
15:          add partial match {(u₁, v_{i₁}), ..., (u_k, v_{i_k}), (u_{k+1}, w)} to OPEN
16:       end if
17:    end for
18: end while
19: terminate
```

---

4   Experimental Results

The experiments described in this section are conducted on three real world data sets, viz. the Enron data set [13], the AIDS data set [14], and the Internet Movie Database [15]. The intention of the experiments is twofold. First, we want to show the power and flexibility of the proposed concept of extended subgraph isomorphism and its applicability to general information retrieval tasks of relational data. Therefore we give several examples of query graphs and their corresponding results. Secondly, an evaluation of the performance of the algorithm is conducted. We are in particular interested in the behavior of our system when parameters in the query graph are changed systematically. For our tests a CPU Intel Core Duo 2GHz is used. The matching framework is implemented in Java.

4.1 Enron Database

The Enron corpus – a large set of e-mail messages – has been made publicly available during the legal investigation concerning the Enron corporation [13]. The Enron corpus contains e-mail messages belonging to 159 users. The underlying data is converted into a graph by representing the senders and addressees as nodes, attributed with their corresponding first and last name, and e-mail address. Note that not only the 159 users are converted into nodes but any person appearing as a sender or addressee in the mail folders of the corpus. Each e-mail that was sent is represented by an edge labeled with the file-path, the subject, the date, and the size. The complete data set consists of 70,400 nodes and 203,865 edges[4].

In Fig. 3 two example queries for the Enron database graph are illustrated. The first query (Fig. 3 (a)) in conjunction with the constraint $U = Z$ corresponds to the question: "which person has written an e-mail to Jennifer Fraser and Ina Rangel on the same day?" That is, there are four variables defined by $A = \{X, Y, U, Z\}$. The second query (Fig. 3 (b)) in conjunction with the constraint $D < E < D + 2h$ is of much higher complexity. The corresponding question is: "which person has first written an e-mail to John Arnold and then within two hours to Caroline Abramo?" Furthermore, we want to find out the subject and date of the two e-mails. We request that there is at least one e-mail sent by Caroline Abramo to John Arnold and we want to get the subject and date of all these e-mails. Hence, there are eight variables $A = \{A, B, C, D, E, F, X, Y\}$.

Our information retrieval system finds 13 possible answers to the first query. One of these answers is, for instance, $A = \{X = John, Y = Arnold, U = 10/4/00, 9.18\,AM, Z = 10/4/00, 9.12\,AM\}$. In order to retrieve all answers it takes our system only about 6 seconds. For the second query, however, 4 minutes are needed to find all possible answers. That is, running time crucially depends on the query graph to be found in the database graph.

In Fig. 4 the run time of processing queries is plotted as a function of the query graph size (Fig. 4 (a)) and as a function of the number of variables used in the query graph (Fig. 4 (b)). The queries of Fig. 4 (a) are defined according to three different scenarios. All scenarios have in common that the number of nodes (senders or addressees) is iteratively incremented by one. That is, starting with a query graph with two nodes, one additional node (and a corresponding edge) is iteratively added to the query graph until it consists of eight nodes totally. Three different query graphs are used for the scenarios plotted in Fig. 4 (b). For each of these query graphs the number of variables is iteratively incremented from zero to ten in steps of one.

It turns out that both the size of the query graph and the number of variables crucially influence the retrieval time. Clearly, the more nodes or variables the

---

[4] Note that nodes and edges are solely of type *person* and *e-mail*, respectively.

Fig. 3. Query graphs to the Enron database graph

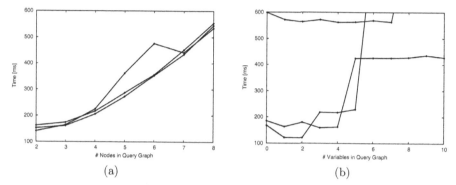

Fig. 4. Retrieval time as a function of the number of nodes (a) and variables (b) in the query graph

query graph contains, the slower the system is. However, query graphs with about eight nodes, which are in fact rather complex subgraphs in real world applications, are typically still processed below one second. Also a high number of variables weakens the system's performance. The proposed graph matching approach, however, copes well in general with up to about seven variables in the query graph which, again, allows quite complex queries.

4.2 AIDS Database

The AIDS data set consists of graphs representing molecular compounds. We construct graphs from the AIDS Antiviral Screen Database of Active Compounds [14]. Our molecule database consists of two classes (*active, inactive*), which represent molecules with activity against HIV or not. The molecules are converted into graphs in a straightforward manner by representing atoms as nodes and the covalent bonds as edges. Hence, there is only one type for both nodes (*atom*) and edges (*bond*). Nodes are labeled with the number of the corresponding chemical symbol and edges by the valence of the linkage. The complete data set consists of 42,689 molecular compounds which serve as a database graph.

In order to test our information retrieval system on the AIDS data set, we define 20 molecular compounds as query graphs. Three of these compounds are illustrated in Fig. 5. Note that different shades of grey represent different chemical symbols, i.e. node labels. For the sake of readability the type of nodes and edges is not indicated.

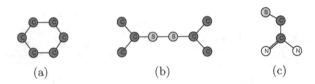

Fig. 5. Four examples of user defined molecular compounds

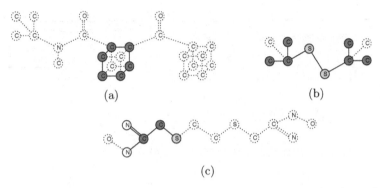

Fig. 6. Three molecular compounds of the original database graph where the user defined query graphs from Figure 5 are found as substructures

Given a certain query graph, our information retrieval system outputs all molecular compounds that contain the considered substructure. In Fig. 6 examples of successful matches for each query graph from Fig. 5 are shown. Note that these matches are not necessarily unique. In Fig. 6 (a) the illustrated match is one out of 32 different possibilities (16 permuations per carbon cube), Fig. 6 (b) shows one match from a total of nine possibilities, and in Fig. 6 (c) there are two possibilities for a positive match. Depending on the algorithm's termination criterion, only one or all possible solutions are returned by the system.

In Fig. 7 the total number of matches of the individual query graphs with the database graph and the total running time for performing the retrieval task are plotted as functions of the 20 individual substructures. Note that all substructures (with the exception of query graph 5 which is illustrated in Fig.5 (b)) are processed within at most 12 minutes. Most query graphs are processed in about three minutes. Clearly, the number of matches crucially depends on the structure and labels of the query graph, i.e. it varies from 30 up to 30,924 matches achieved with complex and rather simple query graphs, respectively.

As seen on the preceding data set, our novel information retrieval framework is able to cope with more complex query graphs than simple subgraph isomorphism tests. Fig. 8 (a), for instance, shows a query graph with answer variables $A = \{X, Y\}$. This query graph with constraints $X = Y, X \neq C, Y \neq C$ represents the question if there exist molecular compounds with circular substructures consisting of four carbon atoms and two opposing non-carbon atoms with the same chemical symbol in the database graph. In this example, besides the output of the original molecular compound from the database graph that contains

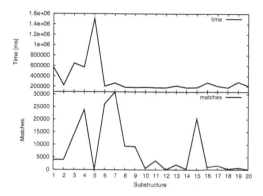

Fig. 7. The number of matches in the database graph and the total run time for each of the 20 query graphs

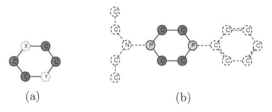

(a)   (b)

Fig. 8. Query graph and example solution from the database graph

the required substructure (Fig. 8 (b)), the answer variables A are linked by $X = Y = \mathbf{P}$.

4.3   Internet Movie Database

The Internet Movie Database (IMDb) [15] maintains a large collection of movie and television information, which is publicly available. We use a subset of the whole dataset with 10,000 movie entities and 73,927 actor entities. Both types of entities are represented by nodes. Hence, there are two types of nodes (*movie* and *actor*). The movie nodes are attributed with the corresponding genre, the title, a description, the year, the language, the type, and the country. Actor nodes are labeled with the first and last name of the corresponding actor. The 83,927 nodes are linked by 107,969 directed edges which represent relationships between actors and movies, i.e. the edges are labeled with the role a certain actor plays in the movie. Consequently, role is the sole edge type.

In Fig. 9 two example queries for the IMDb graph are illustrated. The first query (Fig. 9 (a)) represents the question: "which actors play in which roles in the movie with the title "Huff" released in 2004?" The answer variables are given by $A = \{U, X, Y\}$ and there are no constraints. The second query (Fig. 9 (b)) corresponds to the question: "in which movies the actors *Todd Berger* and *James Duval* play together?" In addition to the title of the movie, we are interested in the roles both actors play, and where and when these movies were shot.

(a)   (b)

Fig. 9. Query graphs to the IMDb database graph

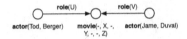

Fig. 10. Misspellings in the query graph

Hence, the answer variables are given by $A = \{U, V, X, Y, Z\}$ and there are no constraints.

For the first query our information retrieval system returns a list with 14 actors playing in the film "Huff" with their corresponding role, i.e., for instance, $A = \{U = Craig\ Huffstodt, Y = Azaria, X = Hank\}$. For this retrieval task our system uses less than a second. For the second query only one answer is returned; $A = \{U = Darryl\ Donaldson, V = Himself, X = \sharp 1\ Fan:\ A\ Darkomentary, Y = 2005, Z = USA\}$. For this retrieval operation the system also uses less than a second.

The proposed system described so far does not return any information from the database graph whenever no match is found. However, in some cases this behavior may be undesirable. Let us assume, for instance, a query graph as shown in Fig 10 is given. Obviously, this query represents the same query as defined in Fig. 9 (b). Note, however, that there are two misspellings in the actors' names (*Tod* and *Jame* instead of *Todd* and *James*). The graph matching framework presented so far merely returns the answer **no** as it finds no match in the database graph[5].

However, we can easily endow the novel graph isomorphism framework presented so far with a certain tolerance to errors. To this end we make use of graph edit distance [16]. Graph edit distance defines the dissimilarity of two graphs by the minimal amount of distortion that is needed to transform one graph into the other. Hence, in cases when no perfect match of the query graph to the database graph is possible, the query is minimally modified such that a match becomes possible. The modifications used in this paper are given by deletions of nodes and edges and substitutions of node and edge labels in the query graph. Furthermore, we also allow violations of constraints imposed on the variables.

Using our system in this extended fashion, approximate results will be returned whenever no perfect match is found. That is, for the misspelled query graph illustrated in Fig. 10 the same answer is returned as for the query graph

---

[5] One can think of similar situations in the two other data sets (e.g. different valences of covalent bonds in a molecular compound).

given in Fig. 9 (b). An additional comment is added making the user aware of the two label substitutions performed on the original query.

5 Conclusions

Data mining is an important area of study in both industry and the scientific community. In the present paper we focus on mining data represented by graphs. Graph based representation is particularly interesting because of its suitability for mining relational data. In order to find out whether a query graph is contained in a large database graph, the concept of subgraph isomorphism can be employed. However, since conventional subgraph isomorphism does not allow us to use don't care symbols, variables, and constraints on the query graph, the concept is rather limited with respect to general information retrieval tasks. Therefore, we provide a substantial extension of subgraph isomorphism which overcomes these limitations. The resulting graph matching framework is characterized by its high flexibility and power. With several experiments on three real world data sets, we prove the applicability of our approach in graph based data mining in quite different application fields. Though the problem of subgraph isomorphism is NP-complete, our framework accomplishes the matchings quite efficiently in practice. By means of the extension with graph edit distance we are able to endow the presented framework with a certain error tolerance such that approximate matches become possible. This is in particular interesting when it is not possible or not desirable to define a query graph which perfectly matches the database graph.

A possible extension of the presented methodology is the generalization of our framework to the domain of hypergraphs [17]. Such an extension, which can be done in a straightforward way, would result in greatly enhanced modelling capabilities. That is, representing n-ary relationships among n entities in the graph is rather difficult with binary edges. However, with hyperedges, which connect n nodes simultaneously, such representations can be easily done.

References

1. Perner, P., Rosenfeld, A.: MLDM 2003. LNCS, vol. 2734. Springer, Heidelberg (2003)
2. Perner, P., Imiya, A. (eds.): MLDM 2005. LNCS (LNAI), vol. 3587. Springer, Heidelberg (2005)
3. Perner, P. (ed.): ICDM 2006. LNCS (LNAI), vol. 4065. Springer, Heidelberg (2006)
4. Perner, P. (ed.): MLDM 2007. LNCS (LNAI), vol. 4571. Springer, Heidelberg (2007)
5. Conte, D., Foggia, P., Sansone, C., Vento, M.: Thirty years of graph matching in pattern recognition. Int. Journal of Pattern Recognition and Artificial Intelligence 18(3), 265–298 (2004)
6. Kandel, A., Bunke, H., Last, M. (eds.): Applied Graph Theory in Computer Vision and Pattern Recognition. Studies in Computational Intelligence, vol. 52. Springer, Heidelberg (2007)

7. Cook, D., Holder, L. (eds.): Mining Graph Data. Wiley-Interscience, Chichester (2007)
8. Blau, H., Immerman, N., Jensen, D.: A visual query language for relational knowledge discovery. Technical report, University of Massachusetts (2001)
9. Marcus, S., Moy, M., Coffman, T.: Social Network Analysis. In: Cook, D., Holder, L. (eds.) Mining Graph Data, pp. 443–467. Wiley-Interscience, Chichester (2007)
10. Ullman, J.: An algorithm for subgraph isomorphism. Journal of the Association for Computing Machinery 23(1), 31–42 (1976)
11. Cordella, L., Foggia, P., Sansone, C., Vento, M.: A (sub)graph isomorphism algorithm for matching large graphs. IEEE Trans. on Pattern Analysis and Machine Intelligence 26(20), 1367–1372 (2004)
12. Larrosa, J., Valiente, G.: Constraint satisfaction algorithms for graph pattern matching. Mathematical Structures in Computer Science 12(4), 403–422 (2002)
13. Klimt, B., Yang, Y.: Introducing the Enron corpus. In: Proc. First Conference on Email and Anti-Spam, CEAS (Electronic Proceedings)(2004)
14. DTP, D.T.P.: Aids antiviral screen (2004), http://dtp.nci.nih.gov/docs/aids/aids_data.html
15. The Internet Movie Database, http://www.imdb.com
16. Bunke, H., Allermann, G.: Inexact graph matching for structural pattern recognition. Pattern Recognition Letters 1, 245–253 (1983)
17. Bunke, H., Dickinson, P., Kraetzl, M.: Theoretical and algorithmic framework for hypergraph matching. In: Roli, F., Vitulano, S. (eds.) ICIAP 2005. LNCS, vol. 3617, pp. 463–470. Springer, Heidelberg (2005)

Contrast-Set Mining of
Aircraft Accidents and Incidents

Zohreh Nazeri, Daniel Barbara, Kenneth De Jong,
George Donohue, and Lance Sherry

George Mason University, 4400 University Drive,
Fairfax, Virginia, 22030, USA
{znazeri,dbarbara,kdejong,gdonohue,lsherry}@gmu.edu

Abstract. Identifying patterns of factors associated with aircraft accidents is of high interest to the aviation safety community. However, accident data is not large enough to allow a significant discovery of repeating patterns of the factors. We applied the STUCCO[1] algorithm to analyze aircraft *accident data* in contrast to the aircraft *incident data* in major aviation safety databases and identified factors that are significantly associated with the accidents. The data pertains to accidents and incidents involving commercial flights within the United States. The NTSB accident database was analyzed against four incident databases and the results were compared. We ranked the findings by the *Factor Support Ratio*, a measure introduced in this work.

Keywords: contrast-set mining, aviation safety, data mining, aircraft accident analysis, aircraft incident analysis, knowledge discovery.

1 Introduction

An aircraft *accident* is an occurrence associated with the operation of an aircraft in which people suffer death or injury, and/or in which aircraft receives substantial damage; an *incident* is an occurrence which is not an accident but is a safety hazard and with addition of one or more factors could have resulted in injury or fatality, and/or substantial damage to the aircraft [1]. Previous research on aircraft accidents has focused on studying accident data to determine factors leading to accidents. In his Why-Because Analysis (WBA) [2] to understand involving causal factors to accidents, Ladkin aims to reveal the causal reasoning behind the events and circumstances leading to an accident. He applied his method to individual aircraft accidents to show how it can improve understanding of the factors involved in those accidents [3]. Dimukes [4] studied 19 airline accidents focusing on pilot errors; his study showed characteristics and limitations of human cognition in responding to different situations and suggested accidents are caused by confluence of multiple factors. Van Es [5] studied Air Traffic Management (ATM) related accidents worldwide and showed *flight crew* is a more important factor in ATM-related accidents than *air traffic control* is. He also

---

[1] STUCCO algorithm is developed by S. D. Bay and M. J. Pazzani, University of California, Irvine.

P. Perner (Ed.): ICDM 2008, LNAI 5077, pp. 313–322, 2008.
© Springer-Verlag Berlin Heidelberg 2008

reported no systematic trends were found in the accident dataset when performing a trend analysis. While these studies help understanding individual accidents and their causal factors, the low rate of accidents however, makes it difficult to discover repeating patterns of these factors.

Other research has analyzed larger sets of data available on incidents to determine the causal factors of incidents. Majumdar [6] applied log-linear modeling technique to analyze seven factors involved in loss-of-separation incidents. Hansen and Zhang [7] tested the hypothesis that adverse operating conditions lead to higher incident rates in air traffic control. NASA [8] studies voluntarily submitted incident reports, mostly by pilots, and publishes the results monthly. While studying incident data is helpful to understand incident causal factors, it does not identify the relationship between the incident factors and accidents. Since the ultimate goal of studying aviation safety data is to reduce accidents, in this research, we analyzed both accident and incident data to show the relationships between the two classes of events and to identify factors that are significantly associated with accidents.

2 Data

The data used in the study consists of accidents and incidents pertaining to commercial flights (part-121) from 1995 through 2004. The accidents were obtained from:

- National Transportation Safety Board (NTSB) database, containing reports of all accidents

The incidents were obtained from four major national databases:

- Federal Aviation Administration Accident and Incident Database System (FAA/AIDS), containing reports of incidents investigated and/or documented by the FAA
- National Aeronautics and Space Administration Aviation Safety Reporting System (NASA/ASRS), containing self-reported errors voluntarily submitted mostly by pilots
- FAA Operational Errors and Deviations (OED), containing mandatory reports of Air Traffic Control errors
- FAA System Difficulty Reports (SDRS), containing reports of mechanical problems with the aircraft system or components

Each report in these databases consists of structured fields plus an unstructured narrative explaining the event. In this study we used the structured fields only. The structured fields contain causal and contributory factors which are identified either by the person reporting the event or by a domain expert who has reviewed the report afterwards. Our analysis used these factors.

2.1 Data Constraints

Some constraints imposed by the data need to be considered. All accidents in the United States involving civil aircraft are investigated by the National Transportation

Safety Board (NTSB), an independent organization, and are reported in the NTSB database. Accident data, therefore, can be assumed complete and free of bias. Incident data however, are under-reported and subject to self-reporting bias. To address these constraints, our study analyzes the underlying factors of accidents and incidents *qualitatively* (and not a quantitative analysis such as regression). The historical data on incidents is large enough to represent these factors qualitatively. Also, we consider *all* factors that have been present in an event, regardless of their primary or contributory role in leading to the event. This minimizes the impact of the bias in reporting the factors.

2.2 Data Selection

Since the purpose of the analysis is to identify operational factors under normal conditions, accidents and incidents due to the following causes were filtered out from the data:

- passenger and cabin-crew related problems, such as passengers being injured due to hot coffee spilling on them
- medical and alcohol related events, such as pilot being sick
- terrorism and security events, such as bomb threats
- bird/animal strike, such as aircraft encountering a deer on the runway
- events during the phases of operation when the aircraft is not operating (parked, standing, preflight)

Also, reports pertaining to the Alaska region were excluded since flight environment and procedures in this region are different from the rest of the regions in the United States and require a separate study.

After applying the filters, there were 184 accidents, and the following sets of incidents left in the data for analysis: 2,188 reports in the AIDS dataset, 29,922 reports in the ASRS dataset, 10,493 reports in the OED dataset, and 85,687 reports in the SDRS dataset.

2.3 Data Preparation

We first normalized the data across the databases and then developed an ontology by developing a hierarchy of factors and sub-factors common across the databases.

Normalization of the values was needed so that all databases use the same term to refer to the same factor or condition. For example, the action where pilot executes a maneuver to avoid an object on the runway is referred to by one database as 'ground encounter' and by another as 'object avoidance'.

Eight high-level categories of factors were identified in the data, each containing corresponding sub-factors. These factors and examples of their sub-factors are shown in Table 1. The 'Other' category contains all sub-factors which didn't fit under the other seven categories and were not big enough to have their own separate category.

We transformed the reports into vectors consisting of fields that indicate presence or absence of each of the common factors and sub-factors in the event (accident or incident). We then analyzed these vectors.

Table 1. Common ontology across multiple databases

Factor	Sub-Factor examples
Aircraft	Engine, Flight control system, Landing gear
Airport	Snow not removed from runway, Poor Lighting, Confusing marking
Air Traffic Control	Communication with pilot, Complying with procedures
Company	Procedures, Management, Training
Maintenance	Compliance, Inspection
Pilot	Visual lookout, Altitude deviation, Decision/Judgment
Weather	Wind, Thunderstorm, Ice
Other	Factors not in the other categories; FAA oversight, Visibility

3 Analysis

We applied the STUCCO algorithm [9] to perform four sets of analyses. In each analysis, the accident vectors were paired with incident vectors from one of the four

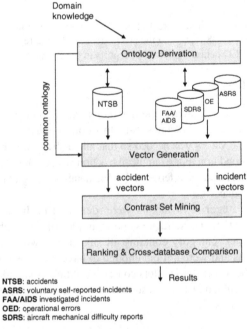

NTSB: accidents
ASRS: voluntary self-reported incidents
FAA/AIDS investigated incidents
OED: operational errors
SDRS: aircraft mechanical difficulty reports

Fig. 1. Analysis of accident data in contrast to incident data

incident databases. Each analysis identified patterns of factors which are significantly associated with accidents (or with incidents). We ranked the findings of each analysis, using the *Factor Support Ratio* measure described below. Final results of the four analyses were compared at the end. Figure 1, depicts the analysis process.

3.1 Algorithm

The STUCCO algorithm finds conjunctions of attribute-value pairs that are significantly different across multiple classes. In the case of our data, there are two classes: accidents and incidents. Attribute-values are binary values (1 or 0) for the factors in each event vector, implying presence or absence of the factors in that event. Figure 2 shows the algorithm used in our study.

In an A-Priori-like process [10], the factors and their children are examined for their support in each class. For each factors-set, *deviation* is calculated as absolute value of the difference between accident support and incident support for the

```
Input accident and incident vectors
C = set of factors in the input vectors
D = set of deviations, initially empty

1. While C is not empty
2.    Scan input data and count support ∀c ∈ C
         supp_acc = (accidents containing the
                        factor/total accidents)*100
         supp_inc = (incidents containing the
                        factor/total incidents)*100
3.    For each factor-set c ∈ C :
4.       If (count_acc > min cell frequency AND
            count_inc > min cell frequency)
5.          If ( |supp_acc - supp_inc| > dev_min )
               then factor-set is large
6.             If (Chi Square test passed)
                  then factor-set is significant
                        Add factor-set to candidates D
7.             Generate children (factor-set, C)
8.             For each child
                  If (supp_acc> dev_min OR supp_inc> dev_min)
                  Then add child to C'
9.    C = C'

10. Rank candidates in D by Factor Support Ratio
```

Legend:
supp_acc= accident support
supp_inc= incident support
dev = deviation

Fig. 2. STUCCO algorithm used in this research

factor-set. Factor-sets with a *deviation* of more than a minimum threshold are tested for the statistical significance of their distribution over the two classes. Chi Square test is used to perform the test. Factor-sets whose test results in a p-value of more than 0.05 are rejected, the rest are added to the list of candidates whose children will be generated and go through a similar test. The contingency table shown in Table 2 is used for the Chi Square test.

The step shown in Figure 2 are slightly different than the original algorithm discussed in [9]. The difference is in step 7 in Figure 2. In this step, the original algorithm generates children for a factor-set if the factor-set is both *large* and *significant*. Here the significance criterion is relaxed for child generation. (Note that this criterion is relaxed only for child generation, passing the significance test is still required for a factor-set to be added to the candidates.) The reason for this modification is to allow for discovery of factors that might not be individually associated with accidents, but if combined together they could be significant accident factors. Discovery of such cases is one of the objectives of the analysis. This modified step generated two additional significant two-factor-sets whose individual factors did not pass the significance test individually.

Table 2. Contingency table used for Chi Square significance test

	accidents	incidents
factor-set true	accidents containing the factor-set	incidents containing the factor-set
factor-set false	accidents not containing the factor-set	incidents not containing the factor-set

3.2 Ranking

Once significant factor-sets are identified by the algorithm, we rank them by their *Factor Support Ratio* measure. We calculate the Factor Support Ratio for each factor-set as the probability of the factor-set given an accident, divided by the probability of that factor-set given an incident, or the ratio of the factor-set's support in accident dataset over its support in the incident dataset (1) where F = factor-set, acc= accident, inc=incident, P(F|acc) = probability of factor-set given an accident, #Facc = number of accidents containing factor F, #acc = total number of accidents.

$$Support\ Ratio\ =\ \frac{P(F\,|\,acc)}{P(F\,|\,inc)}$$

$$=\ \frac{P(acc\,|\,F)P(F)\,/\,P(acc)}{P(inc\,|\,F)P(F)\,/\,P(inc)}$$

$$=\ \frac{\#Facc\,/\#acc}{\#Finc\,/\#inc}\ =\ \frac{\%Facc}{\%Finc}$$

$$Support\ Ratio\ =\ \frac{Support_{accident}}{Support_{incident}}\,.$$

(1)

The information conveyed by the Support Ratio measure about the factor-set is different than that of the *deviation* that is used in the algorithm. Deviation is the difference between the factor-set's accident and incident supports. Support Ratio is the probability of a factor-set being involved in an accident divided by its probability of being involved in an incident. To see the significance of this distinction, consider factor-sets A and B and their corresponding measures in Table 3.

Table 3. Support Ratio vs. deviation

factor-set	accident supp	incident support	Dev	Support Ratio
A	60%	50%	10%	1.2
B	11%	1%	10%	11

Both factor-sets A and B have a deviation of 10% between their accident support and incident support. However, in the case of factor-set B, the support in accidents is 11 times more than in incidents. This can be interpreted as: occurrence of factor-set B in an accident is 11 times more likely than its occurrence in an incident. This is a more distinctive distribution than that of factor-set A which has a Support Ratio of 1.2. We can use this measure to compare factor-sets A and B, and say factor-set A is more likely to be involved in accidents than factor-set B.

4 Results

Results of the analyses were reviewed with domain experts, some results were consistent with previous research findings and some were interesting in the sense that previous studies had not identified them. Highlights of the results are discussed below.

Company factors - factors such as mistakes by the company (or airline) personnel, and inadequate or lack of procedures by the company for performing a task – were consistently the highest ranked category of factors associated with accidents among the eight high-level categories of factors. This was an interesting result. Although previous studies had shown these factors contributed to accidents, their significance relative to other factors was not shown. Our research conducted a holistic study of the factors across multiple databases and in addition to identifying the factors associated with the accidents we could rank the factors in the order of their significance.

The next highest ranked accident factors were *Air Traffic Control (ATC)* followed by the *pilot* factors. Among the *ATC* factors, *communications* sub-factor had the highest rank of association with accidents. And among the *pilot* factors, *visual lookout* had the highest rank. Identification of *ATC communications* and *pilot visual lookout* as accident factors was consistent with previous findings. The interesting finding was that *ATC* factors which are less frequent than *pilot* factors were ranked higher. This implies that although *ATC* factors are less frequent but once they occur there is a high risk of having an accident (as opposed to an incident). *Pilot* factors are more frequent than other factors in accidents but they are also more frequent in incidents, which makes their Support Ratio lower and ranks them after the *company* and *ATC* factors.

Another interesting finding was association of *aircraft* factors with incidents. *Aircraft* factors are mechanical problems with the aircraft system or components, such as landing gears and flight control systems. The results showed these factors are more likely to be involved in incidents except when combined with other factors such as severe weather or pilot errors.

In Table 4 we show the results grouped by the factor category. These results are associations that were consistently identified by multiple analyses. Additional associations were identified by each individual analysis.

Table 4. Selected results of the analyses

Factor Category	Associations
Pilot	(pilot, airport, other) → accident (pilot, weather)→ accident (pilot) → accident
ATC	(ATC, pilot, airport, other) → accident (ATC, airport, company) → accident (ATC) → accident
Aircraft	(aircraft, weather) → accident (aircraft) → **incident**
Company	(company, maintenance, other) → accident (company, maintenance) → accident (company) → accident

Ranking of the results also showed that likelihood of a factor being involved in an accident rises as more factors co-occur with it. This means when multiple factors are present, there is a higher likelihood of having an accident (as opposed to having an incident). Tables 6 and 7 show some examples. For example in Table 6, the Support Ratio for combination of *pilot+airport* factors is 7.2 compared to the Support Ratio of 3.9 for the *pilot* factors, signifying that *pilot* factors combined with *airport* factors are more likely to result in accidents than the *pilot* factors alone.

Table 5. Ranking of results from NASA database analysis

factor-sets in NASA database	Support ratio
pilot, aircraft, company, other	3.7
pilot, company, other	3.6
pilot, aircraft, weather	2.9
pilot, airport, other	2.3
pilot, weather	1.9

Table 6. Ranking of results from FAA database analysis

factor-sets in FAA database	Support ratio
pilot, airport, other	14.3
pilot, aircraft	9.7
pilot, airport	7.2
pilot, weather	4.3
pilot	3.9

Note that the Support Ratio measure cannot be used for cross-database comparison of factor-sets. Factor-sets within a dataset can be compared using their Support Ratios since total numbers of accidents and incidents are the same in calculation of the Support Ratios.

5 Summary and Future Work

By applying contrast-set mining to the aviation safety data, we were able to analyze aircraft accident data in contrast to the incident data and identify patterns of factors which are associated with the accidents. Our ranking measure, the Factor Support Ratio, allowed ranking of the findings and identification of relative significance of the factors in contributing to accidents, compared to other factors.

This work analyzed aircraft accidents and incident pertaining to commercial flights within the United States. The methodology used here could be applied to the general aviation as well. The analysis could be extended to include world-wide safety events. In a future work, other data attributes, such as severity of the event, phase of flight, and type of aircraft could be included in the study to obtain more specific results.

Acknowledgements

Zohreh Nazeri wishes to thank the following individuals for their help: Christina Hunt, manager, at the FAA Aviation Safety Information Analysis and Sharing (ASIAS), provided much of the data needed for the study and patiently answered any questions about the data. Linda Connell, director, and her team, at the Aviation Safety Reporting System (ASRS) program at NASA, helped with downloading the ASRS data from their on-line query interface and explained the data. My colleagues at the MITRE Corporation generously shared their knowledge with me in multiple discussions.

References

1. Federal Aviation Administration, Air Traffic Organization, Aircraft Accident and Incident Notification, Investigation, and Reporting. Order 8020.16,
   http://www.faa.gov/airports_airtraffic/air_traffic/
   publications/at_orders/media/AAI.pdf

2. Ladkin, P.: ATM Related Accidents. Eurocontrol (2006),
   `http://www.eurocontrol.int/corporate/public/standrd_page/`
   `cb_safety.html`
3. Ladkin, P.: Causal Reasoning About Aircraft Accidents. In: 19th International Conference on Computer Safety, Reliability and Security (SAFECOMP), Rotterdam, The Netherlands (2000)
4. Dimukes, K.: The Limits of Expertise: The Misunderstanding Role of Pilot Error in Airline Accidents. ASPA/ICAO regional seminar (2005)
5. Van Es, G.: Review of Air Traffic Management-Related Accidents Worldwide: 1980-2001. In: Fifteenth Annual Aviation Safety Seminar (EASS), Geneva, Switzerland (2003)
6. Majumdar, A., Dupuy, M.D., Ochieng, W.O.: A framework for the Development of Safety Indicators for New Zealand Airspace: the Categorical Analysis of Factors Affecting Loss of Separation Incidents. In: Transportation Research Board (TRB) annual conference (2006)
7. Hansen, M., Zhang, Y.: Safety Efficiency: Link between Operational Performance and Operation Errors in the national Airspace System. Transportation Research Record, journal of Transportation Research Board 1888, 15 (2004)
8. National Aeronautics and Space Administration, Air Traffic Management System (2007),
   `http://quest.arc.nasa.gov/aero/virtual/demo/ATM/tutorial/`
   `tutorial1.html`
9. Bay, S.D., Pazzani, M.J.: Detecting Change in Categorical Data: Mining Contrast Sets. In: Fifth ACM SIGKDD International Conference on Knowledge Discovery and Data Mining. The Association for Computing Machinery, New York (1999)
10. Agrawal, R., Srikant, R.: Fast Algorithm for Mining Association Rules. In: Twentieth International Conference on Very Large Databases, VLDB (1994)

Using Data Mining to Build Integrated Discrete Event Simulations

David A. Holland

Computer Sciences Corporation
1550 Crystal Drive, Suite 1300
Arlington, Virginia 22201
United States
dholland6@csc.com
http://www.csc.com

Abstract. Building a system from disparate software requires analysis to establish commonality of code. The ability of a data mining tool to extract repeating functional structures is the first step to reduce exploration, save development time, and re-use software components. This case study looks specifically at the application of graph-based data mining algorithms to code re-factoring. After writing a module to obtain a graph representation of a discrete event model, we built a tool around the University of Washington's SUBDUE package to find recurring patterns of logic. This resulted in cleaner code and increased awareness of code re-use.

Keywords: subgraph isomorphism, minimum description length, discrete event modeling, code re-factoring.

1 Introduction

Organizations characterized by a high degree of command and control to accomplish a mission depend on hierarchical social structures. The United States military is one example of such an organization. The successful execution of instructions in an operational context demands that subordinates be aware of the flow of information as it flows from the top of the unit down to the rank-and-file and that they be ready to carry out such instructions. Information technology systems are often verified and validated with this organizational reality in view.

Software development, however, does not neatly fit into such a category, at least not initially. Computer programming is as much an art as it is a science, and program code often develops in an organic, ad hoc way. Documenting an artifact that looks more like a Jackson Pollock painting than an engineering design using a formal method such can be daunting.

1.1 Executive Summary of IDEF0

Since the 1960's, Information Technology practitioners have developed a host of notations for describing the architecture of systems. These languages have

P. Perner (Ed.): ICDM 2008, LNAI 5077, pp. 323–329, 2008.

developed in specific communities and can be both characterized and recognized as such. Some of the better known notations include the Unified Modeling Language (UML), the Structured Analysis and Design Technique (SADT), and Viewpoint-Oriented Requirements Definition (VORD).

As communities have become more specialized, these languages have had offspring. One descendant of SADT is the IEEE Standard Integrated Definition family of languages. IDEF0 is a functional modeling language that was developed by the United States Air Force in the 1970s. In the IDEF0 philosophy, a function is conceived of as an activity that produces some outcome. This activity can have one or more inputs that enter into it. Inputs are depicted as entering from the left, and outputs as leaving toward the right.

Additionally, functions can be influenced by controls, such as government regulations or guidance, and can be enabled by mechanisms, which are generally understood as means to accomplishing functions. Controls are depicted as arrows from above, mechanisms as arrows from below.

1.2 Introduction to Discrete Event Simulation Software and Extend

To ensure validation of logistics models, the United States Navy relies on discrete event (DE) simulation software. The Navy has chosen the Extend Simulation Suite to capture ship-based logistical processes in order to arrive at reasonable estimates of time required to accomplish specific tasks on a mobile sea base. Extend is a visual, icon-based language that engineers in general find user-friendly. It has a simple drag-and-drop interface that lets one quickly build a DE model.

Extend uses an event-pending while loop to process objects. Each icon encapsulates a function, such as 'cause the object to wait for a specified period of time.' These functions are linked to each other by means of connectors that associate a single output of one function with one or more outputs of one or more functions. As objects move through the system, functions fire events which are then placed on a queue as pending. The simulation halts when all pending events have been processed.

Behind each Extend function is code written in a proprietary language called Mod-L. Mod-L closely resembles C and is flexible enough to add custom behaviours to the discrete event. The refactoring of Extend icons (functions) is really just a specific instance of the code refactoring of high-level programming languages as it is commonly understood. Most modern integrated development environments (such as Microsoft Visual Studio) now have a code re-factoring feature built into them.

As an Extend model grows, the user may be faced with increased complexity. The software allows for the encapsulation of groups of functions and their connectors into hierarchical blocks, or 'h-blocks.' All connectors that impinge on the encapsulated functions are then re-routed to impinge on the 'black box' that represents the set of functions.

Unfortunately for managers and policy makers, there is nothing in the Extend software that forces developers to be tidy. This is where a language such as

IDEF0 becomes useful (especially if it is mandated by your client!). IDEF0 requires that there be no fewer than three and no more than nine activities represented in a diagram. This means that frequenly occuring patterns of code must be refactored and re-used, like a toolkit. It also means documentation that is easier to understand.

1.3   Representation of DE Models as Graphs

An Extend model can quite naturally be represented mathematically as a graph data structure. Each vertex in the graph represents a function and each edge represents a 'connector' between functions.[1] Each node is assigned a label that is given by the type of Extend icon that was used to generate it. Representing the model in this way affords opportunities for exploiting graph-based data mining algorithms to accomplish tasks such as finding commonality of code and code refactoring.

2   Problems with DE Developer Artifacts

Most of the people who design ships are classical engineers first. They are well-versed in the formulas and techniques of shipbuilding, but may not be familiar with information technology in general or computer programming in particular. The discrete event models developed by shipyards are not necessarily done in a disciplined way, using a top-down requirements definition process followed by object-oriented, modular design.

Moreover, the complexity of designing a single ship, much more the complexity of designing processes for mobile sea bases, usually calls for the involvement of multiple contractual personnel from multiple industrial players. Each contractor may have his or her own notational conventions, further obfuscating the view of a single system design.

The graph compression technique embodied in the Minimum Description Length heuristic, combined with the subgraph isomorphism algorithm, provides a way to mitigate some of the complexity introduced by these factors.

3   Data Mining as a Strategy in Code Refactoring

With the rise of the Internet at the turn of the 21st century and the concomitant explosion in computer storage capacity, there has been a dramatic increase in research on mining sub-structures from networked systems and artifacts. As data mining algorithms have matured, they have been applied to non-traditional domains such as web documents. Among the algorithms that have been developed, three stand out as significant advances in run-time reduction and applicability to non-traditional domains: gSpan, FSG, and SUBDUE. Each of these algorithms

---

[1] In Extend, a connector is essentially a function input or output. In raw source code, these would be the arguments and return values of the functions.

use various strategies to reduce the run-time of finding subgraphs in larger graphs from exponential run-time to $O(nlogN)$ or near-linear.

In the case of SUBDUE (Holder and Cook), researchers discovered that using a Minimum Description Length heuristic can provide more significant insight into the structural content of a graph than previous algorithms that looked only at identifying frequently-occurring subgraphs. Minimum Description Length is a method which aims to reduce the number of bits required to describe a data structure by finding regularity in the data structure.

More precisely, if M is a model (such as a graph) and H is a hypothesis that describes some part of M, then L(M—H) is the length (in bits) of the model description using H as a substitute for the regularity of the model. The minimum description length minimizes L(H) + L(M—H).

SUBDUE includes a C++ implementation of subgraph isomorphism. Our approach was first to run the MDL compression, then apply subgraph isomorphism to identify recurring instances of the graph with the minimum description length.

A brief discussion of the subgraph isomorphism algorithm follows. Here, a graph is defined as a set of vertices V and edges E. We are interested in determining whether a smaller graph G' is a subgraph of a larger graph G. We define two functions l and f. f is a surjection that maps vertices of G onto vertices of G'. l and l' are functions that map vertices and pairs of vertices (edges) to labels for G and G', respectively.

For G' = {V',E'} to be a subgraph of G = {V,E}, the following three conditions must hold:

$$\forall u \in V, (l(u) = l'(f(u))) \tag{1}$$

$$\forall u, v \in V, (u, v) \in E \Leftrightarrow (f(u), f(v)) \in E' \tag{2}$$

$$\forall (u, v) \in E, (l(u, v) = l'(f(u), f(v))) \tag{3}$$

The first condition says that mapped vertices across G and G' have the same names. The second says that a pair of vertices is an edge in graph G if and only if it is an edge in graph G'. And the third condition states that mapped edges across G and G' have the same name.

4   Client Solution

Computer Sciences Corporation built a wrapper (named the Model Integration Application, or MIA) around the SUBDUE package in order to apply MDL and subgraph isomorphism to models of shipboard operations. These models could be characterized as having large numbers of icons occurring repeatedly in groups of similar structure. The shipboard operations, for example, entail the distribution of equipment to soldiers at a number of stations along an assembly line. Four stewards, positioned at regular intervals, would hand off bundles of equipment to soldiers queuing up at these stations. Prior to our analysis, structures such as these were not grouped accordingly but were spread out as individual icons across the entire drawing board. All icons depicting the batching were exposed at the same level. Not surprisingly, this layout was both crowded and confusing.

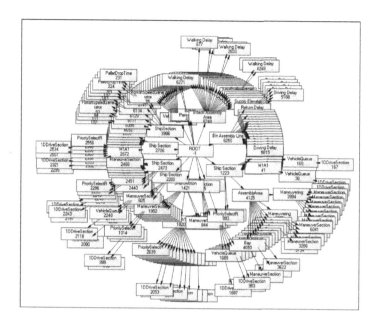

Fig. 1. DE model before MIA processing. Each box represents an Extend H-block.

In mapping these icons and connectors to IDEF0, we made some simplifying assumptions. The client directed us to ignore the controls and mechanisms. This meant that all incident connectors mapped to either an IDEF0 input or output.

Starting with the initial model, we made several passes on the code using the following three-step procedure. First, we applied the subdue.exe (MDL) program to find the subgraphs that most compressed the model. Then, we used these exemplar subgraphs as inputs to sgiso.exe to find the actual instances. The MIA then performed the substitution of each instance with a single node.

By demonstrating the use of the MIA tool and making the Extend developers aware of the benefits of re-factoring code into modules, we were able to persuade them to be more consistent in the use of making so-called 'hierarchical blocks' in Extend, which is essentially the re-factoring of code. As a result, the reader could more easily see the purpose of the stewards along the assembly line, having fewer, 'higher-level' symbols communicating the purpose of that section of the discrete event model. After transforming the code into a more hierarchical format, we exported the the graph data structure to an IDEF0 document, which was then read into System Architect. Using this (mandated) System Architect format, our customer could see, at a glance, the internal structure of the Discrete Event simulation.

4.1   Visualization of Transformed Discrete Event Models

Extend models are hierarchical in nature and resemble the familiar tree data structure, with a root at the top (visible when the model is first loaded) and

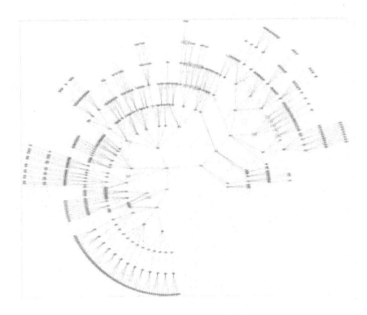

Fig. 2. After MIA processing. Where common patterns have been found, they have been placed in their own diagram and pushed out from the center.

leaves at the edges of the tree (Extend icons that perform some simulation function). Our client asked for a view from 10,000 feet of the DE model we were documenting. In Figure One, we show a representation of the uncompressed ship operations model. Note that many of the interior nodes have many (more than nine) branches. Figure Two shows the model after it has been run through the refactoring tool.

5   Conclusion

The purpose of this paper was to suggest the application of a pair of algorithms to the problem of cleaning messy source code, rather than to refine an algorithm or propose a new one. Discrete event models are only one domain in which this technique could be applied. Code written in C++, C, C#, and Java also come to mind as candidates. Despite the arrival of COTS tools for generating code by machine, there are still many projects where human error, laziness, inexperience, and lack of discipline call for tools that can clean code and make it easier to understand.

References

1. Holder, L.B., Ketkar, N.S., Cook, D.J.: Subdue: Compression-Based Frequent Pattern Discovery in Graph Data. In: Proceedings of the ACM KDD Workshop on Open-Source Data Mining (2005)

2. Grünewald, P.: A Tutorial on the Minimum Description Length Principle, from Advances in Minimum Description Length: Theory and Applications. MIT Press, Cambridge (2004)
3. Kotonya, G., Sommerville, I.: Requirements Engineering: Processes and Techniques. John Wiley and Sons, New York (1998)

Control Charts of Workflows

Calin Ciufudean[1], Constantin Filote[1], and Dumitru Amarandei[2]

[1] Electrical Engineering and Computer Science Department
"Stefan cel Mare" University, University str. 9,
720225 Suceava, Romania
[2] Mechatronics Department
"Stefan cel Mare" University, University str. 9,
720225 Suceava, Romania
calin@eed.usv.ro, filote@eed.usv.ro, mitica@fim.usv.ro

Abstract. This paper focuses on the control of the performance characteristics of workflows modeled with stochastic Petri nets (SPN's). This goal is achieved by focusing on a new model for Artificial Social Systems (ASS's) behaviors, and by introducing equivalent transfer functions for SPN's. ASS's exist in practically every multi-agent system, and play a major role in the performance and effectiveness of the agents. This is the reason why we introduce a more suggestive model for ASS's. To model these systems, a class of Petri nets is adopted, and briefly introduced in the paper. This class allows representing the flow of physical resources and control information data of the ASS's components. In the analysis of SPN we use simulations in respect to timing parameters in a generalized semi-Markov process (GSMP). By using existing results on perturbation (e.g., delays in supply with raw materials, derangements of equipments, etc.) analysis and by extending them to new physical interpretations we address unbiased sensitivity estimators correlated with practical solutions in order to attenuate the perturbations.

Keywords: Artificial Social Systems, Equivalent Transfer Functions, Stochastic Petri Nets, Control Charts of Workflows.

1 Introduction

An Artificial Social System (ASS) is a set of restrictions on agent's behaviour in a multi-agent environment [1]. ASS allows agents to coexist in a shared environment and pursue their respective goals in the presence of other agents. A multi-agent system consists of several agents, where at given point, each agent is in one of several states. In each of its states, an agent can perform several actions. The actions an agent performs at a given point may affect the way that the state of this agent and the state of other agents will change. A system of dependent automata consists of two or more agents, each of which may be in one of a finite number of different local states. We denote the set of local states of an agent i by P_i. The set $(P_1, P_2, ..., P_n)$ of states of the different agents is called system's configuration. The set of possible actions an agent i can perform is a function of the local state. For every state $p \in P_i$ there is a set $A_i(p)$ of

P. Perner (Ed.): ICDM 2008, LNAI 5077, pp. 330–344, 2008.
© Springer-Verlag Berlin Heidelberg 2008

actions that i can perform when in local state p. The row actions $(a_1, ..., a_n)$ denote the actions the different agents perform at a given point and is called their joint action there. An agent's next state is a function of the system's current configuration and the joint action performed by the agents. A goal for an agent is identified with one of its states. That is the reason why an agent has plans how to attain its goal.

A plan for agent i in a dependent automata is a function U(p) that associates with every state p of agent i a particular action $a \in A_i(p)$. A plan [2] is said to guarantee the attainment of a particular goal starting from an initial state, in a given dependent automata system, if by following this plan the agent will attain the goal, regardless of what the other agent will do, and what are the initial states of the other agents. A dependent automata system is said to be social if, for every initial state p_o and goal state p_g, it is computationally feasible for an agent to devise, on-line, an efficient plan that guarantees to attain the goal p_g state when starting in the initial state p_o. For a proper behavior, a dependent automata system is modeled with a social law. Formally, a social law Q for a given dependent automata system consists of functions $(A_1, A_2, ..., A_N)$, satisfying $A_i(p) \subset A_i(p)$ for every agent i and state $p \in P_i$. Intuitively, a social law will restrict the set of actions an agent is "allowed" to perform at any given state. Given a dependent automata system S and a social law Q for S, if we replace the functions A_i of S by the restricted functions A_i, we obtain new dependent automata system. We denote this new system by S^Q. In S^Q the agents can behave only in a manner compatible with the social law S [3], [4]. In controlling the actions, or strategies, available to an agent, the social law plays a dual role. By reducing the set of strategies available to a given agent, the social system may limit the number of goals the agent is able to attain. By restricting the behaviors of the other agents, however, the social system may make it possible for the agent to attain more goals and in some cases these goals will be attainable using more efficient plans than in the absence of the social system. A semantic definition of artificial social systems gives us the ability to reason about such systems. For example, the manufacturer of the agents (e.g., robots) that are to function in the social system will need to reason about whether its creation will indeed be equipped with the hardware and the software necessary to follow the rules. In order to be able to reason properly, we need a mathematical model and a description language [8], [9]. We chose the stochastic Petri nets model in order to model and simulate real conditions encountered in constructions workflow planning. We shall name on further accounts this model as Stochastic Artificial Social System. Petri nets have been recognized as a powerful tool for modeling discrete event systems. State explosion, a typical problem for SPN's, is solved here by introducing the equivalent transfer functions for transitions of SPN's. Data networks, viewed as discrete systems, are analyzed with such models. In the Petri nets theory, mathematical tools are available for analysis of the qualitative properties including deadlock-freeness, boundedness, reversibility, s.a. [1]. However simulation remains the effective for performance evaluation. Perturbation (e.g., delays in supply with raw materials, derangements of equipments, etc.) analysis has been developed for evaluating sensitivity measures by using simulations [2]. A generalized semi-Markov process (GSMP) is the usual model for the stochastic processes of discrete-event simulations, and most existing perturbation analysis methods are based on the GSMP framework. Since GSMP's and stochastic Petri nets (SPN's) have been proven to have the same modeling power [3], existing perturbation analysis methods are expected to apply to SPN's. Petri nets models considered here are SPN's with random transition firing times and the sensitivity estimators can be obtained from a simulation run. Our perturbation analysis is based on work of [5] and [6] which provides unbiased gradient estimators for a broad class of GSMP's. In this study, unbiased estimators are applied by using an appropriate SPN representation. Under correct conditioning, the unbiased estimators are easily confirmed by the simulation run of the SPN representation. This confirms the importance of underlying stochastic process. Practical solutions are shown in the paper, in order to give a concrete utilization of the theoretical model realized with SPN. The remainder of this paper is organized as follows. Section 2 presets SPN`s under consideration, section 3 presents some basic equivalent transfer function used for simplifying the complexity of SPN's, section 4 presents unbiased estimators for general stochastic Petri nets, sections 5 and 6 apply the theoretical approach to a gueuing network, respectively to a construction system perturbation analysis, and explicates some practical correlations between theory and practical implementation, and conclusions underline the approaches presented in this paper and establish future work.

2 Stochastic Petri Nets

In an ordinary Petri net $PN = (P, T, F, M_0)$, where P and T are two disjointed sets of nodes named, respectively, places and transitions. $F \subseteq (PXT) \cup (TXP)$ is a set of directed arcs. $M_0: P \rightarrow N$ is the initial marking. Two transitions t_i and t_j are said to be in conflict if they have at least one common input place. A transition t is said to be conflict free if it is not in conflict with any other transition. A transition may fire if it is enabled. A transition $t \in T$ is said to be enable at marking M if for all $p \in *t$, $M(p) \geq 1$. The SPN's considered here are ordinary Petri nets with timed transitions. Timed transitions can be in conflict therefore we say that a marking is stable if no conflict transitions are enabled. In the following we assume that the initial marking is a stable marking. We note by (M,T) a stable marking reachable from M by firing t. The new stable marking $M*$ is obtained from M according to some routing probability. The basic idea is that in order to guarantee that a stable marking can be reached; we must ensure that the respective circuit contains at least one timed transition. A SPN can be defined by the following elements [4]:

T_t Set of timed transitions
$M_s(M,t)$ Set of stable markings reachable from M by firing transition t
$p(M*, M,t)$ Probability of reaching a stable marking $M*$ from M when t fires.
Obviously, we have: $p(M*,M,t) = 0$ if $M* \notin Ms(M,t)$.
$F_t(.)$ Distribution function of the firing time of t
The GSMP representation of the SPN can be characterized by the following parameters:
$X(t,k)$ Independent random variables, where $t \in T_t$, and $k \in N$. Each $X(t,k)$ has distribution F_t and corresponds to the time of the kth firing of transition t.
$U(t,k)$ Random variables on [0,1]. Each $U(t,k)$ corresponds to the routing indicator at the kth completion of t.

$r_n(t)$ Remaining firing time of transition t at S_n

S(t,k)   Independent uniform random variables on [0,1], where $t \in T_t$, $k \in N$. Each U(t,k) corresponds to the routing indicator at the kth completion of t.

t_n n^{th} completed timed transition

M_n Stable marking reached at the firing of t_n

S_n Completion time of t_n

τ_n Holding time of marking M_{n-1}

V(t,n)   Number of instances of t among t_1 , ..., t_n.

The dynamic behavior of an SPN can be explained in the following way: at the initial marking M_0, set $r_n(t) = X(t,1)$, $\forall \ t \in T_t(M_0)$ and set V(t,0) = 0, $\forall \ t \in T_t$. All other parameters t_{n+1}, τ_{n+1}, s_{n+1}, V(t,n+1), M_{n+1}, r_{n+1} can be determined recursively as usually done in discrete event simulation. Recursive equations are given in [5]. The following routing mechanism is used in GSMP:

$$M_{n+1} = \varnothing(M_n, t_{n+1}, U(t_{n+1}, V(t_{n+1}, n+1))) \tag{1}$$

Where $\varnothing$ is a mapping such that $P(\varnothing(M,t,U) = M^*) = P(M^*, M, t)$.

3 Basic Equivalent Transfer Functions for SPN's

In a timed PN let F: T→R be a vector whose component is a firing time delay with an extended distribution function. By extended distribution functions, we mean that exponential distribution functions are allowed for concurrent transitions. Two transitions are said to be concurrent at marking m if and only if firing either does not disable the other. The firing rule for an SPN provides that when two or more transitions are enabled, the transitions whose associated time delays is statistically the minimum fires. According to the transition-firing rule in PN, when a transition t_k has only one input place p_i, and p_i is marked with at least one token, t_k is enabled. The enabled transition can fire. The firing of t_k removes one token from the p_i and then deposits one token into each output place p_j. Let P(i,k) be a probability that transition t_k can fire. The process from the enabling to the firing of t_k requires a time delay, τ_k. This delay τ_k of a transition can be either a constant or an extended random variable in SPN. P(i,k) and M(s) depend on τ_k as well as the current marking and the time delays of other enabled transitions at that marking. M(s) denote the moment generating function, and is defined as follows:

$$M(s) = \int_{-\infty}^{+\infty} e^{st} F_t(\cdot) \, dt \tag{2}$$

Where s is an extended parameter, and $F_t(\cdot)$ is a probability density function of random variable-transition t (i.e., distribution function of the firing time of transition t $\in T_t$).

A transfer function of a stochastic Petri net [4] is defined as the product P(i,k)·M(s), and is:

$$W_k(s) = P(i,k) \cdot M(s) \tag{3}$$

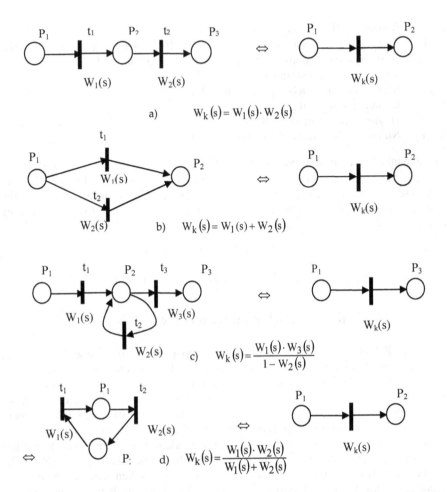

Fig. 1. Equivalent transfer functions for basic structures of SPN

Transition t_k characterized by $P(i,k)$ and τ_k is expressed by a transition characterized by $W_k(s)$. Four fundamental structures can be reduced into a single transition. The reduction rules can be used to simplify some classes of SPN. With these reduction rules we transform PN into finite state machines (in a finite state machine each transition has only one input and output place, and there is one token in such a net). Fig.1, a,b,c, d depict these reduction rules.

The moment generating functions for the state machine SPN which models the construction systems represent the availability of the cells (subsystems) which form the SPN.

4 Perturbation Parameters Modeled with SPN's

Following the approach given in [5], we suppose that the distributions of firing times depend on a parameter Θ. Parameters defined in section 2 are, in the above assumption, functions of Θ. In perturbation analysis the following results hold [6], where performance measures under consideration are of the form $g(M_1, t_1, \tau_1, ...,M_n,t_n,\tau_n)$ and a shorthand notation $g(\Theta)$ is used:

a) For each Θ, $g(\Theta)$ is a.s. continuously differentiable at Θ and the infinitesimal perturbation indicator is:

$$\frac{dg(\theta)}{d\theta} = \sum_{i=1}^{n} \frac{\partial g}{\partial \tau_i} \cdot \frac{d\tau_i}{d\theta}. \tag{4}$$

b) If $d \in [g(\Theta)]/d\Theta$ exists, the following perturbation estimator is unbiased:

$$\sum_{i=1}^{n} \frac{\partial g}{\partial \tau_i} \cdot \frac{d\tau_i}{d\theta} + \sum_{k=1}^{n} f_k(h_k) \cdot G_k \tag{5}$$

$$f_k = \frac{f_{tk+1}(L_k(t_{k+1}))}{F_{tk+1}(L_k(t_{k+1}) + y_k - F_{tk+1}(L_k(t_{k+1})))} \tag{6}$$

$$y_k = \min\{r_k(t) : \forall t \in T(M_k) - \{t_{k+1}\}\} \tag{7}$$

$$\tau_k = \frac{dL_k(t_{k+1})}{d\theta} - \frac{dX(t_{k+1})}{d\theta} \tag{8}$$

$L_k(t)$ is the age of time transition t at S_k; $G_k = g_{pp,k} - g_{DNP,k}$. The sample path $(M_1(\Theta)$, $t_1(\Theta)$, $\tau_1(\Theta)$, ...,$M_n(\Theta)$, $t_n(\Theta)$, $\tau_n(\Theta))$ is the nominal path denoted by NP. The $g_{DNP,k}$ is the performance measure of the kth degenerated nominal path, denoted by DNP_k. It is identical to NP except for the sojourn time of the $(k+1)_{th}$ stable marking in DNP_k. $g_{pp,k}$ is the performance measure of a so-called k^{th} perturbed path, denoted by PP_k. It is identical to DNP_k up to time s_k. At this instant the order of transition t_k and t_{k+1} is reversed, i.e., the firing of t_{k+1} completes just before that of t_k in PP_k. We notice that by definition, DNP_k and PP_k are identical up to s_k. At s_k, the events t_k and t_{k+1} occur almost simultaneously, but t_k occurs first in DNP and t_{k+1} occurs first in PP_k. The commuting condition given in [6] guarantees that the two samples paths became identical after the firing of both t_k and t_{k+1}. Our goal is to introduce a correction mechanism in the structure of the SPN so that the transition t_k and t_{k+1} fire in the desired order, and the routing mechanism given in relation (1) is re-established. We will exemplify this approach on an example, and we will correlate the theoretical assumption with some practical mechanisms in order to verify the approach.

5 Application to a Queuing Network

In Fig.1, we represented a workflow queuing network. The servers are s_1, s_2, and for any of them, if the downstream buffer is full, the customer is blocked until the downstream buffer has one hole. For simplicity of the Petri net model, we consider the perturbation analysis of only one way in the workflow [10]. In the corresponding SPN of the system in Fig.2, the transitions t_1 and t_4 model the arrivals (see Fig.3). Transitions t_3,t_6,t_7,t_9 are used to model the materials departure between constructors.

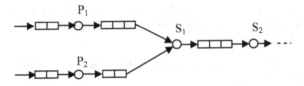

Fig. 2. A data queuing network with finite line capacity

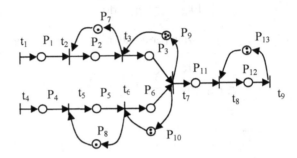

Fig. 3. The SPN model of the queuing network given in Fig.2

The transitions t_2, t_5, t_8 model the service periods in the network. The holding times of the transitions t_2, t_5, t_8 in the SPN are identical to the service times of computers in the workflow. The information transmitted to p_{11} by firing t_7 is determined by u' (routing indicator defined in section 2, see relation (1)) when t_2 fires first and it is determined by u'' when t_5 fires first. Since u' and u'' are independent random variables, the commuting condition given in [6] does not hold (i.e., $\emptyset(\emptyset(M,t_2,u'),t_5,u'')$ can be different from $\emptyset(\emptyset(M,t_5,u''),t_2,u')$. In order to make true the commuting condition we added in Fig.2 the locations p_7, p_8, p_9, p_{10}, and p_{13} and corresponding arcs ensure a Kanban mechanism in the SPN, in order to achieve the desired order in firing transitions t_3 and t_6, and, for p_{13}, a delay in materials transmission to the output. Locations p_7 and p_8 ensure the priorities in servicing of the materials flow arrivals (the arrival of the external raw materials).

For the average delay of demands ($g = \dfrac{4}{n}\sum\limits_{i=1}^{n} M_i(p) \cdot \tau_i$) the perturbation estimator given in (4) is unbiased.

$$\frac{4}{n}\sum_{i=1}^{n}\frac{\partial g}{\partial \tau_i} \cdot \frac{d\tau_i}{d\theta} = \frac{4}{n}\sum_{i=1}^{n} M_{i-1}(p) \cdot \frac{d\tau_i}{d\theta} \qquad (9)$$

$$g = \frac{4}{n}\sum_{i=1}^{n} L(M_{i-1}) \cdot \tau_i \qquad (10)$$

Where $L(M_i) = M_i(p_1) + M_i(p_2) + M_i(p_3) + M_i(p_4) + M_i(p_5) + M_i(p_6) + M_i(p_8)$.

The perturbation estimator is equal to:

$$\frac{4}{n}\sum_{i=1}^{n} L(M_{i-1}) \cdot \frac{d\tau_i}{d\theta} \tag{11}$$

Assuming that firing times are exponentially distributed with mean equal to: Θ for t_1, t_2, t_4, and t_6; 1 for t_3; 0,86 for t_5; 0,75 for t_8, 0,9 fot t_9, we consider the average customer delay (Θ). The mean value of the gradient evaluated at $\Theta = 1.22$ and at $\Theta = 1.24$ is close to the central finite difference: $(E[g(1.24)] - E[g(1.20)]) / 0,04 = -10.27$. This result is acceptable, and we notice that additional values can be obtained by modifying the net structure as discussed before, and as it is drawn in Fig.3, by modifying the marking in the places p_7 and p_8. We notice that we can simplify the structure of the SPN in Fig.3 using the approach presented in chapter 3 (see Fig.1). This approach is useful when we deal with complex Petri nets, and we want to simplify these structures by reducing them to finite state machine, in order to analyze them properly. For the Petri net in Fig.3, we may have the following equivalent schemes, and correspondingly, the equivalent transfer functions. We notice that depending to the specificity of each modeling process, or to the operator skills the reduction procedure can be stoped at a desired level of simplicity.

6 A Flexible Manufacturing System

6.1 The System Description

The manufacturing system considered in this paper consists of two cells linked together by a material system composed of two buffers A and B and a conveyor. Each cell consists of a machine to handle within cell part movement. Pieces enter the system at the load/unload station, where they are released from those two buffers, A and B, and then are sorted in cells (pieces of type "a" in one cell, and pieces of type "b" in the other cell). We notice that in the buffer A are pieces of types "a", "b", and others, where the number of pieces "a" is greater than the number of pieces "b". In the buffer B there are pieces of types "a", "b", and others, where the number of pieces "b" is greater than the number of pieces "a". The conveyor moves pieces between the load/unload station ad the various cells. The sorted piece leaves the system, and an

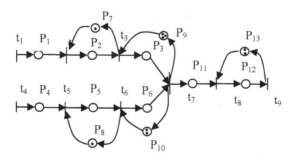

$\Leftrightarrow$

Fig. 4. An example of aplying the equivalent transfer function for minimizing the size of a SPN

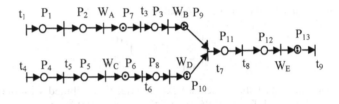

a) Where:

$$W_A(s)=\frac{W_{t_2}(s)\cdot W_{t_3}(s)}{W_{t_2}(s)+W_{t_3}(s)}, \quad W_B(s)=\frac{W_{t_3}(s)\cdot W_{t_7}(s)}{W_{t_3}(s)+W_{t_7}(s)}, \quad W_C(s)=\frac{W_{t_5}(s)\cdot W_{t_6}(s)}{W_{t_5}(s)+W_{t_6}(s)},$$

$$W_D(s)=\frac{W_{t_6}(s)\cdot W_{t_7}(s)}{W_{t_6}(s)+W_{t_7}(s)}, \quad W_E(s)=\frac{W_{t_8}(s)\cdot W_{t_9}(s)}{W_{t_8}(s)+W_{t_9}(s)}$$

⇔

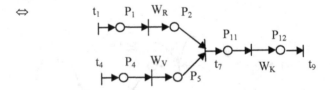

b) Where:

$$W_F(s)=W_{t_2}(s)\cdot W_A(s), \quad W_G(s)=W_{t_3}(s)\cdot W_B(s), \quad W_H(s)=W_{t_5}(s)\cdot W_C(s),$$

$$W_J(s)=W_{t_6}(s)\cdot W_D(s), \quad W_K(s)=W_{t_8}(s)\cdot W_E(s)$$

⇔

c) Where: $W_L(s)=W_{t_3}(s)\cdot W_F(s), \quad W_M(s)=W_{t_6}(s)\cdot W_H(s)$

⇔

d) Where : $W_R(s)=W_L(s)\cdot W_G(s), \quad W_V(s)=W_M(s)\cdot W_J(s)$

Fig. 4. (*continued*)

unsorted piece enters in the system, respectively in one of those two buffers A or B. The conveyor along with the central storage incorporates a sufficiently large buffer space, so that it can be thought of as possessing infinite storage capacity. Thus, if a piece routed to a particular cell finds that the cell is full, it is refused entry and is routed back to the centralized storage area. If a piece routed by conveyor is of a different type of the required types to be sorted, respectively "a", and "b", then that piece is rejected out of the system. We notice that once a piece is blocked from entry in a cell, the conveyor does not stop service; instead it proceed with its operation on the other pieces waiting for transport. At the system level, we assume that the cells are functionally equivalent, so that each cell can provide the necessary processing for a piece. Hence, one cell is sufficient to maintain production (at a reduced throughput). We say that the manufacturing system is available (or, operational) if the conveyor and at least one of the cells are available. A cell is available if its machine is available. Over a specified period of operation, owing to the randomly occurring subsystem failures and subsequent repairs, the cellular construction system (CCS) will function in different configurations and exhibit varying levels of performance over the random residence times in these configurations. The logical model of our manufacturing system is showed in Fig.5.

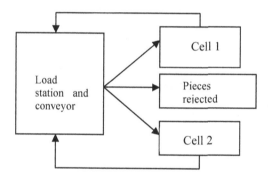

Fig. 5. Logical model for a manufacturing system

6.2 A Markov Model for Evaluating the Availability of the System

For the flexible manufacturing system depicted in Fig.1, we assume that the machines are failure-prone, while the load/unload station and the conveyor are extremely reliable. Assuming the failure times and the repair times to be exponentially distributed, we can formulate the state process as a continuous time Markov chain (CTMC). The state process is given by $\{X(u), u \geq 0\}$ with state space $S = \{(ij), i \in \{0,1,2\}, j \in \{0,1\}\}$, where i denotes the number of machine working, and j denotes the status of the material handling system (load station and conveyor): up (1), and down (0). We consider the state independent (or, time dependent) failure case and the operation dependent failure case separately.

Time dependent failures: In this case, the component fails irrespective of whether the system is operational or not. All failure states are recoverable. Let r_a and r_m denote the repair rates of the material handling system, and a machine respectively. The state

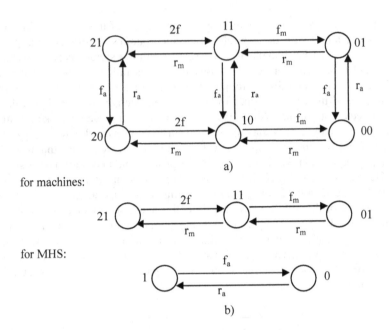

for machines:

for MHS:

Fig. 6. State process of a CCS with time-dependent failures, (a) State process for a state-independent failure model, (b) Decomposed failure/repair process

process is shown in Fig.6.a.   Because the failure/repair behavior of the system components are independent, the state process can be decomposed into two CTMC's as shown in Fig.6.b. Analytically, the state process is expressed by relations: $S_0 = \{(21), (11)\}$ and $S_F = \{(20),(10), (00)\}$. For each failure state in S_f no production is possible since the Material Handling System (MHS) or both the machines are down.

In Fig.6.b the failure/repair behaviour of each resource type (machines or MHS) is described by a unique Markov chain. Thus, the transient state probabilities, $p_{ij}(t)$ can be obtained from relation:

$$p_{ij}(t) = p_i(t)p_j(t) \qquad (12)$$

Where $p_i(t)$ is the probability that i machines are working at time t for $i = 0,1,2$. The probability $p_i(t)$ is obtained by solving (separately) the failure/repair model of the machines. $P_j(t)$ is the probability that j MHS (load/unload station and conveyor) are working at instant t , for $j = 0,1$. Let f_a and f_m denote the failure rates of the MHS and of a machine respectively.

Operation-dependent failures: Assume that when the system is functional, the resources are all fully utilized. Since failures occur only when the system is operational, the state space is: $S = \{(21), (11), (20), (10), (01)\}$, with $S_0 = \{(21), (11)\}$, $S_f = \{(20), (10), (01)\}$. The Markov chain model is shown in Fig.7. Transitions representing failure will be allowed only when the resource is busy. Transitions rates can however be computed as the product of the failure rates and percentage utilization of the resource, and T_k^{ij} represents the average utilization of the kth resource in the state (ij).

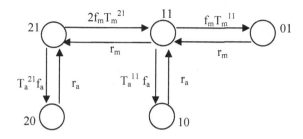

Fig. 7. State process of a CCS with state-dependent failures

6.3 Numerical Study

For the CCS presented in this paper, in the table 1 are given the failure/repair data of the system components We notice that $T_k^{ij} = 1$ since the utilization in each operational state is 100% for all i, j, k, i = {0,1,2}, j = {0,1}, k = 4. The other notations used in table 1 are: f is the exponential failure rate of resources, r is the exponential repair rate of resources, N_p is the required minimum number of operational machines in cell p, p = {1,2}, and n_p is the total number of machines in cell p.

Table 1. Data for the numerical study

	r	f	N_p	n_p	T_k^{ij}
Machines	1	0,05	1	2	1
MHS	0,2	0,001	1	1	1

From Fig.6 and Fig.7 we calculate the corresponding infinitesimal generators and after that, the probability vector of CTMC. With (11) we calculate the availability of CCS. The computational results are summarized in Table 2 for the state process given in Fig.6 (CCS with time-dependent failures), and respectively in Table 3 for the state process given in Fig.7 (CCS with state-dependent failures). We consider the system operation over an interval of 24 hours (three consecutive shifts).

Table 2. Computational results for the CCS in Fig.6

Time hour	Machines	MHS	System Availability
0	1.0000	1.0000	1.0000
1	0.9800	0.9548	0.9217
4	0.9470	0.8645	0.7789
8	0.9335	0.8061	0.7025
12	0.9330	0.7810	0.6758
16	0.9331	0.7701	0.6655
20	0.9330	0.7654	0.6623
24	0.9328	0.7648	0.6617

Table 3. Computational results for the CCS in Fig.7

Time hour	Machines	MHS	System Availability
0	1.0000	1.0000	1.0000
1	0.9780	0.9528	0.9201
4	0.9450	0.8628	0.7762
8	0.9315	0.8039	0.7008
12	0.9310	0.7798	0.6739
16	0.9320	0.7688	0.6632
20	0.9318	0.7639	0.6598
24	0.9320	0.7636	0.6583

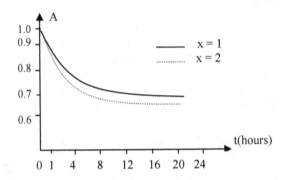

Fig. 8. Availability analysis of the CCS

The results of the availability analysis of the construction system are illustrated in Fig.6, which depicts the availability of the system as a function of the time. The numbers $x = 1, 2$ indicate the system in Fig.5, respectively Fig.7. One can see from Fig.8 that the layout with CCS with time-dependent failures is superior to that with CCS with state-dependent failure.

7 Conclusion

In this paper we analysed the control charts estimators in discrete event systems modelled with stochastic Petri nets (SPN`s). The approach presented in this paper (e.g. Stochastic Artificial Social Systems) can be used to analyze SPN`s that model complex dynamic system interactions. Unbiased gradient estimators proposed in [4], [6] were used for the sensitivity analysis of the GSMP representation and some practical solutions for attenuating the perturbations influences were indicated. The proposed procedure was imagined for a data network perturbation analysis. We estimate that this methodology can be applied to modelling and analysis of manufacturing systems, job scheduling in a chain management system, such as flexible manufacturing systems. Future research will focus on differential and fluid Petri nets in order to estimate throughput of complex systems.

An analytical technique for the availability evaluation of the flexible manufacturing systems was presented. The novelty of the approach is that the construction of large Markov chains is not required. Using a structural decomposition, the construction system is divided into cells. We can simplify the structure of the SPN using the approach presented in chapter 3 (see Fig.1). This approach is useful when we deal with complex Petri nets, and we need to simplify these structures (e.g. graphs) in order to analyze them properly. For each cell a Markov model was derived and the probability was determined of at least Ni working machines in cell i, for i = 1,2,..,n and j working material handling system (MHS) at time t, where N_i and j satisfy the system production capacity requirements. The model presented in this paper can be extended to include other components, e.g., tools, control systems. The results reported here can form the basis of several enhancements, such as conducting performance studies of complex systems with multiple part types.

References

[1] Murata, T.: Petri nets: Properties, analysis and applications. Proc. IEEE 77, 541–580 (1989)

[2] Baccelli, F., Liu, Z.: Comparison properties of stochastic decision free Petri nets. IEEE Trans. on Autom. Contr. 37(12), 1905–1920 (1992)

[3] Haas, P.J., Shendler, G.S.: Stochastic Petri nets: Modeling power and limit theorems. Probability Eng. Inform. Sci. 5, 477–498 (1991)

[4] Fu, M.C., Hu, J.Q.: Extensions and generalizations of smoothed perturbation analysis in a generalized semi-Markov process framework. IEEE Trans. Automat. Contr. 37, 1483–1500 (1992)

[5] Archetti, F., Gaivoronski, A., Sciomachen, A.: Sensitivity analysis and optimisation of stochastic Petri nets. Journal of Discrete Event Dynamic Systems: Theory Appl. 3, 5–37 (1993)

[6] Chiang, S.I., Kuo, C.T., Meerkov, S.M.: DT-Bottlenecks in serial production lines: Theory and Applications. IEEE Trans. Autom. Contr. 16(5), 567–580 (2000)

[7] Xie, X.: Perturbation analysis of stochastic Petri nets. IEEE Trans. on Autom. Control 43(1), 76–80 (1998)

[8] Yee, S.T., Ventura, J.A.: Phase-type approximation of Petri nets for analysis of manufacturing systems. IEEE Trans. on Rob. and Autom. 3, 318–322 (2000)

[9] Recalde, L., Teruel, E., Silva, M.: Modeling and analysis of sequential processes that cooperate through buffers. IEEE Trans. on Rob. and Autom. 14(2), 267–277 (1998)

[10] Ciufudean, C., Popescu, D.: Modeling Digital Signal Perturbation with Stochastic Petri Nets. Advances in Electrical and Computer Engineering 4(II) 1(21), 71–75 (2004)

[11] Ciufudean, C., Larionescu, A.: Safety criteria for production lines modeled with Petri nets. Advances in Electrical and Computer Engineering 2(9) 2(18) 15–20 (2002)

[12] Ciufudean, C.: Modeling the reliability of the interaction man-machine in railway transport. The Annals of the "Stefan cel Mare" University of Suceava 13, 80–84 (2000)

[13] Ciufudean, C., Graur, A.: Availability of Fluid Stochastic Event Graphs. In: International Conference on Control and Automation, ICCA 2005, Budapest, Hungary, June 26-29, 2005, pp. 35–39 (2005) ISBN0-7803-9138-1, IEEE Catalog Number: 05EX1076C

[14] Ciufudean, C., Petrescu, C., Filote, C.: Performance Evaluation of Distributed Systems. In: International Conference on Control and Automation, ICCA 2005, Budapest, Hungary, June 26-29, 2005, pp. 21–25 (2005) ISBN0-7803-9138-1, IEEE Catalog Number: 05EX1076C

[15] Ciufudean, C., Filote, C., Popescu, D.: Worflows in Constructions Modelled with Stochastic Artificial Petri Nets. In: Proc. of The 23rd International Symposium on Automation and Robotics in Construction, ISARC 2006, Tokyo, Japan, October 3-5, pp. 773–778 (2006)

[16] Ciufudean, C., Satco, B., Filote, C.: Reliability Markov Chains for Security Data Transmitter Analysis. In: Proc. of The Second International Conference on Availability, Reliability and Security, ARES 2007, April 10–13, pp. 886–892. Vienna University of Technology, Austria (2007)

[17] Ciufudean, C., Filote, C., Amarandei, D.: Measuring the Performance of Distributed Systems with Discrete Event Formalisms. In: Proc. of The 2nd IEEE –IAS Seminar for Advanced Industrial Control Applications, SAICA 2007, Madrid, Spain, November 5-6, pp. 263–267 (2007)

[18] Zuo, M.J., Liu, B., Murthy, D.N.P.: Replacement-repair policy for multi-state deteriorating products under warranty. European Journal of Operational Research 123, 519–530 (2000)

[19] Hopkins, M.: Strategies for determining causes of events, Technical Report R-306. UCLA Cognitive Systems Laboratory (2002)

[20] Hopkins, M.: A proof of the conjuctive cause conjecture in causes and explanations, Technical Report R-306. UCLA Cognitive Systems Laboratory (2002)

Maximum Margin Active Learning for Sequence Labeling with Different Length*

Haibin Cheng[1], Ruofei Zhang[2], Yefei Peng[2], Jianchang Mao[2], and Pang-Ning Tan[1]

[1] CSE Department, Michigan State University
East Lansing, MI 48824
{chenghai,ptan}@msu.edu
[2] Yahoo, Inc.
2821 Mission College Blvd, Santa Clara, CA 95054
{rzhang,ypeng,jmao}@yahoo-inc.com

Abstract. Sequence labeling problem is commonly encountered in many natural language and query processing tasks. SVM^{struct} is a supervised learning algorithm that provides a flexible and effective way to solve this problem. However, a large amount of training examples is often required to train SVM^{struct}, which can be costly for many applications that generate long and complex sequence data. This paper proposes an active learning technique to select the most informative subset of unlabeled sequences for annotation by choosing sequences that have largest uncertainty in their prediction. A unique aspect of active learning for sequence labeling is that it should take into consideration the effort spent on labeling sequences, which depends on the sequence length. A new active learning technique is proposed to use dynamic programming to identify the best subset of sequences to be annotated, taking into account both the uncertainty and labeling effort. Experiment results show that our SVM^{struct} active learning technique can significantly reduce the number of sequences to be labeled while outperforming other existing techniques.

Keywords: Active Learning, Struct Support Vector Machine, Uncertainty, Sequence Labeling, Natural Language Processing, Subphrase Generation.

1 Introduction

Sequence labeling is the task of mapping an ordered list of inputs to a sequence of output tags. It has many practical applications in natural language processing such as named entity recognition, part-of-speech tagging, shallow parsing, and text chunking. Another potential application, which is investigated in this study, is the subphrase generation problem. The goal of subphrase generation in query processing is to find subphrases in a query that maximally preserve the user's intent. Unlike the classification of record-based data, sequence labeling depends not only on the features extracted from the input sequence but also on its previous output tags. Many algorithms have been proposed in the literature to address this problem, including Conditional Random Field [8], Hidden

---

* The work was performed when the first author worked as a summer intern at Yahoo, Inc.

P. Perner (Ed.): ICDM 2008, LNAI 5077, pp. 345–359, 2008.

Markov Model [12] and Maximum Entropy Markov Model [9]. More recently, a maximum margin method known as Struct Support Vector Machine (SVM^{struct}) [19] was proposed to solve this problem. SVM^{struct} generalizes multi-class Support Vector Machine learning to complex data with features extracted from both inputs and outputs. An empirical study in [11] has demonstrated that SVM^{struct} outperforms other existing methods in the sequence labeling task.

A known problem in supervised learning tasks such as sequence labeling is the difficulty of acquiring labeled examples. The size of training data available is often limited because labeling examples can be very expensive. Labeling a sequence is also more challenging because the output tag depends on both the input and previous output tags. As a result, the tags of a sequence must be determined as a whole, rather than individually for each input element. Active learning may help to address this problem by selecting a small subset of examples for labeling from the large pool of unlabeled sequences available. By selecting the most informative examples, active learning can significantly reduce the required size of training data while maintaining comparable level of performance. However, the definition of "informative" varies for different algorithms and applications. One commonly used method is to select examples with largest uncertainties. In this paper, we treat each sequence as a whole for labeling and propose two strategies to measure the uncertainty of sequences under the SVM^{struct} framework, refered as simple uncertaincy (SU) and most-possible-constraint-violation method ($MPSV$).

Another challenge of active learning for sequences is that we must consider the effort needed to annotate different sequences. Clearly, labeling a long sequence takes more effort than labeling a short sequence. For example, in named entity recognition problem there may be a large variation in the length of sentences from one to another, which will make the effort required for labeling different sequences different. Besides, the effort for labeling sequences from different domain knowledge may also be different. For example, in subphrase generation problem, queries submitted by users coming from different specialty such as biology, which will make the labeling difficult for labeler with different background such as computer science. However, such kind of complexity in sequence is hard to quantify, in this paper we only use sequence length as a measure of effort. Furthermore, previous active learning system in sequence labeling assumes the input to be the number of sequences we want to select. However, this may not be suitable for the active learning in sequence labeling since the number of sequences did not really reflect the effort we need to put on labeling because of the sequence length problem. This paper defines the effort that can be spent in labeling sequence as the total length of all selected sequences instead of the total number of of all selected sequences. We propose a dynamic programming approach that can select the best combination of sequences for labeling which maximize the total uncertainty while restricting the effort that can be spent.

The rest of the paper is organized as follows. In Section 2, we introduce the background of our work including the sequence labeling problem, active learning and SVM^{struct} algorithm for sequence labeling. In Section 2.2, we propose to use SVM^{struct} for sequence active learning. In Section 4, we present the problem of active learning in labeling sequences with different length and propose to solve it by dynamic programming. Section 5 shows some experiment results and we made our conclusion in Section 6.

2 Background

In this section, we will introduce some background on sequence labeling, active learning and Struct Support Vector Machine.

2.1 Sequence Labeling Problem

Sequence labeling is a common problem with many applications in many areas such as named entity recognition [15], POS tagging [16], text chunking [16], etc. In our work, we also investigate the problem of subphrase generation, as a sequence labeling problem. The left panel of Figure 1 shows an example of named entity recognition problem, which labels text elements as predefined categories such as the names of persons, organizations or locations. The right panel of Figure 1 shows another example of subphrase generation problem. Label "0" means that the query word is dropped from the original query and "1" means "keep". As a result, the remaining part of the given query "where can I buy DVD player online?" becomes "buy DVD player". **Definition 1** and **Definition 2** give the formal definition of sequence and sequence labeling problem.

Definition 1. [Sequence]: A sequence x is an ordered list of elements $x = (x^1, x^2, ..., x^t)$.

Definition 2. [Sequence Labeling]: Given a sequence of inputs x, the sequence labeling problem is trying to label it with a sequence of tags $y = (y^1, y^2, ..., y^t)$, where each tag y^i belongs to a tag set D with $|D|$ tags.

One simple way to solve the sequence labeling problem is to use traditional classification algorithm such as SVM [2], which treats each element in the sequence as one example. However, it requires the features extracted only depend on the inputs x, which is not true in sequence labeling problem. The features extracted for sequence labeling not only depends on the inputs x, but also depends on the outputs y. The feature vector for a sequence (x, y) is represented as a joint feature mapping vector $\phi(x, y)$. The definition of ϕ depends on the nature of different applications. One example feature for the subphrase matching problem would be "previous word is dropped $\rightarrow$ current word is kept", which represents the transition from previous tag "0' to current tag "1".

Now assume that we have a training sequence set $\mathcal{X} = \{x_1, x_2, ..., x_n\}$ with its corresponding tag sequence set $\mathcal{Y} = \{y_2, y_2, ..., y_n\}$. We are interested in learning a

```
Joey,    an employee in    Yahoo  ,    has been
[Person] []   []       []  [Organization][]   []
living in    California for 10  years.
  []   []      [Location]  []   []  []
```
```
Query : where can I buy DVD player online?
          0     0  0 1   1    1      0
Subphrase  : buy DVD player
```

Fig. 1. Example of named entity recognition(left) and subphrase generation problem(right)

mapping function $f : \mathcal{X} \to \mathcal{Y}$. Instead of learning f directly, the strategy is to transform the problem into learning a discrimination function F over the joint mapping of input and output:

$$\mathcal{X} \times \mathcal{Y} \to \mathcal{R}$$

Given a test sequence x, its prediction is achieved by maximizing F over the response variable. The generalized form of the hypotheses f becomes:

$$f(x; w) = arg \max_{y \in \mathcal{Y}} F(x, y; w) \tag{1}$$

where w is the parameters to be learned. Using the joint feature vector $\phi(x, y)$, it can be further formulated as:

$$f(x; w) = arg \max_{y \in \mathcal{Y}} F(\phi(x, y), w) \tag{2}$$

Note that many existing methods for sequence labeling problem can be explained in the above framework. For example, the function form F that are maximized in the above prediction function represents the conditional probability $P(y|x)$ in conditional random filed [8], Hidden Markov Models [12] and Maximum Entropy Markov Models [9]. The detailed difference between these methods are illustrated in [11] and not discussed in our paper. In this work we will mainly focus on SVM^{struct} with prediction function:

$$f(x; w) = arg \max_{y \in \mathcal{Y}} w^T \phi(x, y) \tag{3}$$

SVM^{struct} has been proved to outperform all the other methods according to a recent empirical study in [11].

2.2 Active Learning in Sequence Labeling

Active learning is a process to actively select a subset of unlabeled data to query for labeling. There are many different frameworks of active learning. Among those, stream based active learning [4] selectively queries the examples from a stream for labels. The advantage of stream based active learning methods is that it can make fast and instant decision. However, stream based active learning only considers one example at a time and fails to take the underlying distribution of the whole unlabeled data into consideration. In our work, we are mostly concerned with pool based active learning methods, which selects the best examples from the entire pool of unlabeled data. Pool based active learning methods have been used to reduce the number of training data required to obtain an certain level of performance [1] [15] or to improve the overall performance [3].

The most challenging problem in active learning is to choose an appropriate measure as criteria for selecting the best examples from a pool of data. Many criteria have been developed for this purpose. Divergence based methods such as query by committee method [4] select examples with the largest disagreement between different models and aims to minimize the classification variance. Error-reduction-based active learning attempts to select the examples with minimal expected error for querying to minimize classification error over the test data [13]. Another big family of active learning is uncertainty based methods, which use model confidence as a criterion for selecting best examples and thus differ for different models. For example, Jing et al. use entropy based methods [7] to select unlabeled examples for the application of image retrieval. [18] propose three margin based methods in Support Vector Machine to select examples for querying which reduce the version space as much as possible. The underlying distribution of the unlabeled data is also investigated to choose the most representative examples [10].

Active learning on simple data has been well studied, however, there is not much work for more complex data set such as sequences. Active learning for sequence labeling is even more important because it is very expensive to label a long sequence. One challenging problem in sequence active learning is that we must select the whole sequence to query for labeling since the labeling of its elements depends on the context information. Previous methods summarized above can only be used to select one element in the sequence which can not be labeled without context information. [15] proposes a multi-Criteria-based active learning for the problem of named entity recognition using Support Vector Machine. However, they assume that the features depend only on the input sequence and are independent of the output tag sequence. One work that considers the input-output joint feature map is by [17], which utilizes conditional random field as the underlying model for sequence active learning. To the best of our knowledge, non work has been conducted using Struct Support Vector Machine, which has shown its potential improvement over CRF and other sequence learning algorithm such as HMMs [12] and Maximum Entropy [11]. Another common weakness of previous work on active learning for sequence labeling is that they did not take into account the difference in labeling effort for sequences different length. Labeling long sequence usually requires much more efforts than short sequence in terms of the time spent by the labeler. In this paper, we propose some simple but effective measurement for sequence active learning using Struct Support Vector Machine [19] as well as some dynamic programming algorithm to solve the length difference problem.

In this section, we reviewed some previous work on active learning for simple data and sequence data. In the next section, we will present previous work on SVM^{Struct} algorithm for structure prediction.

2.3 Struct Support Vector Machine for Sequence Labeling

The sequence labeling problem can be solved by multi-class Support Vector Machine [20] by treating each tag sequence as a class. For a tag sequence $\boldsymbol{y} = (y^1, y^2, ..., y^t)$ with t elements and $|D|$ possible tags for each element $y^i, i = 1, \cdots, t$, the possible number of classes is $|D|^t$. When the sequence length t is large, the huge number of classes makes the multi-class Support Vector Machine infeasible. Given a set of training sequences $(\mathcal{X}, \mathcal{Y}) = \{(\boldsymbol{x}_1, \boldsymbol{y}_1), (\boldsymbol{x}_2, \boldsymbol{y}_2), ..., (\boldsymbol{x}_n, \boldsymbol{y}_n)\}$, Struct Support Vector Machine solves this problem by exploring the underlying structure with $\mathcal{Y}$. In Struct Support Vector Machine, the margin of each training example $(\boldsymbol{x}_i, \boldsymbol{y}_i)$ is defined as:

$$\forall i : \boldsymbol{r}_i = \boldsymbol{w}^T \phi(\boldsymbol{x}_i, \boldsymbol{y}_i) - \max_{y \in \mathcal{Y} \setminus y_i} \boldsymbol{w}^T \phi(\boldsymbol{x}_i, \boldsymbol{y}) \qquad (4)$$

By maximizing the $\min_i r_i$ and fixing the functional margin ($\min_i r_i \geq 1$), we find a unique solution of $\boldsymbol{w}$. Thus the hard margin of SVM^{struct} learns the parameter vector $\boldsymbol{w}$ in the training phase by solving the following optimization function:

$$\min_{\boldsymbol{w}} \frac{1}{2}\|\mathbf{w}\|_2^2 \tag{5}$$
$$\text{s. t } \forall i, \forall \boldsymbol{y} \in \mathcal{Y} \backslash \boldsymbol{y}_i :$$
$$\boldsymbol{w}^T \phi(\boldsymbol{x}_i, \boldsymbol{y}_i) - \max_{\boldsymbol{y} \in \mathcal{Y} \backslash \boldsymbol{y}_i} \boldsymbol{w}^T \phi(\boldsymbol{x}_i, \boldsymbol{y}) \geq 1,$$

The nonlinear constraint in the above equation is equivalent to a set of linear constraints:

$$\forall i, \forall \boldsymbol{y} \in \mathcal{Y} \backslash \boldsymbol{y}_i : \boldsymbol{w}^T \phi(\boldsymbol{x}_i, \boldsymbol{y}_i) - \boldsymbol{w}^T \phi(\boldsymbol{x}_i, \boldsymbol{y}) \geq 1 \tag{6}$$

which makes the objective function into the form:

$$\min_{\boldsymbol{w}} \frac{1}{2}\|\mathbf{w}\|_2^2 \tag{7}$$
$$\text{s. t } \forall i, \forall \boldsymbol{y} \in \mathcal{Y} \backslash \boldsymbol{y}_i :$$
$$\boldsymbol{w}^T \phi(\boldsymbol{x}_i, \boldsymbol{y}_i) - \boldsymbol{w}^T \phi(\boldsymbol{x}_i, \boldsymbol{y}) \geq 1$$

The above solution assumes the training set is separable. Similar to standard SVM, a slack variable ξ_i is introduced for each sequence $\boldsymbol{x}_i$ to allow errors. Another weakness of the above solution is the assumption of zero-one classification loss, which is infeasible for sequence labeling problem where $|\mathcal{Y}|$ is large. To allow arbitrary loss function $\Delta(\boldsymbol{y}_i, \boldsymbol{y})$, one way is to rescale the margin. By taking error relaxation and loss function into consideration, the final optimization problem is formulated as:

$$\min_{\boldsymbol{w}, \boldsymbol{\xi}} \frac{1}{2}\|\mathbf{w}\|_2^2 + C \sum_{i=1}^{n} \xi_i \tag{8}$$
$$\text{s. t } \forall i, \forall \boldsymbol{y} \in \mathcal{Y} \backslash \boldsymbol{y}_i :$$
$$\boldsymbol{w}^T \phi(\boldsymbol{x}_i, \boldsymbol{y}_i) - \boldsymbol{w}^T \phi(\boldsymbol{x}_i, \boldsymbol{y}) \geq \Delta(\boldsymbol{y}_i, \boldsymbol{y}) - \xi_i,$$
$$\xi_i \geq 0$$

where $\Delta(\boldsymbol{y}_i, \boldsymbol{y})$ is the loss function which is calculated as the number of different tags between $\boldsymbol{y}_i$ and $\boldsymbol{y}$ in our paper. Since the number of constraints is $n|\mathcal{Y}|$, which is large for sequence labeling problem. SVM^{struct} [19] solves this problem in polynomial time by keeping a small working set of constraints and in each iteration adding the most violated constraint as following:

$$\max_{\boldsymbol{y} \in \mathcal{Y}} \Delta(\boldsymbol{y}_i, \boldsymbol{y}) - (\boldsymbol{w}^T \phi(\boldsymbol{x}_i, \boldsymbol{y}_i) - \boldsymbol{w}^T \phi(\boldsymbol{x}_i, \boldsymbol{y})) \tag{9}$$

After learning the parameter w, the tag sequence $\widehat{\boldsymbol{y}}$ for a test sequence $\boldsymbol{x}$ is predicted by solving the following argmax function using Viterbi search algorithm [5]:

$$\widehat{\boldsymbol{y}} = arg \max_{\boldsymbol{y} \in \mathcal{Y}} \boldsymbol{w}^T \phi(\boldsymbol{x}, \boldsymbol{y}) \tag{10}$$

SVM^{struct} is a flexible and effective solution for the sequence labeling problem and has been proved empirically to outperform other sequence labeling algorithm such as CRF [8], HMM [12] and Maximum Entropy [9]. In this paper, we will investigate the usage of SVM^{struct} in active learning setting.

3   Active Learning by SVM^{Struct}

In the last section, we have introduced the background of SVM^{Struct} for sequence labeling problem and some previous work on sequence labeling and active learning. In this section, we will propose to use SVM^{Struct} for active learning. From the previous work on active learning [7] [18], measurement of uncertainty has played an important role in selecting the most valuable examples from a pool of unlabeled data. In the framework of Support Vector Machine[18], three methods have been proposed to measure the uncertainty of simple data, which are referred as simple margin, MaxMin margin and ratio margin. Simple margin measures the uncertainty of an simple example x by its distance to the hyperplane w calculated as:

$$|w \bullet \varphi(x)| \tag{11}$$

As illustrated in Figure 2, examples lying closer to the hyperplane are assigned with larger uncertainty score. This is consistent with the intuition that examples close to the hyperplane are classified with lower confidence. These examples are considered as valuable examples since they have higher probability to be misclassified and thus more informative to be selected for further training.

However, labeling an element in a sequence by itself is almost infeasible in most sequence labeling applications because of the requirement for context information. In most situations we have to consider a whole sequence as an unit for uncertainty measurement and active selection. Given a pool of unlabeled sequences, $\mathcal{U} = \{s_1, s_2, ..., s_m\}$, the goal of active learning in sequence labeling is to select the most valuable sequences from the pool. Similar to regular Support Vector Machine, a straightforward way to

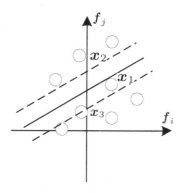

Fig. 2. Simple margin method will select unlabeled data x_1 for querying, which lies closest to the hyperplane

measure the uncertainty of a sequence s is by its prediction score. The prediction score $w^T\phi(s,y)$ measures the certainty of labeling test sequence s using the tag sequence y. The simple uncertainty for sequence s is then calculated in SVM^{struct} as:

$$UC(s) = \exp(-\max_{y\in\mathcal{Y}} w^T\phi(s,y)) \tag{12}$$

which is based on the negative value of the prediction score given by formula (10). Note that the features in sequence labeling not only depend on the input sequence s, but also depends on the output y. As a result, we must run Viterbi algorithm to get the uncertainty score for each sequence in the pool of unlabeled sequences U. Finally, the sequences with larger uncertainty are selected as valuable examples to add to the training set for further learning. We refer this method as simple uncertainty(SU) in this paper.

One drawback of the simple prediction score is its ignorance of the underlying score distribution among different classes and only use the maximum score as a measure of certainty. Here we propose another method which defines the uncertainty of a sequence x as:

$$UC(s) = \exp(-\max_{\substack{y_1,y_2\in\mathcal{Y} \\ y_1\neq y_2}} (w^T\phi(s,y_1) - w^T\phi(s,y_2))) \tag{13}$$

which can be further formulated as:

$$UC(s) = \exp(\min_{y\in\mathcal{Y}} w^T\phi(s,y) - \max_{y\in\mathcal{Y}} w^T\phi(s,y))) \tag{14}$$

We measure the uncertainty of an sequence s as the difference between the minimum prediction score and the maximum prediction score, which is actually the most possible violated constraint for a sequence s that can be added into the optimization problem. We refer this method as the most-possible-constraint-violation method (MPSV) in this paper.

The two methods SU and MPSV proposed here are used to calculate the uncertainty for each test sequence s. The test sequences with maximum uncertainty score are selected as the most informative sequences. These sequences are submitted to the labeler to query for labeling and further added into the training set. The detailed algorithm is presented in Figure 3. However, here we treat each sequence the same disregard with their length. Since labeling sequences with different length requires different effort, we will propose a dynamic programming algorithm to solve it in the next section.

4   Active Learning for Sequence Labeling with Different Length

In previous sections, we have introduced some strategies to select the most valuable sequences from a pool of unlabeled sequences for querying in SVM^{struct} based on the measurement of uncertainty for a sequence. Long sequences tend to have larger value in terms of prediction score as in formula (10) and thus smaller score as in formula (12) than short sequences. One simple way to solve this problem of comparing sequences with different length is by normalization. However, we still did not consider the different effort needed in labeling sequences with different length. Actually, some other factors should also be concerned in measuring labeling effort such as context knowledge difficulty. However, we ignore those factors due to the difficulty in quantitizing context knowledge. For example, given two queries "hotel in LA" and "car loan for people who have filed bankruptcy" in subphrase generation problem, the second query with length of 8 takes longer time for the labeler to label than the first query with length of only 3. This problem is even more severe in named entity recognition problem since there is huge difference in the length of sentences. Furthermore, existing sequence active learning [15] framework usually selects a predefined number of sequences, which is not appropriate since the efforts to be spent in labeling is restricted by the total length of selected sequences. To address these problems, we make the following two assumptions:

- The effort for a labeler that can be spent for labeling a set of sequences is defined as the total number of elements in the sequence set instead of the total number of sequences.
- The effort needed to label a sequence s is related and only related to the length of the sequence.

The first assumption changes the output for the previous sequence active learning system and is able to measure the effort in labeling effectively. The second assumption gives a simple definition to measure the efforts needed in labeling sequences with different length. These two assumptions brings out another concern about the choice between one longer sequence and several shorter sequences.

Here we formulate the problem as follows. Given a pool of unlabeled sequences $\mathcal{U} = \{s_1, s_2, ..., s_m\}$ with uncertainty score $\mathcal{F} = \{f_1, f_2, ..., f_m\}$, we define the effort needed for corresponding sequences as:

$$\mathcal{L} = \{l_1, l_2, ..., l_m\}$$

where $l_i = |s_i|, i = 1, 2, \cdots, m$. We also define the effort that can be spent in labeling sequences by the labeler as L. The goal is to utilize as much effort that can be spent as possible while maximizing the total uncertainty, which leads to the following objective function:

$$\max(\sum_{i=1}^{m} e_i f_i) \tag{15}$$

$$s.t. \sum_{i=1}^{m} e_i l_i \leq L$$

where $\mathcal{E} = \{e_1, e_2, ..., e_m\}$ is the indication vector with:

$$e_i = \begin{cases} 0 & \text{sequence } i \text{ is not selected.} \\ 1 & \text{sequence } i \text{ is selected.} \end{cases} \tag{16}$$

This is a NP-hard problem[6], which is approximately solved by dynamic programming algorithm in this paper. We define $K(i, j)$ as the maximum total uncertainty that can

---

SVM^{struct} Active Learning Algorithm for Sequences with Different Length:

Input: A small set of training sequences $(\mathcal{X}, \mathcal{Y}) = \{(\boldsymbol{x}_1, \boldsymbol{y}_1), (\boldsymbol{x}_2, \boldsymbol{y}_2), ..., (\boldsymbol{x}_n, , \boldsymbol{y}_n)\}$, a large pool of unlabeled sequences $\mathcal{U} = \{\boldsymbol{s}_1, \boldsymbol{s}_2, ..., \boldsymbol{s}_m\}$ and a predefined number of words that can be labeled L

Output: A vector $\mathcal{E} = \{e_1, e_2, ..., e_m\}$ with $e_i = 1$ indicating the selection of sequence i.

Method:

1. Learn the parameter vector $\boldsymbol{w}$ by the standard SVM^{struct} algorithm with training data $(\mathcal{X}, \mathcal{Y})$.

2. Calculate the uncertainty scores $\mathcal{F} = \{\boldsymbol{f}_1, \boldsymbol{f}_2, ..., \boldsymbol{f}_m\}$ for each sequence in the pool of unlabeled sequences $\mathcal{U}$ by $\mathcal{E} = \{e_1, e_2, ..., e_m\}$ Viterbi Search according to:

$$^1\boldsymbol{f}_i = \exp(-\max_{\boldsymbol{y} \in \mathcal{Y}} \boldsymbol{w}^T \phi(\boldsymbol{s}_i, \boldsymbol{y}))$$

$$^2\boldsymbol{f}_i = \exp(\min_{\boldsymbol{y} \in \mathcal{Y}} \boldsymbol{w}^T \phi(\boldsymbol{s}_i, \boldsymbol{y}) - \max_{\boldsymbol{y} \in \mathcal{Y}} \boldsymbol{w}^T \phi(\boldsymbol{s}_i, \boldsymbol{y})))$$

3. Solve formula (16) by dynamic programing to learn the indication vector $\mathcal{E} = \{e_1, e_2, ..., e_m\}$ and send sequence $\boldsymbol{s}_i$ to query for labeling if $e_i = 1$.

Fig. 3. The Active Learning Algorithm for Sequences with Different Length

be achieved with total number of elements less than or equal to j using sequences up to i. The recursive function is defined as:

$$K(0, j) = 0$$
$$K(i, 0) = 0$$
$$K(i, j) = K(i - 1, j), \text{if } \boldsymbol{l}_i > j$$
$$K(i, j) = \max(K(i - 1, j), \boldsymbol{f}_i + K(i - 1, j - \boldsymbol{l}_i)), \text{if } \boldsymbol{l}_i \leq j \qquad (17)$$

$K(m, L)$ is final uncertainty we get for m input sequences and desired effort L. This algorithm is a generalized algorithm that can be applied into any sequence active learning framework with different algorithms or different definitions of uncertainty scores. In this paper, we apply this algorithm into the SVM^{struct} active learning framework, which is described in Figure 3.

In this section, we give a clear definition about the effort needed to label a sequence. We also redefine the effort that can be spent for labeling as the total number of elements instead of the number of sequences. Our sequence active learning algorithm utilizes the SVM^{struct} framework and takes the effort needed in sequence labeling into account to select a subset of sequences with maximum uncertainty from a pool of unlabeled sequences. In the next section, we will conduct some experiments to evaluate our methods.

5 Experimental Result

In this section, we will conduct some experiments on real data sets to prove the effectiveness of our active learning algorithm.

5.1   Experiment Setup

We applied our algorithm to three data sets in our experiment. The first two data sets come from named entity recognition shared task of CoNLL-2002[14]. One is Spanish data (**ESP**), which is a collection of news wire articles made available by the Spanish EFE News Agency. Another is Dutch data **NED**, which consist of four editions of the Belgian newspaper "De Morgen" of 2000. The task is to label each word in the sentence using some predefined entity tags such as person names (PER), organizations (ORG), locations (LOC) and miscellaneous names (MISC) with a B ahead of them denoting the first item of a phrase and an I any non-initial word. The third data we are using is collected from the query subphrase matching project (**QSPM**) of Yahoo Sponsor Search. Given a query by a typical search engine user, the goal is to generate subphrases that preserve the user intent as well as match the bidded terms submitted by advertisers. There are two tags: "KEEP"("1") and "DROP"("0") for each position.

For each position in the sequence, we extract its context features such as "current word is", "previous word is", "next word is" and so on. We also used tag transition features such as "previous tag to current tag". Some word features such as prefix and suffix are also used based on the language of the data such as "th" for English data. We did not employ any feature selection methods in our experiments. For the **DER** data, the Part-Of-Speech tags are also utilized as grammatical features. All our experiments were conducted on a Linux server with 7.2GHz CPU and 15GB RAM.

5.2   Overall Performance

In this experiment,we compare the most-possible-constraint-violations method (**MPSV**) and simple uncertainty(**SU**) method with the random method. To alleviate the length problem in sequence active learning, we select a subset of sequences from the training data, which has the same length. For each data set, we run four experiments, each on a different length selected from the training data. For **NED** data, we select all the sequences with length 12,13,14 and 15 in each experiment. For **ESP** data, we select all the sequences with length 42, 43, 44, 45 in each experiment. For the **QSPM** data, we select all the sequences with length 3, 4, 5, 6. For the **NED** and **MPSV** data set, we select 400 sequences at each length. The first 10 are used for initial training. The pool of the remaining 390 sequences is for active selection. Each time we select 15 sequences and the result is reported as the average error rate of different length. For the **QSPM** data, we select 1930 sequences at each length. The first 10 sequences are used for initial training. The pool of the remaining 1920 sequences is for active selection. Each time we select 60 sequences and the result is reported as the average error rate of different length on the test set.

Figure 4 shows the results for the three methods in the three data sets **ESP**,**NED** and **QSPM**. The x-axis denotes the number of unlabeled sequences selected to query for labeling. The y-axis represents the average error rate, which is calculated in the word level as follows:

$$Error\ Rate_{\{Word\ Level\}} = \frac{Total\ number\ of\ correctly\ tagged\ words}{Total\ number\ of\ words} \qquad (18)$$

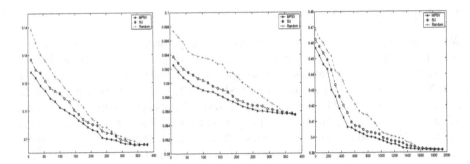

Fig. 4. The average error for ESP data set by three active learning uncertainty measurements

We observe from the Figure 4 that both **MPSV** and **SU** methods outperform random approach on all three data sets. Also **MPSV** performs better than **SU**, which means that MPSV is a better way to measure uncertainty for SVM^{struct}. Furthermore, the gap between the **MPSV** and other two methods seems very large when the number of selected sequences is small. It means that **MPSV** serves as a good criteria that only a small number of sequences are needed to get good performance. In this experiment, all the sequences are of the same length to compare three methods and we are aiming to select a predefined number of sequences. In the next section, we conduct experiments on sequences with efferent length utilizing the dynamic programming algorithm.

5.3   Selecting Sequences with Different Length

In the last section, we have compared two uncertainty measure (**MPSV**) and (**SU**) in SVM^{struct} with random method to select sequences with the same length. In this section, we will conduct experiments on our new active learning system which takes the effort in labeling into consideration. We select 400 sequences randomly from the original data set for **NED** and **ESP** separately with different length. We select 1930 sequences randomly from original **QSPM** data set with different length. Figure 5 shows the histograms of length distribution for the three sample data sets. As we can see, the length of sequences varies from 1 to 102 for **ESP** data set, from 1 to 58 for the **NED** data set and from 2 to 9 for the **QSPM** data. The wide spread of the length distribution elaborates our concern on different effort spent in labeling sequences with different length.

The input here is the percentage of words we want to select from the pool of sequences instead of the number of sequences. The percentage of words to be selected is the effort that can be spent in labeling sequences by the labeler. The baseline here is the previous active learning system, which ranks the sequences in the pool based on the normalized uncertainty score and selects the sequences with highest scores. We compare our dynamic active learning methods with previous active learning methods on the three data sets. Since **MPSV** has shown its improvement over random method and **SU** methods. We use **MPSV** as the uncertainty score measurement for our dynamic active learning algorithm.

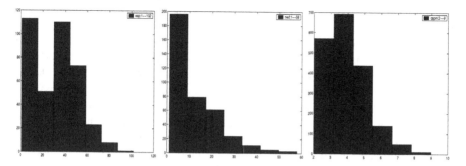

Fig. 5. The length distribution for the three data sets: ESP (left), NED (middle) and QSPM (right)

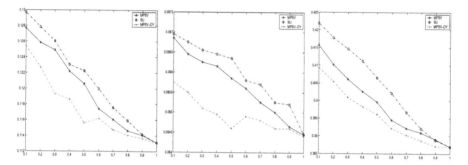

Fig. 6. The average error for the three data sets (EPS,NED,QSPM from left to right) by the dynamic active learning system using **MPSV-DY** as uncertainty measurement and existing active learning methods using **MPSV** and **SU** as uncertainty score

Figure 6 reports the error rate on the ESP, NED and QSPM data sets from left to right comparing our new active learning system with previous active learning system. X-axis is the percentage of words we want to select for labeling which is used to illustrate the effort that can be spent in labeling. Y-axis represents the error rate after querying the selected sequences for labeling and retrain the model with the new labeled sequences added. The result on the esp data shows that our dynamic active learning algorithm with **MPSV-DY** as underlying uncertainty score outperforms previous active learning methods using **MPSV** and **SU** as uncertainty score significantly with lower average error rate. It shows that our active learning system is able to select the most informative subset of unlabeled sequences to query for labeling.

6 Conclusion and Future Work

In this paper, we have proposed two measurements of uncertainty in SVM^{struct} for selecting the most informative sequences to query from labeling from a pool of unlabeled sequences. One is the most-possible-constraint-violation method (**MPSV**) and another is simple uncertainty(**SU**) method. We compare our proposed methods with random selection on three real data set from named entity recognition task and subphrase generation task for queries. Our first experiment result on selecting sequences with same length shows that the most-possible-constraint-violation method (**MPSV**) and simple uncertainty(**SU**) outperform the random method significantly. Also **MPSV** outperforms **SU** by considering the underlying class distribution. We also propose a new active learning for sequence labeling using dynamic programming to select the best combination of sequences that maximizes the total uncertainty and restricts labeling effort, which is defined as the total number of elements of the selected sequences. Experiment result shows that it performs better than previous active learning system for sequence labeling. In the future, we will explore the possibility of considering representativeness of a sequence in a pool and selecting a sequence with both high uncertainty and good representativeness.

References

1. Aidan, F.N.: Active learning selection strategies for information extraction
2. Burges, C.J.C.: A tutorial on support vector machines for pattern recognition. Knowledge Discovery and Data Mining, 2 (1998)
3. Cohn, D.A., Atlas, L., Ladner, R.E.: Improving generalization with active learning. Machine Learning 15(2), 201–221 (1994)
4. Dagan, I., Engelson, S.P.: Committee-based sampling for training probabilistic classifiers. In: International Conference on Machine Learning, pp. 150–157 (1995)
5. Forney, G.D.: The viterbi algorithm. Proceedings of the IEEE 61(3), 268–278 (1973)
6. Garey, M.R., Johnson, D.S.: Computers and Intractability; A Guide to the Theory of NP-Completeness. W. H. Freeman & Co., New York (1990)
7. Jing, F., Li, M.J., Zhang, H.J., Zhang, B.: Entropy-based active learning with support vector machines for content-based image retrieval. In: IEEE International Conference on Multimedia and Expo., June 27–30, 2004. Digital Object Identifier, vol. 1, pp. 85–88 (2004)
8. Lafferty, J., McCallum, A., Pereira, F.: Conditional random fields: Probabilistic models for segmenting and labeling sequence data. In: Proc. 18th International Conf. on Machine Learning, pp. 282–289. Morgan Kaufmann, San Francisco (2001)
9. McCallum, A., Freitag, D., Pereira, F.: Maximum entropy Markov models for information extraction and segmentation. In: Proc. 17th International Conf. on Machine Learning, pp. 591–598. Morgan Kaufmann, San Francisco (2000)
10. McCallum, A.K., Nigam, K.: Employing EM in pool-based active learning for text classification. In: Shavlik, J.W. (ed.) Proceedings of 15th International Conference on Machine Learning, Madison, US, pp. 350–358. Morgan Kaufmann Publishers, San Francisco (1998)
11. Nguyen, N., Guo, Y.S.: Comparisons of sequence labeling algorithms and extensions. In: Proceedings of the 24th international conference on Machine learning, pp. 681–688. ACM Press, New York (2007)
12. Rabiner, L.R.: A tutorial on hidden markov models and selected applications in speech recognition. Proceedings of the IEEE 77(2), 257–286 (1989)
13. Roy, N., McCallum, A.: Toward optimal active learning through sampling estimation of error reduction. In: Proc. 18th International Conf. on Machine Learning, pp. 441–448. Morgan Kaufmann, San Francisco (2001)
14. Sang, E.F.T.K.: Introduction to the conll-2002 shared task: Language-independent named entity recognition. In: Proceedings of CoNLL 2002, Taipei, Taiwan, pp. 155–158 (2002)

15. Shen, D., Zhang, J., Su, J., Zhou, G.D., Tan, C.L.: Multi-criteria-based active learning for named entity recognition. In: Proceedings of the 42nd Meeting of the Association for Computational Linguistics, Main Volume, Barcelona, Spain, July 2004, pp. 589–596 (2004)
16. Stegeman, L.: Part-of-speech tagging and chunk parsing of spoken dutch using support vector machines (2006)
17. Symons, C.T., Samatova, N.F., Krishnamurthy, R., Park, B.H., Umar, T., Buttler, D., Critchlow, T., Hysom, D.: Multi-criterion active learning in conditional random fields. In: Proceedings of the 18th IEEE International Conference on Tools with Artificial Intelligence, Washington, DC, USA, pp. 323–331. IEEE Computer Society, Los Alamitos (2006)
18. Tong, S., Koller, D.: Support vector machine active learning with applications to text classification. In: Langley, P. (ed.) Proceedings of ICML 2000, 17th International Conference on Machine Learning, Stanford, US, pp. 999–1006. Morgan Kaufmann Publishers, San Francisco (2000)
19. Tsochantaridis, I., Joachims, T., Hofmann, T., Altun, Y.: Large margin methods for structured and interdependent output variables. J. Mach. Learn. Res. 6, 1453–1484 (2005)
20. Weston, J., Watkins, C.: Multi-class support vector machines. Technical Report CSD-TR-98-04, Department of Computer Science, Royal Holloway, University of London, Egham, TW20 0EX, UK (1998)

An Efficient Similarity Searching Algorithm Based on Clustering for Time Series

Yucai Feng, Tao Jiang, Yingbiao Zhou, and Junkui Li

College of Computer Science and Technology,
Huazhong University of Science and Technology,
Wuhan 430074, China
jiangtao_guido@yahoo.com.cn

Abstract. Indexing large time series databases is crucial for efficient searching of time series queries. In the paper, we propose a novel indexing scheme RQI (**R**ange **Q**uery based on **I**ndex) which includes three filtering methods: *first-k filtering*, indexing lower bounding and upper bounding as well as triangle inequality pruning. The basic idea is calculating wavelet coefficient whose first k coefficients are used to form a MBR (minimal bounding rectangle) based on haar wavelet transform for each time series and then using *point filtering* method; At the same time, lower bounding and upper bounding feature of each time series is calculated, in advance, and stored into index structure. At last, triangle inequality pruning method is used by calculating the distance between time series beforehand. Then we introduce a novel lower bounding distance function SLBS (**S**ymmetrical **L**ower **B**ounding based on **S**egment) and a novel clustering algorithm CSA (**C**lustering based on **S**egment **A**pproximation) in order to further improve the search efficiency of *point filtering* method by keeping a good clustering trait of index structure. Extensive experiments over both synthetic and real datasets show that our technologies provide perfect pruning power and could obtain an order of magnitude performance improvement for time series queries over traditional naive evaluation techniques.

Keywords: Time series, Clustering, Similarity search, Indexing.

1   Introduction

At present, there are more and more time series data owing to its wide application in many domains, such as finance data analysis, Internet traffic analysis, sensor network monitoring, moving object tracking and motion capture. How to get the useful properties of time series data is an important problem. A wide used method is similarity search in time series. Similarity search can be used as the tools of data mining on time series; it also acts as a subroutine for time series mining (TDM) tasks, for example association rules, clustering, classification and pattern detections. However, time series data is a typical high dimension and massive data. How to develop efficient algorithms is the key of success in time series similarity search.

P. Perner (Ed.): ICDM 2008, LNAI 5077, pp. 360–373, 2008.

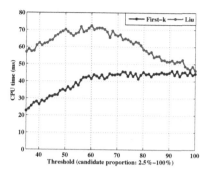

Fig. 1. The comparison of two method spending time. The dataset is from synthetic control chart time series dataset which includes 600 time series. The result is the summation of 50 runs owing to too little time to stat.

Generally, there are three kinds of technologies to improve the efficiency of similarity search, which is lower bounding, time series transform and indexing, as well as quickly efficient algorithms, and moreover, two technologies or three technologies are combined into a scheme for similarity searching.

Indexing large time series databases is crucial for efficient searching of time series queries. Liu et al. [8] proposed a tight lower bounding and upper bounding distance function for L_2-norms. It can be utilized to prune many dissimilar time series and pick up similar time series such that we can save much computing cost and I/O cost. However, their method is not efficient because they take a direct scanning method for each time series. When time series database is very large and mass, the method is less significant. They claims that their method outperforms the traditional method using the first k coefficients of haar wavelet transform [11]. However, our experiment shows that it is not true if the traditional method has a strong pruning capacity and Liu's method doesn't compute two parameters: R_T and S_T [8], in advance. When fetching each time series from database to deal with, we find that spending time of two methods is almost equal as a result of too much I/O cost. When all time series are loaded into memory in advance, Liu's method is slow than *first-k* method (traditional method). Fig. 1 shows the experiment result of two methods.

Motivating to improve on the efficiency of Liu's method, we implement Liu's algorithm in an R-tree index structure and combine *point-filtering* [10] method and triangle inequality pruning method into index method. We make a lot of experiments for our method. Our experiments show that our scheme is more efficient than Liu's method as a result of decreasing the number of post processing time series dramatically. Specifically, in order to further improve the search efficiency of *point filtering* method, we present a novel lower bounding distance function SLBS. Based on SLBS, we present a novel clustering method, namely clustering based on segment approximation (CSA). Building index based on CSA provides a perfect clustering trait which can be made use of further improving the efficiency of *point-filtering* owing to the decrease of searching depth. To summarize, the main contributions of this paper are listed as follows:

1. We propose a novel index scheme RQI which combines three filtering technologies: *point-filtering* method (or *first-k*), tight lower bounding and upper bounding filtering based on index and triangle inequality pruning. Extensive experiments show that our scheme is efficient relative to Liu's algorithm and *first-k* method.

2. We propose a new lower bounding distance function SLBS based on L_2, and then a novel clustering algorithm CSA are introduced based on SLBS. CSA are utilized to optimize index structure which may provide a better searching performance for *point-filtering* method.

3. We make a lot of experiments and prove that our index scheme and clustering algorithm CSA is effective and efficient. Our algorithms represent a perfect pruning capacity and a quick response on similarity search of time series.

The rest of the paper is arranged as follows. Section 2 gives a brief review of the related work. Section 3 demonstrates the background and gives the problem definition. Section 4 is our proposed method. Section 5 evaluates the effectiveness and efficiency of the proposed algorithm vs. Liu's algorithm and traditional algorithm. Finally, Section 6 concludes the paper.

2 Related Work

Since the pioneering work of Agrawal [1] and Faloutsos [2], there emerged many fruit of research in similarity search of time series. In the beginning, many researchers focused on new dimension reduction technologies and new similarity measuring method for time series. The examples of former include DWT (Discrete Wavelet Transform), SVD [6], PAA [11], and APCA [4]. However, these methods have a high percent of false alarms and need a post-processing to guarantee no false dismissals. At the same time, several distance function are proposed, for example L_2-norm (Euclidean distance) [1], DTW and LCSS [9]. Not like L_p-norm, it require that two time series keep the same length, DTW can deal with two time series with different length. However, the time cost in DTW and LCSS is very high.

In order to improve the efficiency of DWT, many researchers proposed lower bounding distance measurement to decrease the number of post-processing time series. These lower bounding distance measure function include LB_Yi [12], LB_Kim [5], LB_Keogh [3] and LB_HUST [7] and so on. LB_Yi [12] is the first DTW lower bounding distance function which makes use of maximum and minimum feature. However, it only can keep a lower filtering efficiency. Afterward, Kim et al. [5] proposed a more approximation lower bounding function, LB_Kim which had used the feature of the first element and the last element, in addition to the maximum and minimum. However, LB_Kim still can not provide a good filtering capacity. In 2002, Keogh et al. [3] brought forward LB_Keogh which is an acknowledged best lower bounding function for DTW and it utilizes local feature of time series and the global restriction for dynamic time warping path. However,

Li et al. [7] found that LB_Keogh is not a symmetrical boundary distance. Therefore, they presented a symmetrical lower bounding LB_HUST which can be made use of clustering time series using DTW. However, all above-mentioned lower bounding functions are aim at DTW. So, Liu et al. [8] put forth a tight lower bounding and upper bounding distance function for L_2 in 2006. However, Liu's algorithm can not acquire a high efficiency when there doesn't exists index structure.

It is very important and necessary to index time series for lower bounding distance measure function. However, there is not much work done in the field. Keogh et al. utilizes LB_PAA [3] and MBR (minimal bounding rectangle) to index time series. Yi et al. [11] used segmented-mean feature to index time series for arbitrary L_p-norms. However, we know that these methods didn't provide a perfect pruning effect. Haar wavelet transform has been used in many domains, for example, time series similarity search [11]. To the best of our knowledge, this is the first work that incorporates tight lower bounding and upper bounding distance function and DWT as well as triangle inequality into index for similarity search in time series database.

3   Problem Statement

3.1   Basic Concept

In the section, we first review some basic concept and useful tool on time series similarity search, which is the main concern of our proposed algorithm.

Definition 1. *Given a query time series $Q(= \{q_1, q_2, \ldots, q_n\})$, a data time series $S(= \{s_1, s_2, \ldots, s_n\})$ and a threshold ε, if their Euclidean distance satisfy the following equation: $D(Q, S) = (\sum_{i=1}^{n}(q_i - r_i)^2)^{1/2} \leq \varepsilon$, we say that S is similar with Q.*

A shortcoming of above definition 1 is that it does not consider the effect of vertical offset to similarity. In fact, vertical offset plays a very important effect in similarity matching on time series. The following figure 2 shows its effect. From Fig. 2, we can find that many candidates are filtered in the case of vertical shift. Therefore, we used the following definition 2 as similarity model of time series. It is more accord with human intuitive cognition to similarity of time series.

Definition 2. *Given a threshold ε, two time series $Q(= \{q_1, q_2, \ldots, q_n\})$ and $S(= \{s_1, s_2, \ldots, s_n\})$ are said to be v-shift similar if $D(S, Q) = (\sum_{i=1}^{n}((s_i - s_A) - (q_i - q_A))^2)^{\frac{1}{2}} \leq \varepsilon$, where $s_A = \frac{1}{n}\sum_{i=1}^{n}s_i$, $q_A = \frac{1}{n}\sum_{i=1}^{n}q_i$.*

Haar wavelet transform is important tools of time series similarity search. As it allow a good approximation with a subset of the first several haar coefficients and it can be computed quickly and easily. The algorithm of haar wavelet transform can find in [11]. In our paper, our similarity matching algorithm is based on haar wavelet transform.

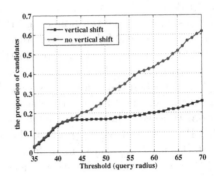

Fig. 2. Candidate proportion comparison using Synthetic Control dataset: vertical shift vs. no vertical shift

3.2   Motivation of the Research

In the section, we explain the motivation of our approach: in particular, why Liu's algorithm is not high efficient and how we improve its efficiency.

In paper [8], Liu et al. introduced a tight lower bounding and upper bounding equation which can be made use of filtering dissimilar time series and picking up most similar time series before post-processing. The main idea of their method embodies the following theorem 1.

Theorem 1. *[8] $T(t_1, t_2, \ldots, t_n)$ and $R(r_1, r_2, \ldots, r_n)$ are the haar wavelet transform of v-shift time series S and Q, there is $\sum_{i=1}^{k}(t_i - r_i)^2 + (T_G - R_G)^2 \leq ||S - Q|| \leq \sum_{i=1}^{k}(t_i - r_i)^2 + (T_G + R_G)^2$ where $T_G^2 = \sum_{i=k+1}^{n} t_i^2$, $R_G^2 = \sum_{i=k+1}^{n} r_i^2$.*

Liu et al. uses the left inequality of theorem 1 to prune the dissimilarity sequences and the right inequality to pick up similarity sequence. However, traditional method makes use of the first k haar coefficients of sequence to prune the dissimilar sequence, which is the inequality $\varepsilon^2 \leq \sum_{i=1}^{k}(t_i - r_i)^2$. In the rest of the paper, we call the traditional method *first-k filtering*.

They claim that using their equation can afford a more quick response time vs. *first-k filtering*. However, we find that it is not true when there doesn't exist index structure. On the one hand, T_G is not a constant which needs $n - k$ square operations and $n - k - 1$ addition operations for each time series, although R_G only requires $n - k$ square operations and $n - k - 1$ addition operations once; On the other hand, our experiments show first-k filtering provides a strong pruning capacity. Therefore, Liu's algorithm can not acquire more quickly response time than *first-k filtering*, especially on large and mass time series data. However, if we build an index structure, computing T_G in advance, and store T_G into index, we can save the time of computing T_G for each time series when comparing the similarity of time series T and R.

4   Our Proposed Method

4.1   Range Query Index Scheme

In the section, we will discuss how to build index structure and our proposed novel indexing scheme based on tight bounding feature which includes three filtering method: *first-k filtering*, indexing lower bounding and upper bounding as well as triangle inequality pruning.

In our scheme, we use R-tree as our high-dimension index method and employ haar wavelet transform to convert v-shift time series into haar coefficients sequence. In order to make use of *first-k filtering*, we take the first even haar coefficients to form a MBR which contains the first $k/2$ points for each haar data sequence (k is an even). Here, we keep odd coefficients of DWT as the first dimension feature, and even coefficients of DWT as the second dimension feature ($f = 2$). Therefore, each two neighbouring coefficients of haar DWT corresponds to an f-point of feature space. In the way, a haar data sequence of $Len(T)$ is mapped to a trail in feature space, consisting of $Len(T)/2$ points: one point corresponding to two adjacent harr coefficients. For example, when k is 4, the MBR corresponding to sequence will contain the first two points. An example about k value, f-point, and MBR of haar sequences is show in table 1.

Table 1. k value, f-point, and MBR for haar sequences

haar sequence	k	f-point($f = 2$)	MBR(min_{f1},min_{f2}, max_{f1},max_{f2})
222.42,18.74,-5.40,17.23,...	4	$(222.42, 18.74)_1$,$(-5.40, 17.23)_2$	(-5.40,17.23,222.42,18.74)
222.70,9.92,-1.91,21.65, 0.90,3.41,...	6	$(222.70, 9.92)_1$,$(-1.91, 21.65)_2$ $(0.90, 3.41)_3$	(-1,91,3.41,222.70,21.65)

Figure 3(a) show the structure of a leaf node and a non-leaf node. Notice that the non-leaf nodes do not need to carry information about *sequence_id*, *pointsArray* and T_G. Here, *pointsArray* is a array variable which stores the first $k/2$ f-points information in order. T_G is the square summary of the last $n - k$ haar coefficients which are calculated in advance. Generally, a MBR contains two f-points in our experiment ($k = 4$). On the one hand, we know that the first four coefficients can prune many dissimilarity sequences; on the other hand, MBR containing too much f-points will bring too false alarms [10]. Of course, you can contain more f-points in a MBR. Fig. 3(b) illuminates the problem. From the figure, we see that two f-points of query time series Q_1 and Q_2, that is fq_1 and fq_2, are in ε-match with the MBR, so all sequences in MBR will be the candidates of Q_1 and Q_2. However, in fact, they are not the candidates of Q_1 and only two corresponding sequences of fp_6 and fp_7 are the candidates of Q_2.

After building the index structure, we will introduce our range query algorithm. It is straightforward and the detailed algorithm is as algorithm RQI

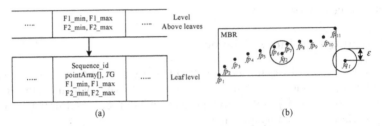

Fig. 3. (a) Index node layout for the last two levels. (b) False alarms caused by storing too many f-points

---

Algorithm. RQI(Range Query based on Index)

Input: Time series database: $TSDB$, Query time series:Q, Threshold: ε.

Output: C_S-similarity sequence set

1. **FOR** each $S \in TSDB$ **DO**
2.   Transform S into v-shift sequence $\tilde{S}$;
3.   Transform $\tilde{S}$ into haar sequence T;
4.   Insert MBR_s and T_G into index;
5. **END**
6. FastSearch($root$, MBR_q, ε);

(**R**ange **Q**uery based on Index) which invokes *FastSearch* algorithm. It has a straightforward recursive implementation. When searching locates in the internal node, we use *point-filter* method which reflects at line 2. When searching arrives at the leaf node, we get the first k coefficients from the array of *points Array*, T_G and R_G by the pointer pointed to data tuples. Then, we prune the dissimilarity sequences through line 5 and gain the similarity sequences through line 6. At last, we invoke a algorithm *TriangleFiltering* at line 7 to further prune dissimilarity sequences and to guarantee no false dismissals.

Algorithm. FastSearch()

Input: *root*-the pointer of root node, MBR_q-a query rectangle, ε-Threshold, and k-the number of the foremost haar coefficients.

Output: C_S-similarity sequence set

1. **IF** ($root-> level > 0$) /* internal node */
2.   **IF**($root-> MBR_s$ intersects query region)
3.   FastSearch($root-> child$, MBR_q, $var\epsilon$);
4. **ELSE** { /* leaf node */
5.   **IF**($\varepsilon^2 \le \sum_{i=1}^{k}(t_i - r_i)^2 + (T_G - R_G)^2$) **RETURN**;
6.   **IF**($\sum_{i=1}^{k}(t_i - r_i)^2 + (T_G + R_G)^2 \le \varepsilon^2$){
         $C_S \leftarrow T_{ID}$; **RETURN**;}
7.   TriangleFiltering(T, $tArray$, $tMatrix$, C_S, Q, ε);
8. }

In the algorithm of *TriangleFiltering*, we make two premises when time series database is very large. The first premise is that the matrix $tMatrix$ is small enough to be contained in main memory. The second premise is that the size of $tArray$ is small enough. Now, we will consider the situation. In fact, it is not needed to store each pair L_2 distances into $tMatrix$ for each pairwise data series. Let $MaxTriangle$ denote the maximum number of time series whose true L_2 distances are kept for triangle inequality pruning and N is the number of time series in the database. Hereafter, we call these time series the reference series. Then, we can find that $MaxTriangle$ is more less than N, namely $MaxTriangle << N$. Thus, the size of the $tMatrix$ is $MaxTriangle * N$. We use a dynamic strategy to fill up the fixed-size $tArray$. At the beginning of running $FastSearch$, there may not be enough reference series for the triangle inequality to be effective. As the reference series are picked and kept, $tArray$ will be gradually filled.

Algorithm. TriangleFiltering(S, $tArray$, $tMatrix$,C_S, Q, ε)

Input: S-current time series, $tArray$-the array of time series with computed true distance to Q, $tmatrix$-precomputed pairwise distance matrix, C_S-resultset, Q-query time series, ε-threshold
1. $maxPruneDist = 0$;
2. **FOR** each time series $R \in tArray$ **DO**
3.   **IF**$((tArray[R].dist - tMatrix[R,S]) \text{¿} maxPruneDist)$
4. $maxPruneDist = tArray[R].dist\text{-}tMatrix[R,S]$;
5.   **END**
6. **IF**$(maxPruneDist > \varepsilon)$**RETURN**;/*pruning by triangle*/
7. **ELSE** {
8. $realDist = L_2(Q,S)$;/*compute true distance*/
9.   insert S and $realDist$ into tArray;
10.   **IF** $(realDist < \varepsilon)$ $C_S \leftarrow$TID;
11.}

4.2 Combining Clustering Algorithm into Index Scheme

In order to further improve pruning capacity of *first-k filtering* in index, we propose an improved strategy to optimize the index structure by clustering time series in advance. We pick up a time series from a cluster until it is empty and then pick up time series from another cluster when we build index structure. By the means, index can gain a best cluster trait, that is, a node of index includes time series more likely from the same cluster.

SLBS(Symmetrical Lower Bounding based on Segment). For a given time series $S = (s_1, s_2, \ldots, s_n)$, we divide up it into N equal segments. Then we take their local maximum and minimum for each segment which consists of w elements. The i-th approximate segment s_i^A of S is defined as: $s_i^A = \{s_i^L, s_i^U\}$, $s_i^L = min\{s_{(i-1)*w+1}, \ldots, s_{i*w}\}$, $s_i^U = max\{s_{(i-1)*w+1}, \ldots, s_{i*w}\}$ where s_i^L and s_i^U are the minimum and maximum values within the subsequence from $s_{(i-1)*w+1}$ to s_{i*w}. An obvious yet important property of lower bounds and upper bounds of s_i^A is the following:

$$s_i^L \leq s_{(i-1)*w+1} \cdots \leq s_{i*w} \leq s_i^U, 1 \leq i \leq N. \tag{1}$$

Therefore, we approximate S by $S^A = \{s_1^A, s_2^A, \ldots, s_N^A\}$ where N denotes the number of segments. We define the distance between s_i^A and q_i^A as $D(s_i^A, q_i^A), 0 \leq i \leq N$. Then we will hold the following equation:

$$D(s_i^A, q_i^A) = \begin{cases} (s_i^L - q_i^U)^2 & (s_i^L > q_i^U) \\ (q_i^L - s_i^U)^2 & (q_i^L > s_i^U) \\ 0 & (otherwise) \end{cases} \tag{2}$$

Based on (2), we can obtain a new **Symmetrical Lower Bounding** based on Segment (**SLBS**) distance measure function as follows:

$$D_{SLBS}(S^A, Q^A) = \sqrt{\sum_{i=1}^{N} w.D(s_i^A, q_i^A)} \tag{3}$$

where w represents the length of each equal segment.

Theorem 2. *For given the approximate segments of S and Q, S^A and Q^A respectively, the following inequality hold: $D_{SLBS}(S^A, Q^A) \leq D_{L2}(S, Q)$.*

Proof. Let us assume the following inequality is true:

$$w.D(s_i^A, q_i^A) \leq \sum_{j=(i-1)*w+1}^{i*w} (s_j - q_j)^2 \tag{4}$$

We prove Theorem 2 by induction. Firstly, we consider the case : $s_i^L > q_i^U$ according to (1) and (2), we have $q_i^L \leq q_{(i-1)*w+1} \cdots \leq q_{i*w} \leq q_i^U < s_i^L \leq s_{(i-1)*w+1} \cdots \leq s_{i*w} \leq s_i^U$, so inequality (4) is correct. For the case $q_i^L > s_i^U$, inequality (4) can be proved by a similar procedure. As the third case, it is obvious. Since, inequality (4) is true, we have $D_{SLBS}(S^A, Q^A) \leq D_{L2}(S, Q)$.

Clustering time series based on SLBS. In above section, we have defined lower bounds and upper bounds of time series segment. Similarly, we can define lower bounds and upper bounds of a cluster time series segment. More formally, given a cluster $C_S = (S_1, S_2, \ldots, S_k)$, the lower bounds of i-th segment of cluster C_S is as: $C_{si}^L = min(s_{i1}^L, s_{i2}^L, \ldots, s_{ik}^L)$ where s_{ik}^L denotes the lower bounds of i-th segment for k-th time series S_k, $1 \leq i \leq N$, the upper bounds of i-th segment of cluster C_S is as : $C_{si}^U = max(s_{i1}^U, s_{i2}^U, \ldots, s_{ik}^U)$ where s_{ik}^U denotes the lower bounds of i-th segment for k-th time series S_k, $1 \leq i \leq N$.

Similar to the $D(s_i^A, q_i^A)$ for time series segment, we define the distance between $C_{s_i^A}$ and $C_{q_i^A}$ as follows:

$$D(C_{s_i^A}, C_{q_i^A}) = \begin{cases} (C_{si}^L - C_{qi}^U)^2 & (C_{si}^L > C_{qi}^U) \\ (C_{qi}^L - C_{si}^U)^2 & (C_{qi}^L > C_{si}^U) \\ 0 & (otherwise) \end{cases} \tag{5}$$

where $C_{s_i^A} = \{C_{si}^L, C_{si}^U\}$, $C_{s_i^A} = \{s_{i1}^A, s_{i2}^A, \ldots, s_{ik}^A\}$ and $C_{q_i^A} = \{q_{i1}^A, q_{i2}^A, \ldots, q_{ik}^A\}$, s_{ik}^A and q_{ik}^A denote the i-th approximate segment of k-th time series S_k and Q_k

in cluster C_S and cluster C_Q respectively. Thus, we have the symmetrical lower bounding based on segment distance measure function between C_S and C_Q as follows:

$$D_{SLBS}(C_S, C_Q) = \sqrt{\sum_{i=1}^{N} w.D(C_{s_i^A}, C_{q_i^A})} \qquad (6)$$

where w represents the length of each equal segment in cluster C_S and C_Q.

Theorem 3. *For given the approximate segments of cluster C_S and C_Q, C_{S^A} and C_{Q^A} respectively, the following inequality hold: $D_{SLBS}(C_{S^A}, C_{Q^A}) \leq D_{SLBS} (S^A, Q^A)$.*

The proof is similar the proof of Theorem 2. Here we omit it for brevity. Having defined (6), we are now ready to introduce a cluster algorithm for time series, namely Clustering based on Segment Approximation (CSA) as follows. We com-

---

Algorithm. CSA

Input: *TSDB*-time series database, D_{clu}-threshold of inter-cluster distance, K-maximal number of clusters

Output: $Clusters = \{C_1, C_2, \ldots, C_k\}$

1. $C_1, C_2, \ldots, C_k \leftarrow \Phi$;
2. **FOR** $S \in TSDB$ **DO**
3. $C_S \leftarrow \{S\}$, $minDist \leftarrow MAXIMUM$, $min \leftarrow 1$;
4.   **FOR** $i = 1$ **TO** K **DO**
5. $dist \leftarrow D_{SLB}(C_S, C_i)$;
6.   **IF** $(dist < D_{clu})$
7. $C_i \leftarrow C_i \cup C_S$,**break**;
8.   **END**
9.   **IF** $(dist < minDist)$
10. $minDist \leftarrow dist$,$min \leftarrow i$;
11.   **END**
12.   **END**
13.   **IF** $(i == K)$
14. $Cluster_{min} \leftarrow Cluster_{min} \cup C_S$
15.   **END**
16. **END**

---

bine CSA into the procedure of index creation. Then, we obtain the algorithm RQIC (**R**ange **Q**uery based on **I**ndex and **C**lustering). It is Identical to Algorithm **RQI** except: Line 1 is replaced with **FOR** $S \in C_i$, $i = 1$ to K **DO**.

5   Experimental Evaluations

5.1   Experimental Setup

In the section, we report the empirical study on the effectiveness and efficiency of our algorithm RQI and RQIC. Two type datasets are used in our experiments.

The first type is Synthetic Control Chart Time Series Dataset (SC datasets) from http://www.cs.ucr.edu/eamonn/TSDMA/index.html, which contains 600 time series, each time series has length of 60. We add 4 zero at the end of each time series so that time series length is the power of 2. The second type is real-world data: stock data from Dow Jones Industrials Index (DJI datasets) which includes 19899 daily close from October 1 in 1928 to December 28 in 2007. You can download from http://finance.yahoo.com/. We use a sliding window to cut the long sequences and every time series' length is 128 or 256. We use the same method to create query time series. We conducted experiments on a Pentium 3 PC with 512M memory. All results are averaged over 20 runs.

In all experiments, filter capacity and CPU time are used to measure the efficiency of the algorithms. Filtering capacity is defined as filtering capacity = the number of time series filtered/the number of time series in database. Note all time series are firstly transformed v-shift time series and then transformed haar wavelet sequences. For *first-k filtering*, the number of time series filtered only refers to those that can be determined whether they are dissimilar to query time series. For other algorithms, it includes the number of all time series excluded the *post-processing* time series. We set the parameter of k to 4.

5.2   Efficiency Test on RQI

In the section, we report the efficiency our experiments using CPU time and filtering capacity two parameters on SC and DJI datasets. The number of DJI time series is 19771 and time series length is 128. CPU time is measured by millisecond. In R-tree, the page size is 512 bytes. The number of reference series is 50 and 200 for SC and DJI dataset respectively in triangle pruning.

The first test is on SC dataset and the results reflect in Fig. 5. From Figure 5(a), we can observe that the performance of RQI is excelled Liu's algorithm in CPU time. Liu's algorithm keeps less change owing to almost the same computing cost and I/O time when using scan method. However, RQI can save much computing cost through computing in advance and keeps the result in R-tree index. CPU time increases gradually and arrives maximum when threshold is 58, then it starts to decrease little by little. This is because that lower bounding filtering capacity descends as the augment of threshold and upper bounding filtering starts to take effect close to threshold 58. From Figure 5(b) we can find that RQI keeps best performance in three methods in filtering capacity. Because RQI uses three filtering methods which include first-k filtering, indexing lower bounding and upper bounding as well as triangle pruning. The worse filtering capacity is not less than 0.36 for our method. However, first-k filtering capacity is close to zero when threshold arrives to 70. In fact, it is not significance when threshold is greater than 70 because candidate proportion excels 70% which reflects in figure 5(a).

The second test is on DJI dataset (length=128) and Fig. 6 and Fig. 4(b) show the results. We can observe that the effectiveness and efficiency keeps better on large dataset by figure 6(a) and figure 6(b). In fact, our algorithm can exert better efficiency only on the large and mass datasets. Here, we intend to make an

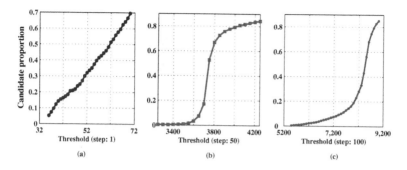

Fig. 4. Candidate proportion for three test: (a) on SC dataset; (b) on DJI dataset (length=128); (c) on DJI dataset (length=256)

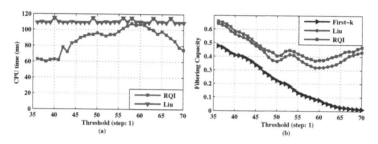

Fig. 5. Test on SC dataset. (a) CPU time comparison: Liu's algorithm and RQI. (b) Filtering capacity comparison: *first-k filtering*, Liu's algorithm and RQI

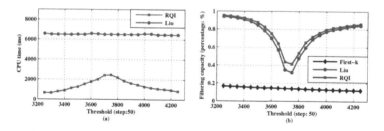

Fig. 6. Test on DJI dataset(length=128). (a) CPU time comparison: Liu's algorithm and RQI. (b) Filtering capacity comparison: *first-k filtering*, Liu's algorithm and RQI.

explanation about the change of candidate proportion which reflects in Fig. 4(b), owing to its correlation to the performance of the second experiment. Figure 4(b) reflects the change of candidate proportion; we see that it keeps a steep rise from threshold 3700 to 3800. Perhaps stock price conforms to the same movement law a certain extent, we conceive. From figure 6(a), we observe that CPU time of RQI is far less than the time of Liu's algorithm and arrives at maximum between thresholds 3700 and 3800 as a result of sudden increasing post processing time

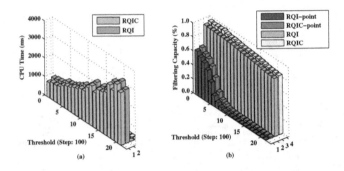

Fig. 7. Efficiency comparison of RQIC and RQI on DJI dataset (length=256) with threshold from 5000 to 7000. (a) CPU time (b) Filtering capacity.

series. On filtering capacity, we observe a similar situation from figure 11. This is derived from two factors. The first factor is that upper bounding filtering takes effect after threshold 3700. The second factor is that many time series would be similar between thresholds 3700 to 3800 based on the observation from figure 6(b). Even the worst situation, our algorithm still keeps better filtering capacity than 40%.

5.3   Efficiency Test on RQIC

In the section, we report the efficiency of RQIC vs. RQI. Here we only use Dow Jones industrial index dataset which includes 19643 time series and the length of time series is 256. From figure 7(a), we can see that CPU time spending in RQIC is less than RQI. The performance progress mainly is derived from point-filtering efficiency which can be observed from figure 7(b). We can obvious find that RQIC algorithm takes a higher point-filtering capacity than RQI through figure 7(b). However, their total filtering capacity is both high and RQIC is little better than RQI. As point-filtering mainly exerts effect between thresholds 5000 to 7000, so we omit other range. Here, RQI-point denotes the point-filtering efficiency of RQI, and RQIC-point is similar.

6   Conclusions

In the paper, a novel index scheme RQI is proposed which takes three pruning strategies, that is, *first-k filtering* which happens at internal node of index, indexing tight lower bounds and upper bounds as well as triangle inequality pruning. RQI keeps a low cost and enables us to prune out most false alarms with no false dismissals. We also introduce a new clustering algorithm CSA which is combined into RQI to form the algorithm RQIC, in order to further improve the pruning capacity of *first-k filtering*. Extensive experiments show RQI and RQIC offer an efficient and scalable searching capacity over large time series data.

References

1. Agrawal, R., Faloutsos, C., Swami, A.: Efficient similarity search in sequence databases. In: Lomet, D.B. (ed.) FODO 1993. LNCS, vol. 730, pp. 69–74. Springer, Heidelberg (1993)
2. Faloutsos, C., Ranganathan, M., Manolopoulos, Y.: Fast subsequence matching in time-series databases. In: Proc. of ACM SIGMOD Conference, pp. 419–429. ACM Press, New York (1994)
3. Keogh, E.: Exact indexing of dynamic time warping. In: Proc. of ACM VLDB Conference, pp. 406–417. ACM Press, New York (2002)
4. Keogh, E., Chakrabarti, K., Mehrotra, S., Pazzani, M.: Locally adaptive dimensionality reduction for indexing large time series databases. In: Proc. of ACM SIGMOD Conference, pp. 151–162. ACM Press, New York (2001)
5. Kim, S. W., Park, S., Chu, W. W.: An index-based approach for similarity search supporting time warping in large sequence databases. In: Proc. of IEEE ICDE Conference, pp. 607–614. IEEE Press, New York (2001)
6. Korn, F., Jagadish, H.V., Faloutsos, C.: Supporting ad hoc queries in large datasets of time sequences. In: Proc. of ACM SIGMOD Conference, pp. 289–300. ACM Press, New York (1997)
7. Li Junkui, Wang Yuanzhen, Li Xinping: LB_HUST: A symmetrical boundary distance for clustering time series. In: The 9th Int'l Conf. on Information Technology, pp. 203–208. IEEE Press, New York (2006)
8. Liu, B., Wang Zhihui, Li Jingtao, Wang W., Shi B.: Tight bounds on the estimation distance using wavelet. In: Yu, J.X., Kitsuregawa, M., Leong, H.-V. (eds.) WAIM 2006. LNCS, vol. 4016, pp. 460–471. Springer, Heidelberg (2006)
9. Vlachos, M., Kollios, G., Gunopulos, D.: Discovering similar multidimensional trajectories. In: Proc. of IEEE ICDE, pp. 673–684. IEEE Press, New York (2002)
10. Moon, Y.-S., Loh, W.-K., Whang, K.-Y.: Duality-based subsequence matching in timeseries databases. In: Proc. of IEEE ICDE, pp. 263–272. IEEE Press, New York (2001)
11. Yi, B.K., Faloutsos, C.: Fast time sequence indexing for arbitrary Lp norms. In: Proc. of ACM VLDB Conference, pp. 385–394. ACM Press, New York (2000)
12. Yi, B.K., Jagadish, H.V., Faloutsos, C.: Efficient retrieval of similar time sequences under time warping. In: Proc. of IEEE ICDE Conference, pp. 23–27. IEEE Press, New York (1998)

Efficient String Mining under Constraints Via the Deferred Frequency Index

David Weese[1] and Marcel H. Schulz[2]

[1] Department of Computer Science, Free University of Berlin, Takustr. 9, 14195 Berlin, Germany
weese@inf.fu-berlin.de
[2] Department of Computational Molecular Biology, Max Planck Institute for Molecular Genetics, Ihnestr. 73, 14195 Berlin, Germany and International Max Planck Research School for Computational Biology and Scientific Computing
marcel.schulz@molgen.mpg.de

Abstract. We propose a general approach for frequency based string mining, which has many applications, e.g. in contrast data mining. Our contribution is a novel algorithm based on a deferred data structure. Despite its simplicity, our approach is up to 4 times faster and uses about half the memory compared to the best-known algorithm of Fischer et al. Applications in various string domains, e.g. natural language, DNA or protein sequences, demonstrate the improvement of our algorithm.

1 Introduction

The storage of data in databases alone does not guarantee that all hidden information is readily available. A promising approach for knowledge discovery in databases is to mine frequent patterns, reviewed in [1]. This general paradigm can be applied in many application domains ranging from mining of customer data to optimize marketing strategies [2], and language identification [3], to finding protein fingerprints or binding motifs in biological sequences [4,5]. The latter is important in Computational Biology, where a gene is regulated by proteins, so-called transcription factors, that bind to its promoter sequence. A common approach taken, is to contrast promoter sequences of genes that are believed to be regulated by the same factor, with promoters of unrelated genes to detect the transcription factor's binding motif. The rationale behind it, is to find sequence motifs that are representative (frequent) for one set of sequences and absent (infrequent) in another, often called discriminatory or contrast data mining [6,7]. Here the Frequency of a motif is defined as the number of distinct sequences in a set that contain the motif at least once. In this paper we propose an approach that can efficiently solve any frequency based string mining problem including the problem introduced above.

1.1 Related Work

There have been several approaches in the context of mining substrings with frequency constraints. Raedt and co-workers introduced the first $\mathcal{O}(n^2)$ algorithm,

P. Perner (Ed.): ICDM 2008, LNAI 5077, pp. 374–388, 2008.

for databases of size n, in 2002 based on the level-wise Apriori algorithm [8]. This algorithm is not suitable for large databases due to repeated scanning of the whole database. Chan and others [9], as well as Lee et al. [10], suggested indexing the database with a suffix tree. Still, suffix trees can be nicely replaced by linear arrays [11], which was utilized by Fischer and colleagues [7] to devise a more efficient algorithm than that of Raedt et al. and Lee et.al [8,10]. One year later, an improvement to their previous algorithm, and the first optimal $\mathcal{O}(n)$ time algorithm was presented by Fischer and the same co-authors [12]. It was established as the fastest known algorithm for the problem, due to optimal time frequency calculation for substring indices via range minimum queries [13].

1.2 Motivation

Fischer and colleagues achieved the optimality [12] at the expense of complicating the algorithm and adding another $\Theta(n)$ space. In addition, both algorithms of Fischer et al. need to sort the whole suffix array and build additional arrays independent of the constraints of the problem. Hence, an interesting approach is to improve upon the frequency calculation of the algorithms [7,8,9,10], while retaining the problem-specific search space pruning. Indeed, we introduce an approach which combines both. We take advantage of partially constructed suffix trees, to design a problem-oriented algorithm like the one of Raedt et al. and Chan et al. [8,9]. Additionally, we utilize a clever solution for the frequency calculation, which comes as a by-product of the sorting procedure without any additional space overhead. On top of that, our approach is surprisingly simple and we show that it is always faster than the optimal algorithm of Fischer and colleagues over a broad range of pattern domains and for different types of frequency string mining problems.

2 Preliminaries

We consider strings over the finite ordered alphabet Σ and use the term *pattern* synonymously. Σ^* is the set of all possible strings over Σ. A string ϕ is a sequence of letters $\phi[1] \ldots \phi[n]$, where each $\phi[i] \in \Sigma$. $\phi\psi$ is the concatenation of two strings ϕ and ψ. $|\phi|$ denotes the length of the string ϕ and $\phi[i..j]$ is a substring of ϕ from position i to j. If $\psi \in \Sigma^*$ is a substring of ϕ, we write $\psi \preceq \phi$, and $\psi \prec \phi$ if $\psi \neq \phi$ holds in addition. For a non-empty set of strings $\Phi \subseteq \Sigma^*$, lcp(Φ) gives the *longest common prefix* of all strings in Φ. If Φ contains exactly 1 string ϕ, lcp(Φ) returns ϕ. A database $\mathcal{D} \subseteq \Sigma^*$ has $|\mathcal{D}|$ many strings over Σ. The *frequency* and the *support* of a string $\phi \in \Sigma^*$ in $\mathcal{D}$ is defined as follows:

$$\text{freq}(\phi, \mathcal{D}) := |\{d \in \mathcal{D} \mid \phi \preceq d\}|, \quad \text{supp}(\phi, \mathcal{D}) := \frac{\text{freq}(\phi, \mathcal{D})}{|\mathcal{D}|} . \tag{1}$$

For a set of databases $\mathcal{D}_1, \ldots, \mathcal{D}_m$ we define the *frequency vector* of ϕ:

$$\text{freq}(\phi, \mathcal{D}_1, \ldots, \mathcal{D}_m) := \Big(\text{freq}(\phi, \mathcal{D}_1), \ldots, \text{freq}(\phi, \mathcal{D}_m)\Big) . \tag{2}$$

For two vectors $u, v \in \mathbb{N}^m$ we define $u \leq v \Leftrightarrow \forall_{i=1,\ldots,m} u_i \leq v_i$.

Example 1. Suppose we are given two databases $\mathcal{D}_1 = \{abab, babb\}$ and $\mathcal{D}_2 = \{baab, aaab\}$, then $\mathrm{freq}(b, \mathcal{D}_1, \mathcal{D}_2) = (2, 2)$ and $\mathrm{freq}(ba, \mathcal{D}_1, \mathcal{D}_2) = (2, 1)$.

2.1   Predicates

A *frequency predicate* on a set of databases $\mathcal{D}_1, \ldots, \mathcal{D}_m$ is defined as a function that for any frequency vector $v \in \mathbb{N}^m$ evaluates to either *true* or *false* and must be *false* for the null vector. In general, our approach is applicable to the task of finding patterns $\phi \in \Sigma^*$ whose frequencies satisfy a predicate *pred* on a given database set $\mathcal{D}_1, \ldots, \mathcal{D}_m$:

$$Th(pred) = \{\phi \in \Sigma^* \mid pred(\mathrm{freq}(\phi, \mathcal{D}_1, \ldots, \mathcal{D}_m)) \text{ is } true\} \ . \tag{3}$$

In the following, we will consider two specific examples of frequency string mining problems:

Problem 1. Given m databases $\mathcal{D}_1, \ldots, \mathcal{D}_m$ of strings over Σ and m pairs of frequency thresholds $(min_1, max_1), \ldots, (min_m, max_m)$, the *Frequent Pattern Mining Problem* is to return all strings $\phi \in \Sigma^*$ that satisfy $min_i \leq \mathrm{freq}(\phi, \mathcal{D}_i) \leq max_i$, for all $1 \leq i \leq m$.

This problem has been considered in a series of research papers [7,8,10]. The next problem considers discriminatory strings for two databases $\mathcal{D}_1, \mathcal{D}_2 \in \Sigma^*$. $\mathcal{D}_1$ is usually called positive (foreground) set, where $\mathcal{D}_2$ is the negative (background) set. As a measure of difference the growth-rate from $\mathcal{D}_2$ to $\mathcal{D}_1$ for a string ϕ is defined as

$$\mathrm{growth}_{\mathcal{D}_2 \to \mathcal{D}_1}(\phi) := \begin{cases} \frac{\mathrm{supp}(\phi, \mathcal{D}_1)}{\mathrm{supp}(\phi, \mathcal{D}_2)} , & \text{if } \mathrm{supp}(\phi, \mathcal{D}_2) \neq 0 \\ \infty & , \text{ otherwise} \end{cases} . \tag{4}$$

Problem 2. Given a support condition ρ_s $(\frac{1}{|\mathcal{D}_1|} \leq \rho_s \leq 1)$, and a minimum growth rate $\rho_g > 1$, the *Emerging Substring Mining Problem* is to detect all strings $\phi \in \Sigma^*$ s.t. $\mathrm{supp}(\phi, \mathcal{D}_1) \geq \rho_s$ and $\mathrm{growth}_{\mathcal{D}_2 \to \mathcal{D}_1}(\phi) \geq \rho_g$ [9].

The minimum support rate ρ_s limits the solution space to representative strings of database $\mathcal{D}_1$, where ρ_g is the discrimination threshold. Patterns which satisfy the conditions of Problem 2 are called *Emerging Substrings*. If the growth rate of the pattern is infinite it is called *Jumping Emerging Substring*, because it is a major discriminator between the databases under investigation.

Example 2. We now apply this problem to databases $\mathcal{D}_1$ and $\mathcal{D}_2$ from Example 1 with $\rho_s = 1$ and $\rho_g = 2$ and want to find all strings $\phi \in \Sigma^*$ with $\mathrm{supp}(\phi, \mathcal{D}_1) \geq 1$ and $\mathrm{growth}_{\mathcal{D}_2 \to \mathcal{D}_1}(\phi) \geq 2$. The corresponding frequency predicate *pred* for the *Emerging Substring Mining Problem* is a function that maps the frequency vector $(d_1, d_2) = \mathrm{freq}(\phi, \mathcal{D}_1, \mathcal{D}_2)$ of a string $\phi \in \Sigma^*$ to a truth value as follows:

$$\begin{aligned} pred(d_1, d_2) &:= (d_1 \geq \rho_s \cdot |\mathcal{D}_1|) \wedge (d_1 \cdot |\mathcal{D}_2| \geq \rho_g \cdot d_2 \cdot |\mathcal{D}_1|) \\ &= (d_1 \geq 2) \wedge (d_1 \geq 2d_2) \ . \end{aligned} \tag{5}$$

The set of patterns whose frequencies satisfy *pred* is $Th(pred) = \{\text{bab, ba}\}$. b for example is not an *Emerging Substring*, because $\text{supp}(b, \mathcal{D}_1) = 1$ but $\text{growth}_{\mathcal{D}_2 \to \mathcal{D}_1}(b) = 1 < \rho_g$.

2.2 Monotonicity

We will now introduce the monotonic property of frequency predicates that we use later to restrict the search space of our algorithm. Examples 3 and 4 show that the frequency predicates of Problem 1 and 2 contain a monotonic subpredicate.

Definition 1. *If for a frequency predicate* $pred : \mathbb{N}^m \to \{\text{true,false}\}$ *holds that:*

$$\forall_{u,v \in \mathbb{N}^m, u \leq v} \left(pred(u) \Rightarrow pred(v) \right) , \tag{6}$$

then pred is called monotonic.

Proposition 1. *For a monotonic[1] frequency predicate pred on databases* $\mathcal{D}_1, \ldots, \mathcal{D}_m \subseteq \Sigma^*$ *it holds that:*

$$\forall_{\phi,\psi \in \Sigma^*, \phi \preceq \psi} : \left(pred(\text{freq}(\psi, \mathcal{D}_1, \ldots, \mathcal{D}_m)) \Rightarrow pred(\text{freq}(\phi, \mathcal{D}_1, \ldots, \mathcal{D}_m)) \right) . \tag{7}$$

Proof. Each occurrence of ψ is also an occurrence of ϕ. Thus, $\text{freq}(\psi, \mathcal{D}_1, \ldots, \mathcal{D}_m) \leq \text{freq}(\phi, \mathcal{D}_1, \ldots, \mathcal{D}_m)$ holds. $\qquad \square$

Example 3. As seen in Example 2 the frequency predicate for the *Emerging Substring Mining Problem* is:

$$pred(d_1, d_2) = (d_1 \geq \rho_s \cdot |\mathcal{D}_1|) \wedge (d_1 \cdot |\mathcal{D}_2| \geq \rho_g \cdot d_2 \cdot |\mathcal{D}_1|) . \tag{8}$$

Generally, *pred* is not monotonic as shown in Example 2. Recall that ba is *emerging* although b is not. However, if we consider only the left inequality:

$$pred_{\text{m}}(d_1, d_2) := (d_1 \geq \rho_s \cdot |\mathcal{D}_1|) , \tag{9}$$

$pred_{\text{m}}$ is monotonic, as for all $u, v \in \mathbb{N}^2$, $u \leq v$ holds $u_1 \geq \rho_s \cdot |\mathcal{D}_1| \Rightarrow v_1 \geq \rho_s \cdot |\mathcal{D}_1|$. Obviously it holds that $pred \Rightarrow pred_{\text{m}}$.

Example 4. For the *Frequent Pattern Mining Problem* with

$$pred(d_1, d_2) = (min_1 \leq d_1 \leq max_1) \wedge (min_2 \leq d_2 \leq max_2) \tag{10}$$

analogously

$$pred_{\text{m}}(d_1, d_2) := (min_1 \leq d_1) \wedge (min_2 \leq d_2) \tag{11}$$

is monotonic and $pred \Rightarrow pred_{\text{m}}$ holds.

---

[1] Note that what we call *monotonic* is called *anti-monotonic* in [8,7], as they consider *pattern* predicates instead of *frequency* predicates.

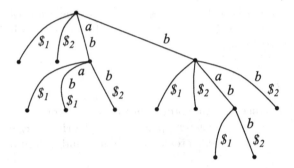

Fig. 1. The resulting generalized suffix tree for database $\mathcal{D}_1 = \{\text{abab}, \text{babb}\}$. As mentioned in the text, to the end of every string from $\mathcal{D}_1$ a unique string marker $\$_1$ and $\$_2$ for the first and second string is appended, respectively.

2.3   Suffix Trees and Suffix Arrays

In this section we will define the *generalized suffix tree* of a database $\mathcal{D} = \{\phi_1, \ldots, \phi_d\}$. To distinguish the suffixes of strings in $\mathcal{D}$, we will use string markers $\$_j$ at the end of each string ϕ_j. String markers are artificial symbols $\$_j$ that must not occur in any string of $\mathcal{D}$ and we implicitly assume $\$_j \in \Sigma$. We define the artificial order $\$_1 < \$_2 < \ldots < \$_d < c$ for any $c \in \Sigma \setminus \{\$_1, \ldots, \$_d\}$.

A generalized suffix tree for a database $\mathcal{D}$ over Σ is a rooted directed tree with edge labels from Σ^*, s.t. every concatenation of symbols from the *root* to a leaf node yields a suffix of $\phi_j\$_j$ for a string $\phi_j \in \mathcal{D}$. Each internal node has at least two children, and no two edges out of the same node are allowed to have edge-labels starting with the same character. By this definition, each node can be mapped one-to-one to the concatenation of symbols from the *root* to itself. The node of a concatenation string α will be denoted by $\bar{\alpha}$.

We will also need the concept of a *generalized suffix array* for a database $\mathcal{D}$ over Σ. Therefore, all strings $\phi_j \in \mathcal{D}$ are concatenated by their string markers $\$_j$ to form conceptually one string $\phi_1\$_1\phi_2\$_2 \ldots \phi_d\$_d$, the *union string* of $\mathcal{D}$. The generalized suffix array stores the starting positions of all lexicographically ordered suffixes of the union string [14].

Generalized suffix trees, generalized suffix arrays, and the union string of a set of databases $\mathcal{D}_1, \ldots, \mathcal{D}_m$ are defined analogously using string markers $\$_1, \$_2, \ldots, \$_{\tilde{d}}$ with $\tilde{d} = \sum_{i=1}^{m} |\mathcal{D}_i|$. Figure 1 shows the generalized suffix tree of the Example 1 database $\mathcal{D}_1$.

3   The Algorithm

This section introduces the *Deferred Frequency Index* (DFI) which is fundamentally based on a generalized lazy suffix tree [15]. The DFI algorithm constructs only the upper part of a generalized suffix tree in a top-down manner. To understand the DFI algorithm we will at first explain the idea of the *write-only, top-down* construction algorithm, abbreviated as *wotd*-algorithm.

3.1   The *wotd*-Algorithm

A lazy suffix tree is a suffix tree whose nodes are created on demand, i.e. when they are visited the first time. For instances where only upper parts of the suffix tree are required, using a lazy suffix tree can be more efficient than constructing the whole suffix tree. Giegerich et al. introduced the first lazy suffix tree data structure [16] that utilizes the *wotd*-algorithm [15,16] for the on-demand node expansion.

The *wotd*-algorithm is a suffix tree construction algorithm that expands a rooted directed tree starting with a tree consisting of only the root node step-by-step to at most the entire suffix tree. We describe a variant of the *wotd*-algorithm to create a generalized suffix tree. Suppose a given non-empty database $\mathcal{D} = \{\phi_1, \ldots, \phi_d\}$ over Σ and a rooted directed tree T. Each node in T is either in *expanded* or *unexpanded* state. In the beginning, T contains only the *unexpanded* root node. Let R be a function that returns for any string $\alpha \in \Sigma^*$ the set of suffixes with string markers of strings in $\mathcal{D}$ so that:

$$R(\alpha) := \{\alpha\beta\$_j \mid \alpha\beta \text{ is a suffix of } \phi_j\} \ . \tag{12}$$

$R(\alpha)$ comprises all suffixes of strings in $\mathcal{D}$ that begin with the string α. In relation to nodes $\bar{\alpha}$ of the generalized suffix tree of $\mathcal{D}$, $R(\alpha)$ contains the concatenated edge labels of paths between the root node and leaf nodes below $\bar{\alpha}$. When an *unexpanded* node $\bar{\alpha}$ of the lazy suffix tree has to be expanded, $R(\alpha)$ can be used to determine the subtree below $\bar{\alpha}$. The node expansion of $\bar{\alpha}$ works as follows: $R(\alpha)$ is divided into groups $R(\alpha c)$ of the same character c that follows α. Let $\alpha c\beta$ be the longest common prefix of $R(\alpha c)$. Out of $\bar{\alpha}$ an edge will be created, labeled with $c\beta$ leading to a node $\overline{\alpha c\beta}$. If $R(\alpha c)$ is a singleton group, the leaf node $\overline{\alpha c\beta}$ is marked as *expanded*. Otherwise, it is a branching node and marked as *unexpanded*. After all groups were processed, $\bar{\alpha}$ will be marked as *expanded*.

Algorithm 1 shows how the whole generalized suffix tree can be constructed recursively starting with *expandNode(root)* on a tree T that contains only the *unexpanded* root node. It can easily be modified to create only an upper part of the suffix tree.

3.2   Monotonic Hull

We now show how to connect arbitrary frequency predicates with the *wotd*-algorithm. To do so, we give a theoretical description of the minimal set of nodes that need to expanded.

Definition 2. *Given frequency predicates pred and $pred_{hull}$. $pred_{hull}$ is called a monotonic hull of pred, if it is monotonic and $pred \Rightarrow pred_{hull}$ holds.*

The most trivial monotonic hull of each frequency predicate *pred* is $pred_{hull} \equiv$ *true*. If we take a look at the generalized suffix tree T of databases $\mathcal{D}_1, \ldots, \mathcal{D}_m$, we make the following observations:

Proposition 2. *Let pred be an arbitrary frequency predicate and $pred_m$ an arbitrary monotonic frequency predicate on $\mathcal{D}_1, \ldots, \mathcal{D}_m$. For all pairs of fathers and sons $\bar{\alpha}$ and $\overline{\alpha\beta}$ in T it holds that:*

Algorithm 1. expandNode($\bar{\alpha}$)

Input   : *unexpanded* node $\bar{\alpha}$
1 Divide $R(\alpha)$ into subsets $R(\alpha c)$ of suffixes starting with character c after α
2 **foreach** $c \in \Sigma$ *and* $R(\alpha c) \neq \emptyset$ **do**
3      $\alpha c\beta \leftarrow \mathrm{lcp}(R(\alpha c))$
4   **if** $|R(\alpha c)| = 1$ **then**   // leaf node
5 | Create the *expanded* node $\overline{\alpha c\beta}$ below $\bar{\alpha}$
6   **else**   // branching node
7 | Create the *unexpanded* node $\overline{\alpha c\beta}$ below $\bar{\alpha}$
8 | expandNode($\overline{\alpha c\beta}$)

9 Mark $\bar{\alpha}$ as *expanded*

---

1. If $pred(\mathrm{freq}(\alpha\beta, \mathcal{D}_1, \dots, \mathcal{D}_m))$ *is* true *then for each string* χ *with* $\alpha \prec \chi \preceq \alpha\beta$ $pred(\mathrm{freq}(\chi, \mathcal{D}_1, \dots, \mathcal{D}_m))$ *is* true.
2. If $pred_m(\mathrm{freq}(\alpha\beta, \mathcal{D}_1, \dots, \mathcal{D}_m))$ *is* true *then* $pred_m(\mathrm{freq}(\alpha, \mathcal{D}_1, \dots, \mathcal{D}_m))$ *is* true.

Proof. The frequency vectors of $\alpha\beta$ *and* χ *with* $\alpha \prec \chi \preceq \alpha\beta$ *must be equal. If not, there would be a branching node between* $\bar{\alpha}$ *and* $\overline{\alpha\beta}$ *which contradicts the assumption* $\bar{\alpha}$ *would be the father of* $\overline{\alpha\beta}$*. Hence 1. holds. 2. is a direct consequence of Proposition 1 as* α *is a substring of* $\alpha\beta$*.* □

In consequence of Proposition 2, it satisfies to evaluate *pred* only on the nodes of T to compute the set *Th(pred)*. For every monotonic hull $pred_{\mathrm{hull}}$ of *pred* the set of nodes, whose frequencies satisfies $pred_{\mathrm{hull}}$, is a directed connected subgraph of T, which if non-empty, contains the root node. Outside of this subgraph there is no node fulfilling *pred*. Our algorithm exclusively traverses this subgraph to compute the set *Th(pred)*. Hence, we are interested in keeping the subgraph as small as possible, leading to the next definition:

Definition 3. $pred_{\mathrm{hull}}$ *is called* the optimal monotonic hull *of pred, if it is a monotonic hull of pred, and for each monotonic hull* $pred'_{\mathrm{hull}}$ *of pred, it holds that* $pred_{\mathrm{hull}} \Rightarrow pred'_{\mathrm{hull}}$*.*

In other words, if $pred_{\mathrm{hull}}$ is optimal, the corresponding subgraph is minimal.

3.3 The Deferred Frequency Index

In the following we will show how the DFI can be built for any given frequency predicate *pred* and a monotonic hull $pred_{\mathrm{hull}}$.

Algorithm 2 starts with *expandNodeWithConstraint(root, pred, $pred_{\mathrm{hull}}$)* on a tree T with α as the *unexpanded* root node. First, *divideAndCountFreq* is called for the current node $\bar{\alpha}$ in line 1. Identically to algorithm 1, the set $R(\alpha)$ is divided

---

Algorithm 2. expandNodeWithConstraint($\bar{\alpha}$, *pred*, *pred*$_{\text{hull}}$)

    Input   : *unexpanded* node $\bar{\alpha}$
1 *Freq* = divideAndCountFreq($\bar{\alpha}$)
2 **foreach** $c \in \Sigma$ **and** $R(\alpha c) \neq \emptyset$ **do**
3      $\alpha c \beta \leftarrow$ lcp($R(\alpha c)$)
4   **if** *pred*(Freq[c]) **then**
5   ⌊ Output strings χ with $\alpha c \preceq \chi \preceq \alpha c \beta$ [4]
6   **if** *pred*$_{\text{hull}}$(Freq[c]) **then**
7   **if** $|R(\alpha c)| = 1$ **then**   // leaf node
8   ⌊ Create the *expanded* node $\overline{\alpha c \beta}$ below $\bar{\alpha}$
9   **else**   // branching node
10   Create the *unexpanded* node $\overline{\alpha c \beta}$ below $\bar{\alpha}$
11   expandNodeWithConstraint($\overline{\alpha c \beta}$, *pred*, *pred*$_{\text{hull}}$)
12 Mark $\bar{\alpha}$ as *expanded*

---

into groups $R(\alpha c)$ of suffixes starting with the same character $c \in \Sigma$ after their prefix α. In addition, an array *Freq*, that stores in *Freq*[c] the frequency vector freq($\alpha c, \mathcal{D}_1, \ldots, \mathcal{D}_m$), is returned. In the next section we explain the implementation details of function *divideAndCountFreq*. The longest common prefix of every non-empty group $R(\alpha c)$ is determined and assigned to $\alpha c \beta$ in line 3. If the predicate *pred* evaluated with the frequency vector *Freq*[c] is *true*, by Proposition 2 all strings χ with $\alpha c \preceq \chi \preceq \alpha c \beta$ belong to *Th*(*pred*) and are output[2]. In line 6 *pred*$_{\text{hull}}$ is evaluated on *Freq*[c]. Only if *true* is returned, the subtree below the node $\overline{\alpha c \beta}$ may contain a node $\bar{\gamma}$ with $\gamma \in$ *Th*(*pred*) and will be expanded recursively. If *false* is returned, the node $\overline{\alpha c \beta}$ is not created, as no further subtree expansion is necessary.

Algorithm 2 is correct and outputs the set *Th*(*pred*) because of the following: For each database substring ϕ there is a path from the *root* ending in a node or on an edge to a node. This node has the same frequency vector as ϕ and will be visited if it satisfies *pred*$_{\text{hull}}$ and output iff it satisfies *pred*. As *pred*$_{\text{hull}}$ is a monotonic hull, no node that satisfies *pred* is left out by the algorithm.

For the *Emerging Substring Mining Problem* and the *Frequent Pattern Mining Problem* one only needs to replace *pred* and *pred*$_{\text{hull}}$ in Algorithm 2 with the predicates deduced in Examples 3 and 4, respectively. The monotonic hulls for these problems are also optimal as Proposition 3 and 4 prove (see Appendix). Figure 2 shows the DFI for the *Emerging Substring Mining Problem* considered in Example 2.

3.4 Algorithm Details

In this section we explain the function *divideAndCountFreq* in detail (Algorithm 3). The sets $R(\alpha)$ are in fact not stored as sets of strings, but as intervals of a generalized suffix array SA. SA is initialized with numbers from 1 to $|S|$, where S

---

[2] In fact, we omit to output strings χ with a trailing $\$_j$ of a string.

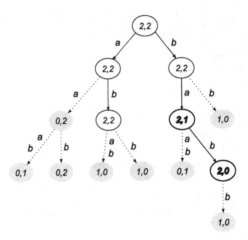

Fig. 2. The generalized suffix tree of our example databases $\mathcal{D}_1$ and $\mathcal{D}_2$. For clarity, the artificial string markers $\$_j$ are omitted. Considering the problem of Example 2, the DFI would construct only the white nodes. Grey nodes are not built. The bold nodes $\overline{ba}$, $\overline{bab}$ represent the *Emerging Substrings*. Each node holds the frequency vector of its corresponding substring (compare the frequencies of $\overline{b}$ and $\overline{ba}$ with Example 1).

is the union string of $\mathcal{D}_1, \ldots, \mathcal{D}_m$. We need a function *getSeqNo* that returns, for each character position i, the sequence number j if $\$_j$ is the next string marker at or to the right of position i in S. We also need a function *getDatabaseNo* that returns, for each sequence number j, the corresponding database number k if $\$_j$ is a string marker of a string in $\mathcal{D}_k$.

When *divideAndCountFreq*($\bar{\alpha}$, *pred*, *pred*$_{hull}$) is called, SA$[l..r]$ contains the start positions of suffixes of S starting with α. Each start position corresponds to a suffix in $R(\alpha)$. Because $\bar{\alpha}$ is *unexpanded*, the suffixes in SA$[l..r]$ have been sorted with counting sort [17] up to the first $|\alpha|$ characters by previous function calls. Because counting sort is stable, the positions in SA$[l..r]$ are in increasing order. Therefore, the corresponding sequence numbers of the positions are stored in contiguous blocks. Counting sort divides $R(\alpha)$ into buckets $R(\alpha c)$ for each character $c \in \Sigma$ (lines 3–4). The frequency of each bucket can simply be counted by counting blocks of equal sequence numbers (line 5).

We keep track of three arrays in the size of the alphabet, i.e. $|\Sigma|$, namely *Bucket*, *Freq* and *Last*. *Bucket* is the original array from counting sort, and *Bucket*$[c]$ counts the occurrences of αc. *Freq* stores frequency vectors, and *Freq*$[c][k]$ determines how often αc occurred in distinct sequences of $\mathcal{D}_k$. *Last* is used to construct *Freq* (lines 5–8).

4   Experiments

To evaluate the performance of our algorithm, we conducted a number of experiments with databases of different characteristics. We used a previously compiled

---

Algorithm 3. divideAndCountFreq($\bar{\alpha}$)

Input   : *unexpanded* node $\bar{\alpha}$
Output : freq($\alpha c, \mathcal{D}_1, \ldots, \mathcal{D}_m$) for each $c \in \Sigma$
Require : SA[$l..r$] stores all suffixes starting with α, suffixes with equal sequence
       numbers are contiguous in this interval
Ensure : suffixes with equal sequence numbers are contiguous in output
       intervals SA[$Bucket[c]..Bucket[c+1] - 1$]

1 Init *Bucket*, *Freq*, *Last* with 0s
   // start to sort the first char after prefix α
2 for $i \leftarrow l$ to r do
3      $c \leftarrow S[SA[i] + |\alpha|]$
4      $Bucket[c] \leftarrow Bucket[c] + 1$
5   if $Last[c] \neq getSeqNo(SA[i])$ then
6          $Last[c] \leftarrow getSeqNo(SA[i])$
7          $k \leftarrow getDatabaseNo(getSeqNo(SA[i]))$
8          $Freq[c][k] \leftarrow Freq[c][k] + 1$

9 Sort suffixes in SA[$l..r$] stable using *Bucket* (Counting sort [17] lines 6–11)
   // now Freq[c] contains the frequency vector freq(αc, 𝒟₁,…,𝒟ₘ)
10 **return** Freq

---

set of human and drosophila core promoters [18], the UniProt [19] proteome
sets of human and mouse, release 12.6, verses of the King James Bible and the
Bible in Basic English, and posts of 5 computer newsgroups from the UCI Machine Learning Repository divided into Windows and non-Windows groups. The
alphabet size $|\Sigma|$ or the sizes of these databases are shown in Table 1.

An experiment consists of two databases $\mathcal{D}_1, \mathcal{D}_2$. These were searched for
Emerging Substrings and for the solution of the *Frequent Pattern Mining Problem* with different values of ρ_s and varying min_1, respectively. As ρ_g and max_2
had no measurable influence on the tested algorithms only the results for $\rho_g = 5$
and $max_2 = \frac{|\mathcal{D}_2|}{2}$ are shown. The results for other values look similar [7]. We
made no other restrictions, i.e. $max_1 = \infty$, $min_2 = 0$.

The theoretically optimal algorithm of Fischer et al. has turned out to be the
hitherto fastest algorithm in practice for the two introduced string mining problems [12]. Hence, we used the implementation of Fischer's algorithm as reference
in our experiments. Both programs were written in C++ and compiled using the
same compiler options. They run under Linux on an Intel Xeon 3.2 GHz with
2 GB of RAM. To reduce influences from the operating system and secondary
storage units, the output was redirected to the null-device, and each experiment
was repeated 5 times. We measured the running time and space consumption of
both algorithms using the GNU tools `time` and `memusage`.

Figure 3 shows, that our approach is in all cases faster than the approach of
Fischer et al. even for small values ρ_s or min_1 when the whole suffix tree needs
to be constructed. As an example, for $\rho_s = 0.2$ the DFI is with 16 seconds on

---

[3] `http://www.o-bible.com/`

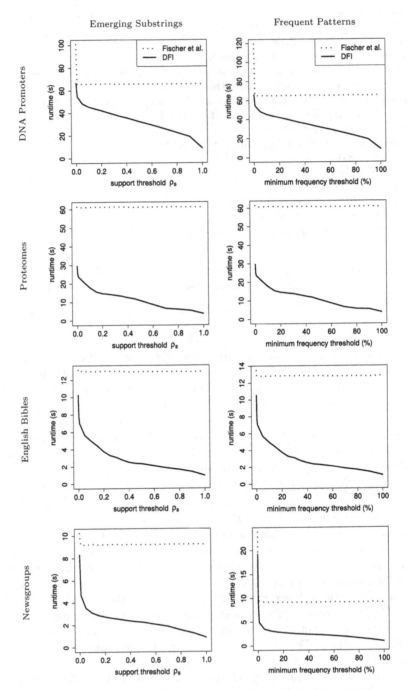

Fig. 3. Runtime comparison of the algorithm of [12] (dotted) and our DFI implementation (solid) for the *Emerging Substring Problem* (left) and the *Frequent Pattern Mining Problem* (right). Experiment details are listed in Table 2.

Table 1. Characteristics and short names for the different databases we used

name	description	$|\Sigma|$	size (mb)	#seqs	source
HProm	human promoters	5	23	15011	Fitzgerald et al. [18]
DProm	drosophila promoters	5	16.7	10914	
HProt	human proteome	24	17.6	40827	Uniprot [19]
MProt	mouse proteome	24	16	35344	
KJB	king james bible	128	4.1	31102	Chinese and English
BBE	bible in basic english	128	4.2	31102	Bible Online[3]
WN	windows newsgroup	128	3.9	2000	Machine Learning
CN	computer newsgroup	128	3.4	3000	Repository [20]

Table 2. Experimental setups and space consumption in MB for various minimum support values

experiment name	$\mathcal{D}_1, \mathcal{D}_2$	Fischer $\rho_s \in [0,1]$	DFI $\rho_s = .001$	DFI $\rho_s = .01$	DFI $\rho_s = .1$
DNA Promoters	HProm , DProm	919.57	531.19	475.82	468.15
Proteomes	HProt , MProt	779.01	365.49	330.95	327.74
English Bibles	KJB , BBE	193.83	109.46	96.96	95.38
Newsgroups	WN , CN	167	104.43	82.25	79.76

the Proteome datasets roughly 4 times faster than the algorithm of Fischer et al. Considering reasonable[4] values of $\rho_s < 0.2$ and $min_1 < 0.2 \cdot |\mathcal{D}_1|$ our algorithm is 1.5–4 times faster in practice. This is surprising, because our algorithm has a worst case running time of $\mathcal{O}(n^2)$ [15], in contrast to the $\mathcal{O}(n)$ algorithm of Fischer and colleagues. Both algorithms have an $\mathcal{O}(n)$ memory consumption, but ours needs only about half of the memory, see Table 2. Fischer's algorithm has an almost constant running time and space consumption as it does not take advantage of the monotonic pruning of the suffix tree like our deferred approach does. The runtime peaks for small values of ρ_s or min_1 are due to the high amount of strings in the solution space that were output.

5 Discussion and Future Work

We presented a new approach to constraint-based string mining that outperforms the best-known algorithm by Fischer et al. [12] in runtime and space consumption as the experiments show. The better running time can be attributed to various factors. Most importantly, the optimal monotonic hull of a frequency

---

[4] Dong and Li [21] report that a minimum support of 1%–20% for finding *Emerging Patterns* could contribute significantly to knowledge discovery.

predicate, is incorporated to prune the search space to a minimum, resulting in the deferred frequency index. Moreover, the frequency information is extracted as a constant time by-product during the suffix tree construction. Our algorithm inherits the good cache locality from the *wotd*-algorithm [15] and in addition uses less memory than Fischer's algorithm.

Depending on the problem at hand, the implementation of our algorithm could be improved. If the DFI should only be used to output the result of *Th(pred)*, the memory consumption of the algorithm could be further reduced. As each node is visited at most once, at any time only nodes of the suffix tree on the path from the *root* to the current node need to be stored. A small alphabet (e.g. DNA) leads to a dense suffix tree with many branching nodes at the top, as observed by Kurtz [22]. In that case, a runtime improvement could be expected by replacing the top of the suffix tree with a q-gram index.

We believe that our constraint oriented algorithm will be useful for the data mining community. Considering constraints during the mining process will play an important role in further algorithmic development, because reducing the solution space of any mining approach to a compact but representative set is one of the open challenges, as mentioned by Han et al. [1]. In the spirit of this observation, the simplicity of our approach opens various avenues of further research. One is to combine *Jumping Emerging Substrings* to build powerful classifiers as was done for *Jumping Emerging Patterns* [23]. This could be achieved by restricting to minimal and highly significant *Jumping Emerging Substrings*. In a recent work [24], a formulation of a similarity pattern predicate composed of an anti-monotonic part was introduced. Our idea can easily be applied, to improve on their approach. Another direction is to extend the algorithm presented here to deal with gap constraints like was done in the work of Ji and colleagues [25]. Our algorithm is freely available at http://www.seqan.de/projects/dfi.html and part of the C++ Sequence Analysis Library SeqAn [26].

Acknowledgements

We thank Knut Reinert who brought the topic to our attention, Clemens Gröpl for helpful discussions, Markus Bauer, Ole Schulz-Trieglaff, and Killian McCutcheon for proofreading.

References

1. Han, J., Cheng, H., Xin, D., Yan, X.: Frequent pattern mining: current status and future directions. Data Min. Knowl. Discov. 15(1), 55–86 (2007)
2. Berry, M.J., Linoff, G.S.: Data Mining Techniques: For Marketing, Sales, and Customer Support, 1st edn., pp. 51–62. John Wiley & Sons, Chichester (1997)
3. Muthusamy, Y.K., Barnard, E., Cole, R.A.: Reviewing automatic language identification. IEEE Sig. Proc. Mag. 11(4), 33–41 (1994)
4. Zhang, M.Q.: Computational analyses of eukaryotic promoters. BMC Bioinformatics 8(Supp 6), S3 (2007)
5. Birzele, F., Kramer, S.: A new representation for protein secondary structure prediction based on frequent patterns. Bioinformatics 22(21), 2628–2634 (2006)

6. Redhead, E., Bailey, T.L.: Discriminative motif discovery in dna and protein sequences using the DEME algorithm. BMC Bioinformatics 8, 385 (2007)
7. Fischer, J., Heun, V., Kramer, S.: Fast frequent string mining using suffix arrays. In: IEEE ICDM 2005, pp. 609–612. IEEE Computer Society Press, Los Alamitos (2005)
8. Raedt, L.D., Jaeger, M., Lee, S.D., Mannila, H.: A theory of inductive query answering. In: IEEE ICDM 2002, pp. 123–130. IEEE Computer Society, Los Alamitos (2002)
9. Chan, S., Kao, B., Yip, C.L., Tang, M.: Mining emerging substrings. In: DASFAA 2003, pp. 119–126. IEEE Computer Society, Los Alamitos (2003)
10. Lee, S.D., Raedt, L.D.: An efficient algorithm for mining string databases under constraints. In: Goethals, B., Siebes, A. (eds.) KDID 2004. LNCS, vol. 3377, pp. 108–129. Springer, Heidelberg (2005)
11. Abouelhoda, M., Kurtz, S., Ohlebusch, E.: Replacing suffix trees with enhanced suffix arrays. Journal of Discrete Algorithms 2, 53–86 (2004)
12. Fischer, J., Heun, V., Kramer, S.: Optimal string mining under frequency constraints. In: Fürnkranz, J., Scheffer, T., Spiliopoulou, M. (eds.) PKDD 2006. LNCS (LNAI), vol. 4213, pp. 139–150. Springer, Heidelberg (2006)
13. Fischer, J., Heun, V.: Theoretical and practical improvements on the RMQ-problem, with applications to LCA and LCE. In: Lewenstein, M., Valiente, G. (eds.) CPM 2006. LNCS, vol. 4009, pp. 36–48. Springer, Heidelberg (2006)
14. Manber, U., Myers, E.: Suffix arrays: a new method for on-line string searches. In: SODA 1990, pp. 319–327. SIAM, Philadelphia (1990)
15. Giegerich, R., Kurtz, S., Stoye, J.: Efficient implementation of lazy suffix trees. Software Pract. Exper. 33(11), 1035–1049 (2003)
16. Giegerich, R., Kurtz, S.: A comparison of imperative and purely functional suffix tree constructions. Sci. Comput. Program. 25, 187–218 (1995)
17. Cormen, T.H., Leiserson, C.E., Rivest, R.L., Stein, C.: 8.2: Counting sort. In: Introduction to Algorithms, 2nd edn., pp. 168–170. MIT Press and McGraw-Hill (2001)
18. Fitzgerald, P.C., Sturgill, D., Shyakhtenko, A., Oliver, B., Vinson, C.: Comparative genomics of drosophila and human core promoters. Genome Biol. 7, R53 (2006)
19. The UniProt Consortium: The Universal Protein Resource (UniProt). Nucl. Acids Res. 36(suppl_1), D190–D195 (2008),
   ftp://ftp.ebi.ac.uk/pub/databases/integr8/uniprot/proteomes
20. Asuncion, A., Newman, D.: UCI machine learning repository (2007),
   http://www.ics.uci.edu/~mlearn/MLRepository.html
21. Dong, G., Li, J.: Efficient mining of emerging patterns: discovering trends and differences. In: KDD 1999, pp. 43–52. ACM, New York (1999)
22. Kurtz, S.: Reducing the space requirement of suffix trees. Software Pract. Exper. 29(13), 1149–1171 (1999)
23. Li, J., Dong, G., Ramamohanarao, K.: Making use of the most expressive jumping emerging patterns for classification. In: PADKK 2000, pp. 220–232. Springer, Heidelberg (2000)
24. Mitasiunaite, I., Boulicaut, J.F.: Looking for monotonicity properties of a similarity constraint on sequences. In: SAC 2006, pp. 546–552. ACM, New York (2006)
25. Ji, X., Bailey, J., Dong, G.: Mining minimal distinguishing subsequence patterns with gap constraints. Knowl. Inf. Syst. 11(3), 259–286 (2007)
26. Döring, A., Weese, D., Rausch, T., Reinert, K.: SeqAn an efficient, generic C++ library for sequence analysis. BMC Bioinformatics 9, 11 (2008)

Appendix

Proposition 3. *Let $\mathcal{D}_1, \mathcal{D}_2$ be two databases, $\rho_s, \rho_g \in \mathbb{R}$, and pred $: \mathbb{N}^2 \to \{true, false\}$ be defined as:*

$$pred(d_1, d_2) = (d_1 \geq \rho_s \cdot |\mathcal{D}_1|) \wedge (d_1 \cdot |\mathcal{D}_2| \geq \rho_g \cdot d_2 \cdot |\mathcal{D}_1|) . \qquad (13)$$

The monotonic hull $pred_{\mathrm{hull}}$ of pred with:

$$pred_{\mathrm{hull}}(d_1, d_2) := (d_1 \geq \rho_s \cdot |\mathcal{D}_1|) \qquad (14)$$

is optimal.

Proof. We assume $pred_{\mathrm{hull}}$ is a non-optimal monotonic hull of *pred*. Then there exists a monotonic hull $pred'_{\mathrm{hull}}$ of *pred* with $pred_{\mathrm{hull}} \not\Rightarrow pred'_{\mathrm{hull}}$. Thus, $d \in \mathbb{N}^2$ exist so that $pred_{\mathrm{hull}}(d_1, d_2)$ is *true* and $pred'_{\mathrm{hull}}(d_1, d_2)$ is *false*. By the contraposition of the monotonicity criterion, $pred'_{\mathrm{hull}}(d_1, 0)$ also is *false*. It holds that $pred(d_1, 0) = pred_{\mathrm{hull}}(d_1, d_2) = true$ and $pred \not\Rightarrow pred'_{\mathrm{hull}}$. This is a contradiction to $pred'_{\mathrm{hull}}$ being a monotonic hull of *pred*. Hence the proposition holds.   □

Proposition 4. *Let $min_1, max_1, min_2, max_2 \in \mathbb{N}$, $(min_1, min_2) \leq (max_1, max_2)$, and pred $: \mathbb{N}^2 \to \{true, false\}$ be defined as:*

$$pred(d_1, d_2) = (min_1 \leq d_1 \leq max_1) \wedge (min_2 \leq d_2 \leq max_2) \qquad (15)$$

The monotonic hull $pred_{\mathrm{hull}}$ of pred with:

$$pred_{\mathrm{hull}}(d_1, d_2) := (min_1 \leq d_1) \wedge (min_2 \leq d_2) \qquad (16)$$

is optimal.

Proof. Analogously holds for a $pred'_{\mathrm{hull}}$ and $d \in \mathbb{N}^2$: $pred_{\mathrm{hull}}(d)$ is *true* and $pred'_{\mathrm{hull}}(d)$ is *false*. Thus it holds that $(min_1, min_2) \leq d$ and $pred'_{\mathrm{hull}}(min_1, min_2)$ also is *false*. It holds that $pred(min_1, min_2) = true$ and $pred \not\Rightarrow pred'_{\mathrm{hull}}$. This is a contradiction to $pred'_{\mathrm{hull}}$ being a monotonic hull of *pred*. Hence the proposition holds.   □

Autonomous Forex Trading Agents

Rui Pedro Barbosa and Orlando Belo

Department of Informatics, University of Minho, 4710-057 Braga, Portugal
{rui.barbosa,obelo}@di.uminho.pt

Abstract. In this paper we describe an infrastructure for implementing hybrid intelligent agents with the ability to trade in the *Forex Market* without requiring human supervision. This infrastructure is composed of three modules. The *"Intuition Module"*, implemented using an **Ensemble Model**, is responsible for performing pattern recognition and predicting the direction of the exchange rate. The *"A Posteriori Knowledge Module"*, implemented using a **Case-Based Reasoning System**, enables the agents to learn from empirical experience and is responsible for suggesting how much to invest in each trade. The *"A Priori Knowledge Module"*, implemented using a **Rule-Based Expert System**, enables the agents to incorporate non-experiential knowledge in their trading decisions. This infrastructure was used to develop an agent capable of trading the USD/JPY currency pair with a 6 hours timeframe. The agent's simulated and live trading results lead us to believe our infrastructure can be of practical interest to the traditional trading community.

Keywords: Forex trading, data mining, hybrid agents, autonomy.

1 Introduction

The *Forex Market* is the largest financial market in the world. In this market currencies are traded against each other, and each pair of currencies is a product that can be traded. For instance, USD/JPY is the price of the United States Dollar expressed in Japanese Yen. At the time of writing of this paper the USD/JPY price is 102.55, meaning we need 102.55 JPY to buy 1 USD. Trading this pair in the *Forex Market* is pretty straightforward: if a trader believes the USD will become more valuable compared to the JPY he buys USD/JPY lots (goes long), and if he thinks the JPY will become more valuable compared to the USD he sells USD/JPY lots (goes short). The profit/loss of each trade can be expressed in pips. A pip is the smallest change in the price of a currency pair. For the USD/JPY pair a pip corresponds to a price movement of 0.01. The actual value of each pip depends on the amount invested. For example, if we buy/sell 100,000 USD/JPY each pip is worth 1,000 JPY (100,000 times 0.01), or 9.75 USD (1,000 divided by 102.55).

While it is very easy to trade in the *Forex Market*, it is actually pretty hard to be profitable doing it. Exchange rate prices are just too unpredictable. This unpredictability is due in part to the fact that the *Forex Market* is a 24 hours a day decentralized market, with many types of participants with different goals. Add to this the non-stop stream of news coming out each day that can affect several pairs at the same time, and

P. Perner (Ed.): ICDM 2008, LNAI 5077, pp. 389–403, 2008.

it becomes clear why the *Forex Market* is one of the hardest financial markets to beat. Trying to overcome these intrinsic hardships of *Forex* trading has led *Forex Market* participants to steadily move from traditional trading to algorithmic trading. This move, often referred to as the "algorithms arms race", is happening at a very fast pace. The adoption of algorithmic trading in the *Forex Market* is expected to grow from 7% by the end of 2006 to 25% by 2010 [1].

With this growing interest in quantitative methods in mind, we will describe an infrastructure for implementing autonomous *Forex* trading agents that makes extensive use of artificial intelligence models. The concept of using artificial intelligence models in trading is not exactly new, as there are already plenty of studies in this field. A special emphasis has been given to the use of neural networks to perform financial time series prediction [4][7][11]. In fact, several studies have shown that neural networks can model financial time series better than traditional mathematical methods [6][8]. Lately, researchers have displayed a growing interest in the development of hybrid intelligent systems for financial prediction [5][9][12]. These studies have shown that hybrid systems can outperform non-hybrid systems.

Even though most studies demonstrate that artificial intelligence models can produce reasonably accurate financial predictions, that in itself will not impress most traditional traders. These studies usually measure a model's performance based on its accuracy (for classification) or the mean squared error (for regression). The problem with this approach, from a trader's point of view, is that higher accuracy does not necessarily translate into higher profit. A single losing trade can wipe out the profit of several accurately predicted trades. A low mean squared error is also far from being a guarantee that a model can produce profitable predictions [3]. Some studies try to tackle this problem by using model predictions on out-of-sample data to simulate trades. This might make for a better study from a traders' point of view, but it is still not a perfect solution. Simulated trades do not account for problems that frequently occur while trading live, such as slippage and partial fills. The effect of these problems on the overall profitability of a trading strategy is not negligible.

In the end, profit and drawdown are the only performance gauges that really matter to the trading community. Any performance claims are also expected to be backed up by a meaningful track record of live trading. With that in mind, our study will be exclusively directed at what the trading community wants. We will describe an infrastructure for implementing trading agents whose main goal is to maximize the profit and to minimize the drawdown while trading live. Each implemented agent is also expected to be able to operate autonomously, placing trades and handling money management without requiring human intervention. The infrastructure is loosely based in the decision process of a traditional trader: it enables the agents to intuitively recognize patterns in financial time series, to remember previous trades and use that empirical knowledge to decide when and how much to invest, and to incorporate knowledge from trading books and trading experts into the trading decisions.

2   Infrastructure

The infrastructure for implementing trading agents is represented in Figure 1. This infrastructure defines two percepts (price changes over a period of time and result of previous trades) and a single action (placement of new trades). The agent's structure is organized in three interconnected modules:

- *Intuition Module* – this module is responsible for predicting if the price of a currency pair will go up or down. This prediction is done by an **Ensemble Model,** which consists of several classification and regression models that try to find hidden patterns in price data.
- *A Posteriori Knowledge Module* – this module uses information from previous trades to suggest when and how much to invest in each trade. This suggestion is done by a **Case-Based Reasoning System.** Each case in this system corresponds to a trade executed by the agent and its final result (profit or loss in pips).
- *A Priori Knowledge Module* – this module is responsible for making the final trading decision, using the prediction from the *Intuition Module* and the suggestion from the *A Posteriori Knowledge Module*. This decision is done by a **Rule-Based Expert System,** which contains several rules regarding when to invest and when to stop a trade. These rules must be provided to the agent because it will not be able to learn them by itself while trading.

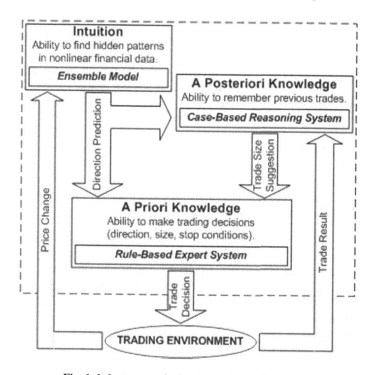

Fig. 1. Infrastructure for implementing trading agents

2.1 The *Intuition Module*

A common definition for intuition is "knowing without reasoning". It is hard to explain how this mental process works and even harder to try to implement its software equivalent. In a loose way we can look at intuition as a complex pattern recognition process [2]. Even if we are oversimplifying a complex concept, that definition perfectly suits our needs. We can easily base our trading agents' intuition in a set of classification and regression models capable of finding hidden patterns in nonlinear financial data.

To implement the pattern recognition mechanism we could follow the typical approach of training a model with as much training data as possible, verifying that it was able to recognize past patterns in some out-of-sample test data and finally use it to predict future prices. If the future price predicted is higher than the current price then we should go long (buy) and if it is lower we should go short (sell). But this strategy has several problems:

- It is susceptible to overfitting – the only way to decide if a model is accurate before using it to make predictions is to analyze its performance with the training and test data. By selecting a model based on this performance we may be selecting a model that performs well with those sets of data, but does not have any ability to generalize from the training data to new unseen data.
- It is "dumb" – once trained the model will not be able to learn anymore. This would not be a problem if the training data contained all the information that the model needs to be aware of. Unfortunately that is never the case when dealing with financial time series, no matter how big the training set is.
- It cannot live up to the expectations – anyone who has ever watched the price action of a major currency pair for a couple of sessions will know that expecting a model to accurately predict a price in the future is probably unrealistic, especially when dealing with smaller timeframes. The price is just too volatile.
- It is optimized for accuracy instead of profit – learning algorithms aim at building models that are accurate. When trading, high accuracy does not necessarily mean high profitability.
- It is not autonomous – because it is susceptible to overfitting and it cannot learn, the model needs to be continuously monitored to make sure it does not need to be retrained or replaced with a better model.

To try to overcome these problems we implemented the *Intuition Module* using an **Ensemble Model**. This **Ensemble** is a weighted voting system composed of several classification and regression models, where the weight of each vote is based on the profitability of each model. The models do not try to predict the price in the future, they simply try to predict what will happen to the price in the future. The prediction of each model thus corresponds to one of two classes: "*the price will go up*" or "*the price will go down*". That is straightforward for classification models, but for regression models we need to convert the price prediction into one of the classes. That is very simple to accomplish: if the predicted price is higher than the current price then the predicted class is "*the price will go up*", otherwise it is "*the price will go down*".

Before each prediction, the available instances (each consisting of the correct class to be predicted and a set of attributes that depends on the model) are divided into two datasets: the test set consisting of the most recent N instances, and the training set consisting of all the instances left. Using these two sets of data the following sequence of steps is applied to each model in the **Ensemble**:

1. The model is retrained using the training set and tested using the test set.
2. For each instance in the test set a trade is simulated (if the model predicts "*the price will go up*" we simulate a buy, otherwise we simulate a short sell). The results from the simulation are used to calculate the *overall profit factor*, *long profit factor* and *short profit factor* of the retrained model:

$$Overall\ PF = \frac{\sum pips\ won}{\sum pips\ lost} - 1 \qquad (1)$$

$$Short\ PF = \frac{\sum pips\ won\ when\ predicting\ down}{\sum pips\ lost\ when\ predicting\ down} - 1 \qquad (2)$$

$$Long\ PF = \frac{\sum pips\ won\ when\ predicting\ up}{\sum pips\ lost\ when\ predicting\ up} - 1 \qquad (3)$$

3. If the *overall profit factor* of the retrained model is higher or equal to the *overall profit factor* of the original model, then the retrained model replaces the original model in the **Ensemble**. Otherwise, the original is kept and the retrained model is discarded.
4. The selected model makes its prediction: if it predicts "*the price will go up*" the weight of its vote is its *long profit factor*; if it predicts "*the price will go down*" the weight of its vote is its *short profit factor*. If the weight is a negative number then it is replaced with zero, which effectively means the model's prediction is ignored.

After all individual models have made their predictions, the ensemble prediction is calculated by adding the votes of all the models that predicted "*the price will go up*" and then subtracting the votes of all the models that predicted "*the price will go down*". If the ensemble prediction is greater than zero then the module's final prediction is "*the price will go up*", otherwise if it is lower than zero the final prediction is "*the price will go down*".

There are several reasons why we decided to perform the predictions using an **Ensemble Model** and the previously described algorithm:

- Some models are more profitable under certain market conditions than others. An **Ensemble Model** can be more profitable than any of its individual models because it can adapt to the market conditions. That is accomplished by continuously updating the weight of the vote of each of the individual models: as a model becomes more profitable its vote becomes more important.
- Some models are better at predicting when the market will go up and others are better at predicting when the market will go down. By using an **Ensemble Model** we can combine the qualities of the best models at predicting long trades and the best models at predicting short trades. That is accomplished by using the models' *long profit factor* and *short profit factor* as their votes' weight.

- An **Ensemble Model** makes our trading strategy resilient to changes in market dynamics. If a single classification or regression model is used for prediction and it starts turning unprofitable, the trading strategy will soon become a disaster. On the other hand, if that model is a part of our **Ensemble Model**, as it becomes unprofitable its vote continuously loses weight, up to a point where its predictions are simply ignored. And since our strategy tries to improve the models by retraining them with more data as it becomes available, it is very likely that the unprofitable model will end up being replaced with a more profitable retrained version of itself.
- Our algorithm optimizes profitability instead of accuracy. Obviously the learning algorithms used to retrain the models still optimize their accuracy, but the decision to actually make the retrained models a part of the **Ensemble Model** is based entirely on their profitability.
- Retraining the models before each prediction is the key to our agents' autonomy. The agents can keep learning even while trading, because new unseen data will eventually become a part of the training set.

Our strategy is not without faults though. The decision to replace an original model with a version of itself trained with more data is based on the simulated profitability displayed with the test data. This means we are selecting models based on their test predictions, which might lead to selecting models that overfit the test data. However, this ends up not being a very serious problem, because our algorithm eventually replaces unprofitable models with more profitable retrained versions of themselves (that might or might not overfit a different set of test data).

2.2 The *A Posteriori Knowledge Module*

Deciding when to buy or short sell a financial instrument is a very important part of successful trading. But there is another equally important decision: how much to invest in each trade. If we have a model that consistently produces profitable predictions we might feel tempted to double our investment per trade. That will in fact double the profit, but will also double the exposure and the drawdown (loosely defined as the maximum loss an investor should expect from a series of trades). Keeping the drawdown low is of vital importance to traditional traders because, no matter how profitable a trading strategy is, a large drawdown can cause a margin call and pretty much remove the trader from the market. So doubling the investment per trade is not the best money management strategy for our trading agents. A better way to increase the profitability without a proportional increase in the risk would be to double the investment in trades with high expected profitability, use the normal investment amount for trades with average expected profitability, and skipping trades with low expected profitability.

In order to determine the expected profitability of a trade we will be looking at the individual predictions of the models that are part of the **Ensemble Model**. Intuitively, we might expect that the probability of a trade being successful will be higher if all the individual models make the same prediction (all predict "*the price will go up*" or all predict "*the price will go down*"), compared to a trade where the models' predictions are mixed (some predict "*the price will go up*" and some predict "*the* *price will go down*"). Empirical evidence demonstrates that those expectations are well founded. Certain combinations of individual predictions really are more profitable than others. Our agents' money management strategy is based on that empirical observation.

We implemented the *A Posteriori Knowledge Module* using a **Case-Based Reasoning System** where each case represents a trade previously executed by the agent. The following information is contained in each case in the database: the predicted class, the trade result (profit or loss in pips) and the individual predictions from the models in the **Ensemble Model**. The agent uses that information to calculate the expected profitability of a trade before it is placed. It then decides if a trade is worth opening, and if so how much should be invested. The following sequence of steps is executed before each trade is placed:

1. The *Intuition Module* makes the ensemble prediction and sends the sequence of individual predictions from the models in the **Ensemble** to the **Case-Based Reasoning System**. This system retrieves from its database all the cases with the same class prediction and the same sequence of individual predictions.
2. If the number of retrieved cases is not higher or equal to a predefined minimum number of cases, the **Case-Based Reasoning System** removes the last prediction in the sequence of individual predictions and retrieves the cases again. This process is repeated until enough cases are retrieved.
3. The **Case-Based Reasoning System** calculates the *overall profit factor* of the retrieved cases using Equation (1). That is the expected profitability of the trade.
4. If the *overall profit factor* is greater or equal to a predefined value the agent doubles the investment; if it is lower than another predefined value the agent skips the trade; otherwise, the regular investment amount is used.

After a trade is executed and closed, a new case is inserted in the **Case-Based Reasoning System** database. Our infrastructure uses the *overall profit factor* of the matching cases in the database to make the money management decision, which is yet another way in which it tries to optimize the profit.

2.3 The *A Priori Knowledge Module*

No matter how "smart" the implemented agents are, there is still some trading knowledge they will not be able to pick up from their empirical trading experiences. For this reason, the module responsible for making the final trading decision consists of a **Rule-Based Expert System** where important rules can be defined by trading experts.

Some of these rules can be quite simple. For example, we may want the agents to skip trades in low liquidity days, such as those around Christmas or New Year's Day, when the already naturally high volatility of the *Forex Market* is exacerbated. Or we may want them to skip trades whenever major economic reports are about to be released, to avoid the characteristic chaotic price movements that happen right after the release. The primary example of such a report is the United States Nonfarm Payrolls, or NFP, released on the first Friday of every month.

Other more important rules are those where the settings for take profit and stop loss orders are defined. These are necessary so that the agents know when to exit each trade. A take profit order is used to close a trade when it reaches a certain number of pips in profit, to guarantee that profit. A stop loss order is used to close a trade when it reaches a certain number of pips of loss, to prevent the loss from widening.

Before each trade the **Rule-Based Expert System** receives the prediction from the *Intuition Module* and the suggested investment amount from the *A Posteriori Knowledge Module*. It then uses the rules defined by the expert traders to make the final decision regarding the trade direction, investment amount and exit conditions. The agents can be made completely autonomous by using a broker's proprietary API to send the final trade decisions directly into the market.

3 Sample Implementation

There are countless ways in which our infrastructure can be used to implement a trading agent. As a sample implementation, we opted for developing an agent with the ability to place a trade every 6 hours, from Sunday 18:00 GMT to Saturday 00:00 GMT, using the USD/JPY currency pair.

3.1 Agent's Intuition

To implement the agent's *Intuition Module* we had to select the classification and regression models to be inserted in its **Ensemble Model**. The first step to accomplish this was to obtain historical price data that could be used to train the models.[1]

We decided to download the data in the form of USD/JPY 6 hours candlesticks. A candlestick is a figure that displays the high, low, open and close price of a financial instrument over a specific period of time. Using higher timeframe candlesticks would probably make more sense, because the higher the timeframe the less noise would be contained in the financial time series. But since the amount of data available was scarce, we needed to use the 6 hours timeframe to be able to obtain enough instances to train the models. We downloaded 4,100 candlesticks, comprising the period from May 2003 to January 2007. These candlesticks were used to calculate the price return over each 6 hours period, which was one of the attributes we inserted in the training instances. Using the return instead of the actual price to train the models is a normal procedure in financial time series prediction, because it is a way of removing the trend from the series. Of the available returns, 4,000 (corresponding to the period from May 2003 to December 2006) were used to train the models and the remaining 100 (corresponding to the month of January 2007) were used to test the models.

Table 1 describes the exact attributes used to train and test each model in the **Ensemble**. While the classification models tried to predict the next class ("*the price will go up in the next 6 hours*" or "*the price will go down in the next 6 hours*"), the regression models tried to predict the price return in the following 6 hours period. That return was then converted to a class (if the predicted return was greater or equal to zero than the class prediction was "*the price will go up in the next 6 hours*", otherwise it was "*the price will go down in the next 6 hours*"). The models were trained with attributes such as the hour, the day of the week and the current class or

---

[1] Our agent's training data sources were *www.dukascopy.com* and *www.oanda.com*

Table 1. Attributes used to train each model in the Ensemble

Model	Attributes	Prediction
Instance-Based K*	hour (nominal), day of week (nominal), last 6 returns moving average, current class	next class
C4.5 Decision Tree	hour (nominal), day of week (nominal), last 6 returns moving average, current class	next class
RIPPER Rule Learner	hour (nominal), day of week (nominal), current class	next class
Best-First Decision Tree	hour (numeric), day of week (numeric), last 6 returns moving average, previous return, current return	next class
Naïve Bayes	hour (nominal), day of week (nominal), current return	next class
Logistic Decision Tree	hour (nominal), last 6 returns moving average, current class	next class
Multilayer Perceptron	hour (nominal), day of week (nominal), last 6 returns moving average, current return	next class
Hidden Naïve Bayes	hour (nominal), current class	next class
K-Nearest Neighbor	hour (nominal), day of week (nominal), last 6 returns moving average, current class	next class
Instance-Based K*	hour (nominal), day of week (nominal), last 6 returns moving average, current class	next return
Support Vector Machine	hour (numeric), day of week (numeric), last 10 returns moving average, last 2 returns moving average, current return	next return

return. We also tried several attributes regularly used in technical analysis by traditional traders, such as moving averages, the *Relative Strength Index*, the *Williams %R* and the *Average Directional Index*, amongst others. Of these, only the moving averages added predictive power to the models. The usefulness of the moving averages was not unexpected, as it had already been demonstrated by several studies in the past [10].

All the models were trained and tested using the *Weka* data mining software.[2]

As previously mentioned, before each trade the *Intuition Module* uses a fixed size dataset to test the models and calculate their simulated profitability. We decided that our agent would use a test set consisting of the most recent 100 instances. The decision to use only 100 instances for testing might seem a bit odd, as most literature regarding supervised learning would recommend the use of at least 30% of the available data. However, there are several reasons why we made our agent use such a small set of test data:

---

[2] *Weka* is an open source data mining software available at www.cs.waikato.ac.nz/ml/weka/

- Usually we would need a lot of test data to make sure a model did not overfit the training data. Our agent does not need that because its predictions are not based in a single model. So even if one of its models overfits the training data, that is not necessarily a problem. Over time the agent is able to ignore models that overfit the data (i.e., models that are unprofitable in out-of-sample trading) and eventually replaces them with retrained versions of themselves. That is the reason why we can save much needed data for training, which would otherwise be required for testing.

- Heteroskedasticity is a key feature of most economic time series. This means that the volatility is clustered: usually a long period of low volatility is followed by a short period of high volatility and this pattern is repeated *ad eternum*. Since the weights of the models' votes are based in their simulated profitability using the test instances, we need to keep the test set small enough that the weights can adapt quickly when the market enters a period of high volatility. In other words, the shorter the test set, the faster the agent can adapt to changes in market dynamics.

- A new instance is available after each trade. This instance becomes a test instance, and the oldest instance in the test set becomes a training instance. This means that, as time goes by, the training set grows while the test set remains the same size and moves like a sliding window. What this implies is that the shorter the test set, the faster the new instances can be used for training. In other words, the shorter the test set, the faster the agent can learn new patterns.

After completing the implementation of the *Intuition Module*, we used its predictions to simulate trades with out-of-sample data corresponding to the period from February 2007 to March 2008. Figure 2 casts some light into the way the *Intuition Module* was able to adapt to the changing market conditions over that period of time. It shows the average long and average short weights of the votes of the 11 models in the **Ensemble**, and the USD/JPY price changes.

As the price trends up the long votes' average weight increases, while the short votes' average weight shrinks, and vice-versa. The periods marked with arrows in the chart are particularly interesting. Over these periods, the average weight for long

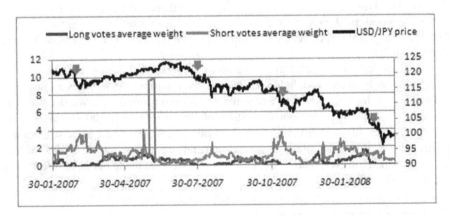

Fig. 2. Average weight of the models' votes

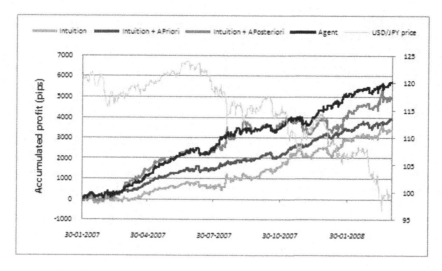

Fig. 3. Performance comparison using different module combinations

votes is very close to zero. What this means is that models predicting that *"the price will go up in the next 6 hours"* are being ignored. So if a single model with **short profit factor** greater than zero predicts *"the price will go down in the next 6 hours"*, then the final ensemble class prediction will automatically be the same, even if the other 10 models predict a price increase. It is this mechanism of selecting the best models according to the market conditions that allows the *Intuition Module* to quickly adapt to changes in the price trend and volatility.

The accumulated profit in pips over the simulation period for the *Intuition Module* is displayed in Figure 3. After an initial period of unprofitable trading, where the weights of the models' votes in the **Ensemble** were adapting to the market conditions, the *Intuition Module* was able to recover and ended up with a profit of 3,330 pips after 1,276 trades. It is fairly obvious that this trading strategy needs improvements: the drawdown is too high (463 pips) and the profit curve is too volatile and erratic.

3.2 Agent's Empirical Knowledge

Implementing the agent's *A Posteriori Knowledge Module* was as easy as defining a couple of user variables regarding money management. After a couple of trial and error tests, we decided to use the following settings:

- double the investment whenever the *overall profit factor* of the cases retrieved is greater or equal to 1;
- skip trades whenever the *overall profit factor* of the cases retrieved is lower or equal to 0;
- require a minimum of 7 retrieved cases before a decision is taken.

The chart in Figure 3 shows the result of combining the *Intuition Module* and the *A Posteriori Knowledge Module* to simulate trades using the out-of-sample data. This combination performed 907 trades, with a final profit of 4,837 pips and a drawdown of 882 pips. Compared to using the *Intuition Module* alone, the profit increased 45% and the drawdown increased 90%. There was also an important reduction of 29% in the number of trades. So the *A Posteriori Knowledge Module* was able to increase the final profit while making fewer trades, but this strategy needs improvement because the drawdown is too high.

3.3 Agent's Expert Knowledge

To implement the agent's *A Priori Knowledge Module* we just had to define the rules in the **Rule-Based Expert System**. We started by defining some rules to avoid low liquidity days. These should not have a big impact in the trading results. But we also added one rule that will certainly have a significant impact: each trade is accompanied by a take profit order of 20 pips. This means that whenever a trade reaches a profit of 20 pips it is automatically closed. In other words, we are capping our maximum profit per trade to 20 pips (40 pips when the investment is doubled). A trade that is not closed with the take profit order will only be closed when the 6 hours period ends and a new trade is open.

Figure 3 shows the results of combining the *Intuition Module* and the *A Priori Knowledge Module* to simulate trades with the out-of-sample data. This strategy netted 3,958 pips of profit, with a drawdown of 360 pips. Compared to using the *Intuition Module* alone there was a 19% increase in the profit and a 22% decrease in the drawdown. The lower drawdown is exactly what we needed, but unfortunately there is also a big profit reduction if we compare these results with the ones obtained with the combination between the *Intuition Module* and the *A Posteriori Knowledge Module*.

3.4 Results

Through simulation, we have shown that each module makes a different contribution to the trading profit and the drawdown. The actual agent consists of all the three modules working together. Figure 3 shows the agent's simulated trading results. The agent was able to take advantage of the *A Posteriori Knowledge Module* ability to increase the profits and the *A Priori Knowledge Module* ability to reduce the drawdown. It obtained a final profit of 5,742 pips with a drawdown of 421 pips. This is, by any standards, an excellent performance.

By looking at Figure 3 it is easy to see that not only is the agent more profitable than any combination of its modules, its profit curve is also the smoothest. We can also see that the agent is not directionally biased: it is profitable no matter if the USD/JPY price is going up or down. It is also important to note that the agent performed acceptably in periods of high volatility (such as the month of August).

Table 2 resumes the trading statistics of both the agent and the module combinations. The first interesting statistic in this table is the fact that the *Intuition Module* can only predict if the price will go up or down with 52.74% accuracy. This percentage might seem too low, but it makes sense when we consider that this module optimizes profitability instead of accuracy. Therefore, even though the module is not very accurate, the profit it obtains from the accurately predicted trades is a lot higher than the losses it suffers from incorrectly predicted trades. Its success rate, i.e., the

Table 2. Trading statistics

Module combination	Accuracy	Success	Profit	Drawdown	Trades
Intuition	52.74%	52.74%	3,330	463	1,276
Intuition + A Posteriori Knowledge	54.47%	54.47%	4,837	882	907
Intuition + A Priori Knowledge	52.74%	64.89%	3,958	360	1,276
Agent	54.41%	66.67%	5,742	421	873

percentage of trades that are closed in profit, is equal to its accuracy because all the trades are closed at the end of the 6 hours period, when a new trade is opened.

The 54.41% accuracy of the agent is higher than the accuracy of its *Intuition Module* because both its *A Posterior Knowledge Module* and its *A Priori Knowledge Module* can make it skip trades that are expected to be unprofitable. That explains why the agent did only 873 trades, against the 1,276 trades that would have been performed by the *Intuition Module* alone. The agent has a 66.67% success rate, which is considerably higher than its accuracy. That is due to the take profit rule in the *A Priori Knowledge Module*. This rule allows the agent to be profitable even if it makes a wrong prediction, just as long as the price moves at least 20 pips in the predicted direction.

While pips are a good way to measure the performance of our Forex trading agent, it might be interesting to see how that performance translates into actual money won or lost. Forex investments are usually leveraged (which means they are done with borrowed funds), so the total profit obtained by the agent will always depend on the size of its trades. Let us assume we have a starting capital of $100,000, and we want our agent to use a low risk trading strategy, with trades of 100,000 USD/JPY. As long as the agent has more than $100,000 in its account its trades will not be leveraged, except when it doubles the investment for trades with high expected profitability. As previously seen, for a USD/JPY price of 102.55, the pip value for a 100,000 USD/JPY trade will be $9.75. Since our agent obtained a total profit of 5,742 pips, its profit in dollars after 14 months of trading is $55,985, or 60%. This is a really good performance, but things get even more interesting if we consider the agent could have used a higher initial leverage. Figure 4 displays the equity curves for a $100,000 account, using different trade sizes.

Amazingly, if the agent used a standard trade size of 2,000,000 USD/JPY, its $100,000 account would have grown to $1,219,690 in 14 months, or around 1,120%. However, it is easy to see why using such high leverage would be too risky in live trading. From November 23rd to December 5th the agent suffered its maximum drawdown of 421 pips. A trade size of 2,000,000 USD/JPY corresponds to $195 per pip, so there was a drawdown of $82,095. This loss is barely noticeable in the equity curve displayed in Figure 4, because it happened at a time when the agent had already a really high account balance. But let us imagine the agent placed its first trade on November 23rd. Its initial balance of $100,000 would then drop $82,095 in 13 days, which would inevitably result in a margin call. The agent would not be able to trade again, and would end up with a loss of over 80%. If the agent was using a more

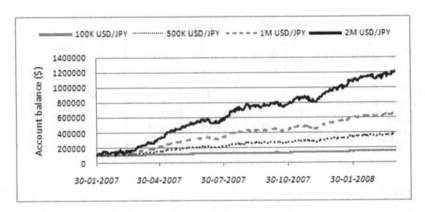

Fig. 4. Equity curves for different trade sizes

reasonable trade size of 500,000 USD/JPY, the maximum drawdown would have been only $20,528, and it would have turned $100,000 into $379,980 in 14 months.

As previously mentioned, simulated results can give us a general idea regarding an agent's ability to be profitable while trading live, but cannot provide any guarantees. There are many details concerning live trading that can have a tremendous impact in the final net profit. The only way to prove that an agent can be profitable is to allow it to create an extensive track record of live trading. In order to accomplish this we integrated our agent with an Electronic Communication Network, where it has been trading autonomously since the middle of June 2007. As expected, the agent's actual live trading results are not as good as the simulated results, with a decrease of around 22% in the total profit. This difference is due to commissions, slippage, partial fills and interest payments, amongst other things. But the agent's results are still very good, with an average profit of 5.1 pips per live trade, which compares with an average profit of 6.6 pips per simulated trade over the same period of time.

Obviously, it is still too soon to reach any conclusions regarding the agent's profitability in the long run, because its live trading track record is too short. But so far our results seem to show that the agent is capable of profiting from inefficiencies in the *Forex Market*. Furthermore, we expect the agent's success rate to increase over time, as its models are trained with more data and more cases are inserted in its **Case-Based Reasoning System**.

4   Final Remarks

In this paper we described an infrastructure for implementing agents with the ability to trade autonomously in the *Forex Market*. The infrastructure is loosely based in traditional trading, i.e., the agents are capable of:

- recognizing patterns in financial time series,
- learning from empirical experience,
- incorporating knowledge obtained from non-experiential sources into the trading strategy.

Each of these capabilities corresponds to a module in the infrastructure, named *Intuition, A Posteriori Knowledge* and *A Priori Knowledge Module*, respectively.

There are multiple ways the infrastructure can be used to implement a trading agent. In this paper we described a sample implementation of an agent capable of trading the USD/JPY currency pair with a 6 hours timeframe. Using simulated trading we were able to demonstrate the positive impact of each of the infrastructure's modules in the trading profit. Live trading results seem to suggest that the agent is indeed capable of being profitable while trading without supervision. However, only after a couple of years will we be able to make any claims regarding the agent's ability to survive and thrive in all market conditions.

A common way to reduce the risk inherent to trading is through diversification. In our case, investment diversification can be easily achieved by simply using the infrastructure to implement a basket of agents trading different uncorrelated currency pairs and using different time frames. Even though the infrastructure was developed with the *Forex Market* in mind, it is obvious it can be used to implement agents capable of trading any other financial instruments, such as stocks or futures. In fact, it would be an excellent idea to implement such agents and to add them to the basket of currency trading agents. We are currently looking into this multi-agent investment strategy. Given the growing interest in algorithmic and quantitative trading, it is our belief that it will be of much interest to the traditional trading community.

References

1. Cole, T.: Foreign Exchange Implications of Algorithmic Trading. In: Foreign Exchange Contact Group, Frankfurt, Germany, May 23 (2007)
2. Thomas, A.: The intuitive algorithm. Affiliated East-West Press (1991) ISBN 8185336652
3. Swingler, K.: Financial Prediction, Some Pointers, Pitfalls, and Common Errors. Stirling University (1994)
4. Franses, P., Griensven, K.: Forecasting Exchange Rates Using Neural Networks for Technical Trading Rules. Erasmus University (1998)
5. Abraham, A.: Analysis of Hybrid Soft and Hard Computing Techniques for Forex Monitoring Systems. Monash University (2002)
6. Dunis, C., Williams, M.: Modelling and Trading the EUR/USD Exchange Rate: Do Neural Network Models Perform Better? Liverpool Business School (2002)
7. Kondratenko, V., Kuperin, Y.: Using Recurrent Neural Networks To Forecasting of Forex. St. Petersburg State University (2003)
8. Kamruzzamana, J., Sarker, R.: Comparing ANN Based Models with ARIMA for Prediction of Forex Rates. Monash University (2003)
9. Abraham, A., Chowdhury, M., Petrovic-Lazarevic, S.: Australian Forex Market Analysis Using Connectionist Models. Monash University (2003)
10. Kamruzzamana, J., Sarker, R.: ANN-Based Forecasting of Foreign Currency Exchange Rates. Monash University (2004)
11. Yu, L., Wang, S., Lai, K.: Adaptive Smoothing Neural Networks in Foreign Exchange Rate Forecasting. Chinese Academy of Sciences (2005)
12. Yu, L., Wang, S., Lai, K.: Designing a Hybrid AI System as a Forex Trading Decision Support Tool. Chinese Academy of Sciences (2005)

An Exploration into the Power of Formal Concept Analysis for Domestic Violence Analysis

Jonas Poelmans[1], Paul Elzinga[2], Stijn Viaene[1,3], and Guido Dedene[1,4]

[1] K.U.Leuven, Faculty of Business and Economics, Naamsestraat 69,
3000 Leuven, Belgium
[2] Police Organisation Amsterdam-Amstelland, James Wattstraat 84,
1000 CG Amsterdam, The Netherlands
[3] Vlerick Leuven Gent Management School, Vlamingenstraat 83,
3000 Leuven, Belgium
[4] Universiteit van Amsterdam Business School, Roetersstraat 11,
1018 WB Amsterdam, The Netherlands
{Jonas.Poelmans,Stijn.Viaene,Guido.Dedene}@econ.kuleuven.be,
Paul.Elzinga@amsterdam.politie.nl

Abstract. The types of police inquiries performed are very diverse in nature and the current data processing architecture is not sufficiently tailored to cope with this diversity. Many information concerning cases is still stored in databases as unstructured text. Formal Concept Analysis is showcased as an exploratory data analysis technique for discovering new knowledge from police reports. It turns out that it provides a powerful framework for exploring the dataset, resulting in essential knowledge for improving current practices. It is shown that the domestic violence definition employed by the police organisation of the Netherlands is not always as clear as it should be, making it hard to use it effectively for classification purposes. In addition, newly discovered knowledge for automatically classifying certain cases as either domestic or non-domestic violence is presented. Moreover, essential techniques for detecting incorrect classifications, performed by police officers, are provided. Finally, some problems encountered because of the sometimes unstructured way of working of police officers are discussed. Both using Formal Concept Analysis for exploratory data analysis and its application on this area are novel enough to make this paper into a valuable contribution to the literature.

Keywords: Formal Concept Analysis (FCA), domestic violence, knowledge discovery in databases, data mining.

1 Introduction

According to the U.S. Office on Violence against Women, domestic violence is a "pattern of abusive behavior in any relationship that is used by one partner to gain or maintain power and control over another intimate partner" [1]. Domestic violence can take the form of physical violence, which includes biting, pushing, maltreating, stabbing or even killing the victim. Physical violence is often accompanied by mental or

P. Perner (Ed.): ICDM 2008, LNAI 5077, pp. 404–416, 2008.

emotional abuse, which includes insults and verbal threats of physical violence to the victim, the self or others including children. Domestic violence occurs all over the world, in various cultures [2] and affects people across society, irrespective of economic status [3].

Domestic violence is one of the top priorities of the police organization of the region Amsterdam-Amstelland in The Netherlands. Of course, in order to pursue an effective policy against offenders, being able to swiftly recognize cases of domestic violence and label reports accordingly is of the utmost importance. Still, this has proven to be problematic. In the past, intensive audits of the police databases related to filed reports have established that many reports tended to be wrongly classified as domestic or as non-domestic violence cases. One of the conclusions was that there was a need for an in-depth investigation of this problem area.

In this paper, it shall be demonstrated that from the unstructured text in police reports, essential knowledge regarding domestic violence can be obtained by using a technique known as Formal Concept Analysis (FCA) [8, 9]. FCA arose twenty-five years ago as a mathematical theory [14]. It has over the years grown into a powerful framework for data analysis, data visualization, [10, 15, 18], information retrieval and text mining [16, 17, 20]. However, FCA has never been used for exploratory data analysis, which is one of the core contributions of this paper. What makes FCA into an especially appealing knowledge discovery in databases technique from a practitioner point of view is the compactness of its information representation and the minimal need for users to tune (hyper-) parameters to distill a useful, actionable picture of the mining exercise.

The remainder of this paper is composed as follows. In section 2, we shall cover the essentials of FCA theory, introducing the pivotal FCA notions of concept and concept lattice. In section 3, the dataset used in our research will be elaborated on. Section 4 then showcases and discusses the results of the application of FCA for exploratory analysis of domestic violence cases using this data set. Finally, section 5 rounds up with conclusions.

2   FCA Essentials

This section introduces the main ideas of Formal Concept Analysis in a very elementary way.

2.1   Concepts and Lattices

The starting point of the analysis is a database table consisting of rows (i.e. objects), columns (i.e. attributes) and crosses (i.e. relationships between objects and attributes). The mathematical structure used to reference such a cross table is called a formal context. An example of a cross table is displayed in table 1. In the latter, reports of domestic violence (i.e. the objects) are related (i.e. the crosses) to a number of terms (i.e. the attributes); here a report is related to a term if the report contains the term. The dataset in table 1 is an excerpt of the one we used in our research. Given a formal context, FCA then derives all concepts from this context and orders them according to a "subconcept-superconcept" relation. This results in a line diagram (a.k.a. lattice).

Table 1. Example of a formal context

	kicking	dad hits me	stabbing	cursing	scratching	maltreating
report 1	X	X				X
report 2			X	X	X	
report 3	X	X	X	X	X	
report 4						X
report 5				X	X	

The notion of "concept" is central to FCA. The extension consists of all objects belonging to the concept, while the intension comprises all attributes shared by those objects. Let us illustrate the notion of concept of a formal context using the data in table 1. Take the attributes that describe report 5, for example. By collecting all reports of this context that share these attributes, we get to a set O consisting of reports 2, 3 and 5. This set O of objects is closely connected to set A consisting of the attributes "cursing" and "scratching." That is, O is the set of all objects sharing all attributes of A, and A is the set of all attributes that are valid descriptions for all the objects contained in O. Each such pair (O, A) is called a formal concept (or concept) of the given context. The set O is called the extent, while A is called the intent of the concept (O, A).

There is a natural hierarchical ordering relation between the concepts of a given context that is called the "subconcept-superconcept" relation. A concept d is called a subconcept of a concept e (or equivalently, e is called a superconcept of a concept d) if the extent of d is a subset of the extent of e (or equivalently, if the intent of d is a superset of the intent of e). For example, the concept with intent "cursing," "scratching" and "stabbing" is a subconcept of a concept with intent "cursing" and "scratching." With reference to table 1, the extent of the latter is composed of reports 2 and 3, while the extent of the former is composed of reports 2, 3 and 5.

The set of all concepts of a formal context combined with the "subconcept-superconcept" relation defined for these concepts gives rise to the mathematical structure of a complete lattice, called the concept lattice of the context. The latter is made accessible to human reasoning by using the representation of a (labeled) line diagram. The line diagram in figure 1, for example, represents the concept lattice of the formal context abstracted from table 1. The circles or nodes in this line diagram represent the formal concepts. The shaded boxes (upward) linked to a node represent the attributes used to name the concept. The non-shaded boxes (downward) linked to the node represent the objects used to name the concept. The information contained in the formal context of table 1 can be distilled from the line diagram in figure 1 by applying the following "reading rule:" An object "g" is described by an attribute "m" if and only if there is an ascending path from the node named by "g" to the node named by "m". For example, report 5 is described by the attributes "cursing" and "scratching".

Retrieving the extension of a formal concept from a line diagram such as the one in figure 1 implies collecting all objects on all paths leading down from the corresponding node. In this example, the objects associated with the third concept in row three are reports 2 and 3. To retrieve the intension of a formal concept one traces all paths leading up from the corresponding node in order to collect all attributes. In this example, the third concept in row three is defined by the attributes "stabbing," "cursing"

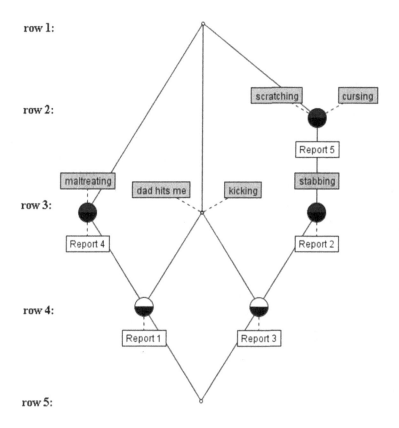

Fig. 1. Line diagram corresponding to the context from table 1

and "scratching". The top and bottom concepts in the lattice are special. The top concept contains all objects in its extension. The bottom concept contains all attributes in its intension. A concept is a subconcept of all concepts that can be reached by travelling upward. This concept will inherit all attributes associated with these superconcepts. Note that the extension of the concept with attributes "kicking" and "dad hits me" is empty. This does not mean that there is no report that contains these attributes. However, it does mean that there is no report containing only these two attributes.

2.2 FCA Software

We used FCA as an unsupervised clustering technique [11, 13]. Police reports containing terms from the same term-clusters were grouped together in concepts. The aim was to make these concepts as pure as possible. When a concept contained domestic and non-domestic violence reports, we investigated these reports and searched them for new attributes that can be used to discriminate between the domestic and non-domestic violence reports from this concept. This process was repeated until a classification was obtained that minimizes the number of false negative cases (i.e. domestic violence cases that were not classified as such).

The tool Concept Explorer [7] was used to visualize the concepts and their relationships.

3 Data Set

The domestic violence definition employed by the police organization of the Netherlands is as follows: "Domestic violence can be characterized as serious *acts of violence* committed by *someone of the domestic sphere of the victim.* Violence includes all forms of physical assault. The domestic sphere includes all partners, ex-partners, family members, relatives and family friends of the victim. Family friends are those persons who have a friendly relationship with the victim and who (regularly) meet the victim in his/her home", [6].

The XPol database – the database of the Amsterdam police organization – contains all documents relating to criminal offences. Documents related to certain types of crimes receive corresponding labels. Immediately after the reporting of a crime, police officers are given the possibility to judge whether or not it is a domestic violence case. If they believe it is a domestic violence case, they can indicate this by assigning the project code "domestic violence" to the report. However, not all domestic violence cases are recognized as such by police officers and by consequence, many documents are wrongly lacking a "domestic violence" label. The in-place case triage system is used to filter out these reports for in-depth manual inspection and classification. For example, going back to the first quarter of 2006, the in-place triage system retrieved 367 of such cases.

The dataset used during the research consists of 4146 police reports describing all violent incidents from the first quarter of 2006. All domestic violence cases from that period are a subset of this dataset. The 367 cases selected by the in-place case triage system are also a subset of this dataset. Unfortunately, many of these 4146 police reports did not contain the reporting of a crime by a victim, which is necessary for establishing domestic violence. Therefore, we only retained the 2288 documents in which the victim reported a crime to a police officer. From these documents, we removed the follow-up reports referring to previous cases. This filtering process resulted in a set of 1794 reports. From these reports, we extracted the person who reported the crime, the suspect, the persons involved in the crime, the witnesses, the project code and the statement made by the victim to the police. These data were used to generate the 1794 html-documents that were used for our research. An example of such a report is displayed in figure 2.

We also have at our disposal a thesaurus – a collection of terms – that was obtained by performing frequency analyses on these police reports. The terms that occurred most often were retrieved and added to the initially empty thesaurus. This resulted in a set of 123 terms.

Our validation set consists of 9147 cases describing all violent incidents from the year 2005 where the victim made a statement to the police. After removing the follow-up reports, 7817 cases were retained. In 2005, the in-place case triage system retrieved 2668 documents that had to be manually classified by police-officers. 1526 of them were classified as domestic violence, while 1142 of them were classified as non-domestic violence. These documents are a subset of our validation set.

Title of incident	Violent incident xxx
Reporting date	26-11-2007
Project code	Domestic violence against seniors (+55)
Crime location	Amsterdam Keizersgracht yyy
Suspect (male) Suspect (18-45yr)	zzz
Address	Amsterdam Keizersgracht yyy
Involved (male) Involved (18-45yr)	Neighbours
Address	Amsterdam Keizersgracht www
Victim (male) Victim (>45jr)	uuu
Address	Amsterdam Keizersgracht vvv

Reporting of the crime

Last night I was attacked by my only son. I was watching television in the living room when he suddenly attacked me with a knife. I felt on the floor. Then he tried to kick me. I tried to escape through the back door while I was yelling for help. I ran to the neighbours for help. They called the emergency services. Meanwhile my son ran away. My leg was bleeding etc.

Fig. 2. Example police report

We intend to verify whether a report can be classified as domestic violence by checking that it contains one or more terms from each of the two components of the domestic violence definition. A case can be labelled as domestic violence if:

1. a criminal offence has occurred. This may range from verbal threats over pushing and kicking to even killing the victim. To verify whether a criminal offence has occurred, the report is searched from terms like "hit", "stab", "kick", etc. These terms are grouped into the term-cluster "acts of violence".

2. and a person of the domestic sphere of the victim is involved in the crime. It should be noted that a report is always written from the point of view of the victim and not from the point of view of the officer. A victim always adds "my", "your", "her" and "his" to the persons involved in the crime. Therefore, the report is searched for terms like "my dad", "my mom", "my son", etc. These terms are grouped into the term-cluster "family members". The report is also searched for terms like "my ex-boyfriend", "my ex-husband", "my ex-wife", etc. These terms are grouped under the term-cluster "ex-partners". Furthermore, the report is searched for terms like "my nephew", "her uncle", "my aunt", "my step-father", "his step-daughter", etc. These terms are grouped under the term-cluster "relatives". Then the report is searched for terms like "family friend", "co-occupant", etc. These terms are grouped under the term-cluster "family friends".

The reports having the attribute "domestic violence" were classified by police officers as domestic violence. The remaining reports were classified as non-domestic violence. This results in the following lattice:

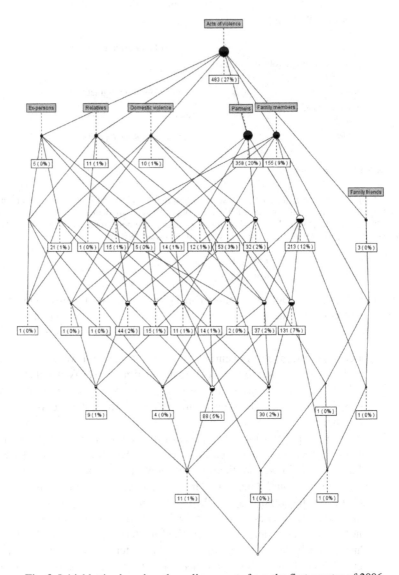

Fig. 3. Initial lattice based on the police reports from the first quarter of 2006

4   Results

By analysing the lattice displayed in figure 3, it became clear that a lattice containing only term-clusters based on the domestic violence definition does not sufficiently

Table 2. Results from lattice in figure 3

	Non-domestic violence	Domestic violence
Acts of violence	483	10

discriminate domestic from non-domestic violence reports. In other words, the definition cannot be used to automatically classify domestic violence cases as such, because many non-domestic violence also contain terms belonging to one or more of these clusters. However, the following interesting knowledge emerged out of this lattice.

Some 36% of the non-domestic violence reports only contain terms from the "acts of violence" cluster, while there are only 10 such domestic violence reports in the dataset. After in-depth manual inspection, these reports turned out to be wrongly classified as domestic violence. Although the lattice can not be used to distinguish domestic violence reports from non-domestic violence reports, it can be used to detect cases that were wrongly classified as domestic violence. Therefore, this lattice was constructed for all domestic violence reports from the year 2005. 116 (5%) of the cases that were classified as domestic violence only contained one or more terms from the "acts of violence" cluster, while there was no person of the domestic sphere of the victim involved. In-depth manual inspection of these reports proves that they were almost all wrongly classified as domestic violence. Furthermore, 49 (2%) of the domestic violence cases did not describe a violent incident. They were also wrongly classified as domestic violence. 7% of the domestic violence cases were thus reclassified as non-domestic violence. By using this lattice, it was also possible to discover the most frequently occurring types of domestic violence cases from the year 2005.

Table 3. Most frequently occurring types of domestic violence in 2005

	% of all domestic violence cases of 2005
Acts of violence and family members and partners	27%
Acts of violence and family members and partners and ex-persons	15%
Acts of violence and family members	13%
Acts of violence and family members and ex-persons	12%
Acts of violence and family members and partners and relatives	6%
Acts of violence and partners	6%

The next step consisted of searching for additional attributes that can be used to distinguish a domestic violence report from non-domestic violence reports and vice versa. First of all, it became apparent that in a large number of the domestic violence cases (137 cases or 28%), the perpetrator and the victim lived at the same address at the time the victim made his/her statement to the police. 128 of these cases were classified as domestic violence. When we studied the 9 non-domestic violence cases we found that the perpetrator and the victim always lived together in the same institution (e.g. a youth institution, a prison, an old folk's home). Of the 21 cases where the perpetrator and the victim lived in the same institution, only 12 were classified as domestic violence. This finding brought about a lively discussion amongst the police officers of the Amsterdam police force. More importantly, it exposed the mismatch between the management's conception of domestic violence and the classification as performed by the police officers. Going back to 2005, 138 cases described incidents between inhabitants of an institution. Police officers classified a substantial part of these as non-domestic violence, while only a limited number were classified as domestic violence. However, according to the board members responsible for the domestic violence policy, all these cases should have been classified as domestic violence. In other words, the definition employed by the management was much broader than the one employed by the police officers performing the classification task.

To classify the remaining cases, we explored the corresponding police reports, in search of new attributes. We found that 42% of these reports (749 cases) did not mention a suspect. However, according to the domestic violence definition (which says that the perpetrator must belong to the domestic sphere of the victim), the offender should be known. By consequence, we assumed that these reports described non-domestic violence cases. However, 30 of them turned out to be domestic violence cases. After in-depth inspection of these reports, we concluded that this was due to unstructured way of working of the police officers. Some officers immediately label a person as a suspect, when the victim mentions this person as a suspect. However, other officers first want to interrogate the suspect. In the latter case, this person is added to the list of persons who were involved in or witnessing the crime. This list of persons might include friends and family members of the victim, bystanders, etc. and can be very large. By consequence it is often very difficult to identify the suspect from this list in filed reports. We asked the proper authorities whether or not there exists a policy that regulates the labelling of persons as suspects. It turned out that such a regulation did not exist. Our research proves that such a regulation is necessary. We also found that a 37% of these reports that lack a suspect contain a description of the suspect (277 cases). These 277 reports were all classified as non-domestic violence.

After the consultation of the proper authorities with regard to this subject, it became clear that the best and the most feasible solution would be to introduce an additional field in police reports that can be used by police officers to record the person who was mentioned by the victim as the offender. This relatively small change makes it easier to identify the suspect(s) for a given case.

According to the literature, domestic violence is a phenomenon that mainly occurs inside the house [4, 5, 6, 21]. Therefore, an attribute called "private locations" was introduced. This term-cluster contained terms like "bathroom", "living room", "bedroom", etc. As was expected, 393 (86%) of the domestic violence cases from the dataset contained one or more terms from this term-cluster. However, 568 (43%) of the non-domestic violence cases also contained one or more terms from this term-cluster. An attribute called "public locations" was also introduced. It was expected that there would be almost no domestic violence case that took place on the street. Surprisingly, this turned out to be incorrect. In about one-fourth of the domestic violence cases there had been an incident on a public location. While studying these police reports, we discovered that this was often the case when ex-partners were involved in the case. It thus became clear that it is not possible to distinguish domestic from non-domestic violence reports by means of the type of locations mentioned in the reports. Combining the clusters "private locations" and "public locations" with clusters like "family members" or "ex-persons" for example did not yield the expected results either. While exploring the domestic violence reports, terms like "divorce", "marriage problems", "relational problems", etc. regularly occurred. Therefore, a new term-cluster called "relational problems" was introduced.

In order to keep the lattice surveyable, we clustered the terms from the clusters "family members", "relatives", "partners", "ex-partners" and "family friends", together in the cluster "persons" and added the newly discovered attributes "same address", "institution", "no suspect", "description of suspect" and "relational problems". This resulted in the following lattice:

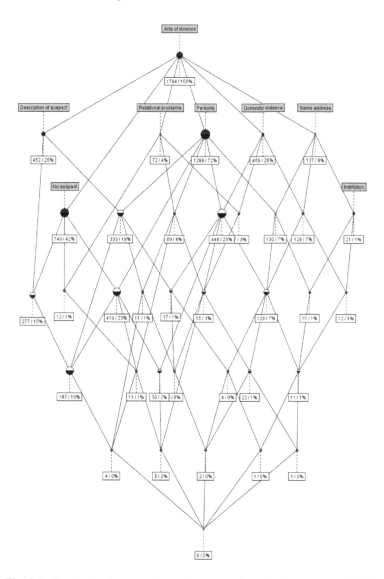

Fig. 4. Refined lattice based on the police reports from the first quarter of 2006

This lattice provides us with essential knowledge needed to discriminate domestic from non-domestic violence reports and vice versa. The most interesting findings are displayed in the following table:

Table 4. Results from lattice in figure 4

	Non-domestic violence	Domestic violence
Acts of violence	483	10
Acts of violence and same address	9	128
Acts of violence and no suspect and description of suspect	277	0
Acts of violence and no suspect	719	30

This lattice was also constructed for the validation set containing police reports from the year 2005. The most interesting findings are displayed in the following table.

Table 5. Discovered knowledge applied on dataset from 2005

	Non-domestic violence	Domestic violence
Acts of violence	2065	116
Acts of violence and same address	56	422
Acts of violence and no suspect and description of suspect	937	27
Acts of violence and no suspect	2579	186

The 56 cases where the perpetrator and the victim lived at the same address and that were not classified as domestic violence contained many cases where the perpetrator and the victim lived in the same institution. The remaining cases turned out to be wrongly classified after in-depth manual inspection. After in-depth manual inspection of the 116 police reports containing only one or more terms from the "acts of violence" cluster and that were classified as domestic violence, they turned out to be wrongly classified.

5 Conclusions

In this paper, the possibilities of using FCA as an exploratory data analysis technique for discovering new knowledge from police reports were explored. The construction of an initial lattice containing term-clusters created by a domain expert on the basis of the domestic violence definition and the incremental refinement of this lattice provided the user with a powerful framework for exploring the dataset.

It was shown that the domestic violence definition is often not applied properly by police officers, but incorrect classifications can be automatically corrected on the basis of this definition. In addition, some essential characteristics that discriminate domestic from non-domestic violence reports were discovered. However, the limitations of the FCA technique also became apparent. Concept lattices could only be used effectively with a maximum of 14 attributes. If more attributes were used, the lattices became too cluttered to be useful for data exploration. Issues for future research include improving the used thesaurus by amongst others taking negated sentences like "my dad didn't hit me" into account.

Acknowledgements

The authors would like to thank the police organization of the region Amsterdam-Amstelland and in particular Deputy Chief Organisation Information and Programme Manager Intelligence Led Policing Reinder Doeleman and Chief Information and Operations Hans Schönfeld for supporting this research.

References

[1] About Domestic Violence. Office on Violence against Women (Retrieved on 22-10-2007), http://www.usdoj.gov/ovw/domviolence.htm
[2] Watts, C., Timmerman, C.: Violence against women: global scope and magnitude. The Lancet 359(9313), 1232–1237. RMID 1155557
[3] Waits, K.: The criminal Justice System's response to Battering: Understanding the problem, forging the solutions. Washington Law Review 60, 267–330 (1984-1985)
[4] Vincent, J.P., Jouriles, E.N.: Domestic violence. In: Guidelines for research-informed practice. Jessica Kingsley Publishers, London, Philadelphia (2000)
[5] Minleer-Black, C.: Domestic violence: Findings from a new British Crime Survey self-completion questionnaire. Home Office Research Study, London (1999)
[6] Keus, R., Kruijff, M.S.: Huiselijk geweld, draaiboek voor de aanpak. Directie Preventie. Jeugd en Sanctiebeleid van de Nederlandse justitie (2000)
[7] Yevtushenko, S.A.: System of data analysis "Concept Explorer". In: Proceedings of the 7th national conference or Artificial Intelligence. KII 2000, Russia, pp. 127–134 (2000)
[8] Ganter, B., Wille, R.: Formal Concept Analysis: Mathematical Foundations. Springer, Heidelberg (1999)
[9] Wille, R.: Restructuring lattice theory: an approach based on hierarchies of concepts. In: Rival, I. (ed.) Ordered sets, pp. 445–470. Reidel, Dordrecht-Boston (1982)
[10] Priss, U.: Formal Concept Analysis in Information Science. In: Cronin, B. (ed.) Annual Review of Information Science and Technology, ASIST, vol. 40 (2005)
[11] Wille, R.: Why can concept lattices support knowledge discovery in databases? Journal of Experimental & Theoretical Artificial Intelligence 14(2), 81–92 (2002)
[12] Stumme, G., Wille, R., Wille, U.: Conceptual Knowledge Discovery in Databases Using Formal Concept Analysis Methods. In: Zytkow, J.M., Quafofou, M. (eds.) PKDD 1998. LNCS, vol. 1510, pp. 450–458. Springer, Heidelberg (1998)
[13] Stumme, G.: Efficient Data Mining Based on Formal Concept Analysis. In: Hameurlain, A., Cicchetti, R., Traunmüller, R. (eds.) DEXA 2002. LNCS, vol. 2453, pp. 3–22. Springer, Heidelberg (2002)
[14] Stumme, G.: Formal Concept Analysis on its Way from Mathematics to Computer Science. In: Proc. 10th Intl. Conf. on Conceptual Structures (ICCS 2002). LNCS. Springer, Heidelberg (2002)
[15] Priss, U.: Lattice-based information Retrieval. Knowledge Organization 27(3), 132–142 (2000)

[16] Godin, R., Gescei, J., Pichet, C.: Design of browsing interface for information retrieval. In: Belkin, N.J., van Rijsbergen, C.J. (eds.) Proc. SIGIR 1989, pp. 32–39 (1989)

[17] Carpineto, C., Romano, G.: Using concept lattices for text retrieval and mining. In: Formal Concept Analysis-State of the Art, Proc. of the first International Conference on Formal Concept Analysis, Berlin. Springer, Heidelberg (2005)

[18] Cole, R., Eklund, P.: Browsing Semi-structured Web Texts Using Formal Concept Analysis. In: Delugach, H.S., Stumme, G. (eds.) ICCS 2001. LNCS (LNAI), vol. 2120, pp. 319–332. Springer, Heidelberg (2001)

[19] Eklund, P., Ducrou, J., Brawn, P.: Concept Lattices for Information Visualization: Can Novice Read Line Diagrams? In: Eklund, P. (ed.) ICFCA 2004. LNCS (LNAI), vol. 2961, pp. 57–73. Springer, Heidelberg (2004)

[20] Priss, U.: A Graphical Interface for Document Retrieval Based on Formal Concept Analysis. In: Santos, E. (ed.) Proc. of the 8th Midwest Artificial Intelligence and Cognitive Science Conference. AAAI Technical Report CF-97-01, pp. 66–70 (1997)

[21] Beke, B.M.W.A., Bottenberg, M.: De vele gezichten van huiselijk geweld. In: opdracht van Programma Bureau Veilig / Gemeente Rotterdam. Uitgeverij SWP, Amsterdam (2003)

Leatherbacks Matching by Automated Image Recognition

Eric J. Pauwels[1], Paul M. de Zeeuw[1], and Danielle M. Bounantony[2]

[1] Centrum Wiskunde & Informatica, Amsterdam, The Netherlands
{Eric.Pauwels,Paul.de.Zeeuw}@cwi.nl
[2] Nicholas School of the Environment and Earth Sciences, Duke University, USA
danielle.buonantony@duke.edu

Abstract. We describe a method that performs automated recognition of individual laetherback turtles within a large nesting population. With only minimal preprocessing required of the user, we prove able to produce unsupervised matching results. The matching is based on the Scale-Invariant Feature Transform by Lowe. A strict condition posed by biologists reads that matches should not be missed (no false negatives). A robust criterion is defined to meet this requirement. Results are reported for a considerable sample of leatherbacks.

1 Introduction

The ability to individually identify sea turtles in the field has been one of the most valuable tools in advancing our understanding of these animals. Marked or identified turtles allow for the measurement of a wide variety of biological and population variables (e.g. reproductive output, longevity, and survival rates). Traditional marking methods have included flipper, transponder, and mutilation tagging. In leatherbacks the pink spot, overlying the pineal gland on the dorsal surface of the head, has been reported as a unique identifier (McDonald and Dutton [6]). Leatherback nesting colonies of Trinidad offer the ideal research location for collecting photos of these spots as it annually supports nesting by 10,000 turtles. The Matura Beach/Fishing Pond nesting colony, located on the east coast of the island accounts for approximately half of all nesting on the island with over 150 turtles nesting per night. The beach is patrolled continuously by a local conservation organization, The Nature Seekers, which enabled most turtles to be detected. Photos are taken only during the laying stage of nesting to preclude disturbance of the turtle.

Identification of leatherbacks by humans involves laborious and tedious browsing through a (growing) photo database. Therefore, we seek to determine whether identification can be automated using image recognition algorithms. The time that can be put in watching colonies is limited, already for this reason the algorithm needs to avoid false negatives at all costs. The latter is an important issue for biologists, the presence of the same leatherback at a different place and different time should not be overlooked. The Scale Invariant Feature Transform

P. Perner (Ed.): ICDM 2008, LNAI 5077, pp. 417–425, 2008.
© Springer-Verlag Berlin Heidelberg 2008

(SIFT, Lowe [4]) appears capable to provide us with automated matches robust to changes in 3D viewpoint and illumination, noise and occlusion.

Biologists are waking up to the possibilities of computer-assisted photo-identification and a number of stand-alone systems are under development (cf. [2,5,7,8]). The method demonstrated in this paper allows for a web based service by which one can query and contribute to a database of images. See [1] for a similar service under development also within the field of biodiversity.

The paper is organized as follows. In Section 2 we decribe the necessary pre-processing of images and the use of Lowe's SIFT features. Section 3 describes how to decide whether we can presume that images of pineal spots are matching or not. In Section 4 we provide statistics derived from a comparison with the groundtruth. There is the option of providing a future webservice, see Section 5.

2   Preprocessing and Feature Transform

2.1   Preprocessing

Cropping. An individual leatherback can uniquely be identified by its pineal spot [6], see the top row of Figure 1. We benefit from the fact that the pineal spot stands out in pink on the dorsal surface of the animal. We search and isolate the "pink spot" by human intervention, i.e. a rectangular region around the spot is selected. An additional advantage of the cropping is the reduction of dimensions which speeds up the subsequent processing. Clearly, the selection procedure introduces some arbitrariness as it is not always obvious to what extent "satellite" spots and marks should be included. However, as long as the main salient parts are retained, the resulting classification appears quite robust, see Section 4. The cropping is the only manual intervention required at this stage and typically takes 5 secs per image.

Contrast enhancing. The cropped colour image is turned into a gray-value image, where the gray-value is computed in such a way that it enhances the contrast between the pink spot and the dark background. This can be done adaptively (i.e. data-driven) by selecting the colour combination that corresponds to the first PCA (principal component analysis) factor. In the current implementation we simply convert a colour image into a gray-scale image by defining the gray value K at each pixel as

$$K = R - 0.5(G + B)$$

where R, G and B are the intensity-values of the red (R), green (G) and blue (B) component.

2.2   Recapitulation on SIFT

To recognize (gray-value) images we use the features produced by the Scale Invariant Feature Transform (SIFT, Lowe [4]). This method selects so-called *keypoints* in an image. These are local points of interest, furnished with loca-tion, best fitting scale, and orientation with respect to the gradient. Along with

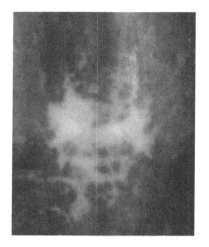

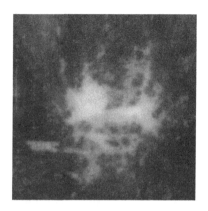

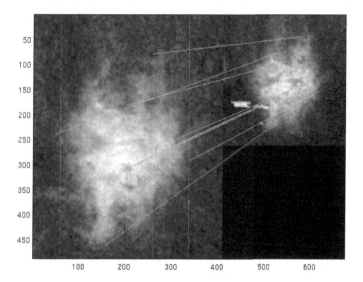

Fig. 1. Top: pineal spots of leatherbacks photographed on different days with different cameras. Bottom: matching keypoints found by SIFT.

each keypoint comes a *keypoint descriptor*, which is a feature vector summarizing local gradient information. The keypoints are selected in a strict manner through a cascade filtering approach. The features are defined such that they appear both invariant to image scaling plus rotation and, to a considerable extent, invariant to change in illumination and 3D camera viewpoint. Moreover, they are well localized in both the spatial and frequency domains, reducing the probability of disruption by occlusion, clutter, or noise. The descriptors prove highly distinctive, which allows a single feature to find its correct match with good probability in a large database of features. Below we describe major stages within the intricate transform, and omit lots of (important) details.

1. **Extrema detection in scale-space.** The image $I(x, y)$ is convolved with a difference-of-Gaussian function which computes the difference of two nearby scales (separated by a constant factor k)

$$D(x, y, \sigma) = (G(x, y, k\sigma) - G(x, y, \sigma)) * I(x, y) \qquad (1)$$

   where

$$G(x, y, \sigma) = \frac{1}{2\pi\sigma^2} e^{-(x^2+y^2)/2\sigma^2}.$$

   This can be computed efficiently and closely approximates the result as if a scale-normalized Laplacian of Gaussian $\sigma^2 \Delta G$ were applied (Lindeberg [3]).

   In order to detect the local maxima and minima of image $D(x, y, \sigma)$, each sample point is compared not just to its eight direct neighbors but also to its nine direct neighbors in a scale above and below. It is selected only if it is larger than all of these neighbors or smaller than all of them. Still, the scale-space difference-of-Gaussian function has a large number of extrema, all candidate keypoints. Fortunately, a coarse sampling of scales suffices.

2. **Keypoint localization in scale-space.** A Taylor expansion (up to quadratic terms) of $D(x, y, \sigma)$ is used to determine an accurate location of the extremum in the coordinates $(x, y, \sigma)^T$. An expansion around the newly found extremum helps to detect low contrast, upon which the candidate keypoint is rejected as it is deemed unstable. Edges occurring in the original $I(x, y)$ provide another source of extrema with poorly defined locations. This situation is detected when the ratio of principal curvatures at an extremum rises above a certain threshold, upon which, again, the candidate keypoint is rejected as it is deemed unstable.

3. **Orientation assignment.** We proceed with the keypoints that have remained. In order to achieve rotation invariance for our keypoint descriptor to be (next stage, stage 4), we want to determine the keypoint orientation. The convolved version of $I(x, y)$ with scale closest to the one of the keypoint is selected for doing so. Magnitude and direction of the gradient are computed pixelwise using simple differences. A histogram is formed from the orientations of sample points within a certain Gaussian-weighted circular window around the keypoint. Obviously, peaks in the histograms correspond to dominant directions. At most two of such directions are taken into account (two directions leading to two different keypoints).

4. **Descriptor assignment.** This stage is similar to the previous one in that orientation histograms are computed. Again the scale of the keypoint determines the level of Gaussian blur for the image. To achieve rotational feature invariance, coordinates are rotated relative to the keypoint orientation as determined in the previous stage. The feature descriptor is computed as a set of orientation histograms over 4×4 sampling regions. Only 8 different orientations are considered, leading to 8 bins in each histogram. This leads to a feature vector / descriptor of $4 \times 4 \times 8 = 128$ elements per keypoint.

Keypoint descriptor matching. Comparing two images I and I' now boils down to comparing their respective sets of keypoints and descriptors. To decide whether an individual keypoint with descriptor in one image matches with a counterpart in the other image is not trivial. A uniform treshold on distance between descriptors is not wise as some descriptors discriminate more easily than others. For a positive match it is not good enough for mutual descriptors to be at close range. Far too many descriptors may apply, hereby invoking lots of false matches. Instead, a match of descriptors is required to *excel*. This is expressed by the criterion explained below. For keypoint p_i in image I one looks for the best matching keypoint p'_j in image I' by searching for the smallest distance $d(\delta_i, \delta'_j)$ between their 128-sized descriptors δ_i and δ'_j. This point match will only be retained if it excels: the (minimum) distance of the first choice should be smaller than a predefined fraction of the second best choice. More formally, p_i in image I is matched to p'_j in I' only if

$$d(\delta_i, \delta'_j) = \min_k d(\delta_i, \delta'_k)$$

and

$$d(\delta_i, \delta'_j) < D_R \min_{k \neq j} d(\delta_i, \delta'_k).$$

Otherwise it is rejected which implies that keypoint p_i has no match in I'. The fraction D_R is called the *distance ratio* by Lowe [4] and is often fixed at a value of 0.6.

3   Matching of Images

Here we explain on what grounds (criteria) we presume the result of matching two images to be positive or negative and how reliable (and why) we want our presumptions to be. We rely on SIFT keypoints and use the accompanying descriptors. One needs to be aware that the matching of images is *not* symmetric: it depends on whether an image is considered a *query* or a *reference* image. The asymmetry is due to the way a match of descriptors has been defined (see the last paragraph of Section 2.2). A case in point is that one keypoint in the "query" image may resemble more than one keypoint in the "reference" image. To come up with a *symmetric* similarity measure we compute the number of bi-directional matches (n_{bi}): i.e. matches are only retained if they persist when swapping the roles of query and reference image. If a point-match is bi-directional the chances of it being erroneous are slim (see the lower part of Figure 1 for examples).

Deformations between different images that occur are due to the use of different cameras at different times by different people. This involves differences in resolution or scale, rotations and translations, changes in illumination (including glare), viewing angles and pollution (see the upper part of Figure 1 for examples). SIFT is apt to deal with such variations. However, since we cannot afford to overlook a genuine match, we relax the value of the distance ratio D_R

to 0.7 (see Section 2.2). The net result of this adjustment is that the number of matching keypoints between the query and reference image will be higher.

The standard way to decide whether two images are similar could be straightforward: compute the number of bi-directional matches n_{bi} and compare it to a predefined threshold. The images are then declared to be either matching or non-matching depending on whether or not n_{bi} exceeds this threshold. Again, as explained before, it is of paramount importance to reduce the risk of overlooking a genuine match. We therefore thread cautiously and introduce *two* thresholds: an upper threshold n_{bi}^{high} and a lower one n_{bi}^{low}. If the number of bi-directional matches (n_{bi}) between two images exceeds n_{bi}^{high} then we presume to have a high quality match between the images and it is kept in the database. If, on the other hand, $n_{bi} < n_{bi}^{low}$ then the images appear dissimilar and the match is rejected. For image pairs that achieve a score in between these two thresholds, this is substantial evidence that the images might be similar but it needs to be backed up by an additional test (introduced below).

The deformation between different images of the same spot is moderate (see above). We therefore assume that if keypoints in the query image are correctly matched to their counterparts in the reference image, the distance between any pair of keypoints in the query image should be the same (up to a scaling) as the distance between the corresponding points in the reference image. This can easily be checked by regressing the distances in the reference images over the corresponding distances in the query image. Data points due to correct point matches will trace out a line, the slope of which reflects the afore-mentioned scaling factor. Mismatches on the other hand, will create outliers.

The proposed additional test can now be summarized as follows: for two images, find all pairs of points p_i (in the query image) and p_i' (in the reference image) which are joined by a bi-directional match. Next, compute the distances between all such points in each image separately. This results in a set of distance $d_{ij} = d(p_i, p_j)$ for the points in the query image, and another set $d_{ij}' = d(p_i', p_j')$ for the corresponding points in the reference image. The latter set of values is regressed on the the the former (using the regression model $y = ax + \epsilon$ which corresponds to a line that passes through the origin). As argued above, the fit of regression model reflects the quality of the match. This is quantified by computing the *mean squared error* (MSE) for the regression:

$$MSE = \frac{1}{n-1} \sum_{i=1}^{n} (y_i - \hat{y}_i)^2$$

where $\hat{y}_i$ is the predicted value based on the regression. If the MSE exceeds a predefined threshold, the regression fit is low indicating the the point matches are erroneous. As a result the images are classified as non-matching. If on the other hand, the regression fit is satisfactory, we conclude that the point matches — although relatively few in numbers — enjoy a consistency that is indicative of true underlying similarity. The image pair is therefore tagged as a potential match, to be verified by a human expert for final validation or rejection. The ones that are retained are presented to the user for a final confirmation or rejection decision.

4   Results

In a first experiment we worked with a database of 613 images that were collected over the period of about six weeks in the Leatherbacks nesting colonies of Trinidad. During the night, groups of around 150 turtles would emerge from the sea to lay and bury their eggs on the beach. During this activity the pineal spot of most animals was photographed twice, usually within the time span of a few minutes. As a consequence, the database comprises lots of individual animals for which we have two photos taken in quick succession and labeled to reflect the fact that they depict the same individual. These pairs are very valuable as they furnish us with a set of genuine matches that can be used to check minimal performance measures (e.g. whether the number of false negatives among these trivial matches is actually zero). In addition to these trivial matches, there are the more interesting repeat encounters where the same individual was photographed on different nights. In the current database 13 such individuals were discovered by manual inspection. The challenge faced by the matching algorithm outlined above therefore amounts to identifying all true matches (i.e. both the trivial and the non-trivial ones) while simultaneously minimizing the number of images that need to be checked manually.

Recall that the matching decision logic involves two thresholds (n_{bi}^{high} and n_{bi}^{low}, see Section 3) for the number n_{bi} of bi-directional matches. In the current experimental set-up we use the values $n_{bi}^{high} = 10$ and $n_{bi}^{low} = 3$. If the number of bi-directional matches between two images exceeds 10 then we presume a high level of similarity and they are automatically accepted as a matching pair. Conversely, if the number of matches is less than 3 then the image pair is automatically rejected. Finally, if $3 \leq n_{bi} \leq 10$ then we compute the square root of the MSE for the regression model. If $\sqrt{MSE}$ exceeds a threshold (which has been set equal to 7% of the data range), then the regression fit is deemed unsatisfactory and also this pair is rejected. If however $\sqrt{MSE}$ is smaller than this threshold value, the image pair is presented to a human supervisor for final approval or rejection.

The algorithm checked $613 \cdot 612/2 = 187,578$ image pairs. The above outlined decision strategy succeeded in recovering all true matches while no genuine matches were overlooked. Notably, the algorithm managed to uncover one additional match which happened to be overlooked by human experts. A total of 73 pairs (i.e. less than 0.04% of all pairs) were singled out by the algorithm for final inspection by a human supervisor. For the biologists involved this algorithm therefore provided highly reliable and welcome assistance.

Overview algorithm and results. Let p_i $(i = 1, \ldots, n_{bi})$ be the keypoints in the query image (Q) that have been bi-directionally matched (using SIFT descriptors) to keypoints p_i' in the reference image (R), i.e. if $m_{AB}()$ denotes the matching function from image A to image B, then $\forall i = 1, \ldots, n_{bi} : m_{QR}(p_i) = (p_i')$ AND $m_{RQ}(p_i') = (p_i)$. Hence, the number of bi-directional matches between images Q and R equals n_{bi}.

Algorithm	
if $n_{bi} > n_{bi}^{high}$	Accept match between images Q and R;
else if $n_{bi} < n_{bi}^{low}$	Reject match between images Q and R;
else	Compute distances $d_{ij} = d(p_i, p_j)$ and $d_{ij}' = d(p_i', p_j')$, regress d_{ij}' over d_{ij} and compute $\sqrt{\text{MSE}}$;
if $\sqrt{\text{MSE}} > q$	Reject match between images Q and R;
else	Present presumed match between Q and R to human supervisor for final confirmation or rejection.

In the current implementation $n_{bi}^{low} = 3$, $n_{bi}^{high} = 10$, and q equals 7% of the d_{ij}' range, i.e. $q = 0.07(\max\{d_{ij}'; j > i\} - \min\{d_{ij}'; j > i\})$. The results for the current database are summarized in the table below.

Nr. of images	613
Nr. of false positives	0
Nr. of false negatives	0
Nr. of pairs processed	187,578
Nr. of pairs retained for manual inspection	73 (i.e. 0.04%)

5   Discussion and Future Directions

Leatherback turtles migrate over large distances and it would therefore be interesting to collect all data in a readily accessible global database. It seems to us that a web-based database running the proposed photo-identification algorithm could be an interesting addition to the current data repositories. Since the only manual work involved is the initial cropping of the pineal spot and, possibly, the acceptance or rejection of a small number of ambiguous matches, organizing this as a web-service would be rather straightforward. This way groups of biologists could easily share and compare data collected at different times and locations. At the same time, it would allow large groups of amateurs to significantly contribute to the scientific enterprise by submitting their own pictures. We believe that this type of web-enabled collective effort will play an increasingly important role in the near future.

Acknowledgments

We thank Scott A. Eckert (WIDECAST, Duke University) for sharing his expert knowledge on leatherback turtles. Danielle M. Bounantony acknowledges the practical support provided by The Nature Seekers, and the financial support received from the Environmental Internship Fund, Andrew W. Mellon Foundation and the Duke Center for Latin American and Caribbean Studies. This work was partially supported by MUSCLE, part of EU's Sixth Framework Programme for Research and Technological Development (FP6).

References

1. de Zeeuw, P.M., Ranguelova, E., Pauwels, E.J.: Towards an Online Image-Based Tree Taxonomy. In: Perner, P. (ed.) ICDM 2007. LNCS (LNAI), vol. 4597, pp. 296–306. Springer, Heidelberg (2007)
2. Hillman, G., et al.: Computer-assisted photo-identification of flukes using blotch and scar patters. In: Proceedings of 15th Biennial Conference on the Biology of Marine Mammals (December 2003)
3. Lindeberg, T.: Scale-space theory: A basic tool for analysing structures at different scales. Journal of Applied Statistics 21(2), 224–270 (1994)
4. Lowe, D.G.: Distinctive image features from scale-invariant keypoints. International Journal of Computer Vision 60(2), 91–110 (2004)
5. Mizroch, S., Beard, J., Lynde, M.: Computer Assisted Photo-Identification of Humpback Whales. In: Hammond, P., Mizroch, S., Donovan, G. (eds.) Individual Recognition of Cetaceans, pp. 63–70. International Whaling Commission, Cambridge (1990)
6. McDonald, D.L., Dutton, P.H.: Use of PIT tags and photoidentification to revise remigration estimates of leatherback turtles (*Dermochelys coriacae*) nesting on St. Croix, U.S. Virgin Islands (1979-1995); Chelonian Conservation and Biology 2 (2), 148–152 (1996)
7. Ranguelova, E., Pauwels, E.J.: Saliency Detection and Matching Strategy for Photo-Identification of Humpback Whales. In: International Conference on Graphics, Vision and Image Processing - GVIP 2005, Cairo, Egypt, December 2005, pp. 81–88 (2005)
8. Van Tienhoven, A., den Hartog, J., Reijns, R., Peddemors, V.: A computer-aided program for pattern-matching of natural marks on the spotted raggedtooth shark (*Carcharias taurus*). Journal of Applied Ecology 44(2), 273–280 (April 2007)

Author Index

Lecture Notes in Artificial Intelligence (LNAI)

Vol. 4845: N. Zhong, J. Liu, Y. Yao, J. Wu, S. Lu, K. Li (Eds.), Web Intelligence Meets Brain Informatics. XI, 516 pages. 2007.

Vol. 4840: L. Paletta, E. Rome (Eds.), Attention in Cognitive Systems. XI, 497 pages. 2007.

Vol. 4830: M.A. Orgun, J. Thornton (Eds.), AI 2007: Advances in Artificial Intelligence. XIX, 841 pages. 2007.

Vol. 4828: M. Randall, H.A. Abbass, J. Wiles (Eds.), Progress in Artificial Life. XII, 402 pages. 2007.

Vol. 4827: A. Gelbukh, Á.F. Kuri Morales (Eds.), MICAI 2007: Advances in Artificial Intelligence. XXIV, 1234 pages. 2007.

Vol. 4826: P. Perner, O. Salvetti (Eds.), Advances in Mass Data Analysis of Signals and Images in Medicine, Biotechnology and Chemistry. X, 183 pages. 2007.

Vol. 4819: T. Washio, Z.-H. Zhou, J.Z. Huang, X. Hu, J. Li, C. Xie, J. He, D. Zou, K.-C. Li, M.M. Freire (Eds.), Emerging Technologies in Knowledge Discovery and Data Mining. XIV, 675 pages. 2007.

Vol. 4811: O. Nasraoui, M. Spiliopoulou, J. Srivastava, B. Mobasher, B. Masand (Eds.), Advances in Web Mining and Web Usage Analysis. XII, 247 pages. 2007.

Vol. 4798: Z. Zhang, J.H. Siekmann (Eds.), Knowledge Science, Engineering and Management. XVI, 669 pages. 2007.

Vol. 4795: F. Schilder, G. Katz, J. Pustejovsky (Eds.), Annotating, Extracting and Reasoning about Time and Events. VII, 141 pages. 2007.

Vol. 4790: N. Dershowitz, A. Voronkov (Eds.), Logic for Programming, Artificial Intelligence, and Reasoning. XIII, 562 pages. 2007.

Vol. 4788: D. Borrajo, L. Castillo, J.M. Corchado (Eds.), Current Topics in Artificial Intelligence. XI, 280 pages. 2007.

Vol. 4775: A. Esposito, M. Faundez-Zanuy, E. Keller, M. Marinaro (Eds.), Verbal and Nonverbal Communication Behaviours. XII, 325 pages. 2007.

Vol. 4772: H. Prade, V.S. Subrahmanian (Eds.), Scalable Uncertainty Management. X, 277 pages. 2007.

Vol. 4766: N. Maudet, S. Parsons, I. Rahwan (Eds.), Argumentation in Multi-Agent Systems. XII, 211 pages. 2007.

Vol. 4760: E. Rome, J. Hertzberg, G. Dorffner (Eds.), Towards Affordance-Based Robot Control. IX, 211 pages. 2008.

Vol. 4755: V. Corruble, M. Takeda, E. Suzuki (Eds.), Discovery Science. XI, 298 pages. 2007.

Vol. 4754: M. Hutter, R.A. Servedio, E. Takimoto (Eds.), Algorithmic Learning Theory. XI, 403 pages. 2007.

Vol. 4737: B. Berendt, A. Hotho, D. Mladenic, G. Semeraro (Eds.), From Web to Social Web: Discovering and Deploying User and Content Profiles. XI, 161 pages. 2007.

Vol. 4733: R. Basili, M.T. Pazienza (Eds.), AI*IA 2007: Artificial Intelligence and Human-Oriented Computing. XVII, 858 pages. 2007.

Vol. 4724: K. Mellouli (Ed.), Symbolic and Quantitative Approaches to Reasoning with Uncertainty. XV, 914 pages. 2007.

Vol. 4722: C. Pelachaud, J.-C. Martin, E. André, G. Chollet, K. Karpouzis, D. Pelé (Eds.), Intelligent Virtual Agents. XV, 425 pages. 2007.

Vol. 4720: B. Konev, F. Wolter (Eds.), Frontiers of Combining Systems. X, 283 pages. 2007.

Vol. 4702: J.N. Kok, J. Koronacki, R. Lopez de Mantaras, S. Matwin, D. Mladenič, A. Skowron (Eds.), Knowledge Discovery in Databases: PKDD 2007. XXIV, 640 pages. 2007.

Vol. 4701: J.N. Kok, J. Koronacki, R. Lopez de Mantaras, S. Matwin, D. Mladenič, A. Skowron (Eds.), Machine Learning: ECML 2007. XXII, 809 pages. 2007.

Vol. 4696: H.-D. Burkhard, G. Lindemann, R. Verbrugge, L.Z. Varga (Eds.), Multi-Agent Systems and Applications V. XIII, 350 pages. 2007.

Vol. 4694: B. Apolloni, R.J. Howlett, L. Jain (Eds.), Knowledge-Based Intelligent Information and Engineering Systems, Part III. XXIX, 1126 pages. 2007.

Vol. 4693: B. Apolloni, R.J. Howlett, L. Jain (Eds.), Knowledge-Based Intelligent Information and Engineering Systems, Part II. XXXII, 1380 pages. 2007.

Vol. 4692: B. Apolloni, R.J. Howlett, L. Jain (Eds.), Knowledge-Based Intelligent Information and Engineering Systems, Part I. LV, 882 pages. 2007.

Vol. 4687: P. Petta, J.P. Müller, M. Klusch, M. Georgeff (Eds.), Multiagent System Technologies. X, 207 pages. 2007.

Vol. 4682: D.-S. Huang, L. Heutte, M. Loog (Eds.), Advanced Intelligent Computing Theories and Applications. XXVII, 1373 pages. 2007.

Vol. 4676: M. Klusch, K.V. Hindriks, M.P. Papazoglou, L. Sterling (Eds.), Cooperative Information Agents XI. XI, 361 pages. 2007.

Vol. 4667: J. Hertzberg, M. Beetz, R. Englert (Eds.), KI 2007: Advances in Artificial Intelligence. IX, 516 pages. 2007.

Vol. 4660: S. Džeroski, L. Todorovski (Eds.), Computational Discovery of Scientific Knowledge. X, 327 pages. 2007.

Vol. 4659: V. Mařík, V. Vyatkin, A.W. Colombo (Eds.), Holonic and Multi-Agent Systems for Manufacturing. VIII, 456 pages. 2007.

Vol. 4651: F. Azevedo, P. Barahona, F. Fages, F. Rossi (Eds.), Recent Advances in Constraints. VIII, 185 pages. 2007.

Vol. 4648: F. Almeida e Costa, L.M. Rocha, E. Costa, I. Harvey, A. Coutinho (Eds.), Advances in Artificial Life. XVIII, 1215 pages. 2007.

Vol. 4635: B. Kokinov, D.C. Richardson, T.R. Roth-Berghofer, L. Vieu (Eds.), Modeling and Using Context. XIV, 574 pages. 2007.

Vol. 4632: R. Alhajj, H. Gao, X. Li, J. Li, O.R. Zaïan (Eds.), Advanced Data Mining and Applications. XV, 634 pages. 2007.

Vol. 4629: V. Matoušek, P. Mautner (Eds.), Text, Speech and Dialogue. XVII, 663 pages. 2007.

Vol. 4626: R.O. Weber, M.M. Richter (Eds.), Case-Based Reasoning Research and Development. XIII, 534 pages. 2007.